ENCYCLOPEDIA OF INORGANIC CHEMISTRY

ENCYCLOPEDIA OF INORGANIC CHEMISTRY

Vol. 3

By
Sananda Chatterjee
DAE, DCA, CMCNet
Scientific Consultant
Institute for Natural Sciences
Kolkata
(West Bengal)

D P H

DISCOVERY PUBLISHING HOUSE PVT. LTD.
NEW DELHI-110 002

Published by:
Tilak Wasan
DISCOVERY PUBLISHING HOUSE PVT. LTD.
4383/4A, Ansari Road, Darya Ganj
New Delhi-110 002 (India)
Phone : +91-11-23279245, 43596064-65
Fax : +91-11-23253475
E-mail : parul.wasan@gmail.com
discoverypublishinghouse@gmail.com
web : www.discoverypublishinggroup.com

***First Edition:* 2012**

ISBN: 978-81-8356-874-6 (Set)

Encyclopedia of Inorganic Chemistry

Printed at:
Shree Balaji Art Press
Delhi

Preface

Chemistry is limitless subject but full of experiments and surprises, which is interesting indeed, especially for a new student, though some mathematical and theoretical calculations are there, in spite of all these, chemical world is world of lure.

The origin of chemistry is very deep-rooted and comes from a very early time, the time of the mediaeval history rather from the time of the Alchemists.

Egyptians, whom are thought to be the father of the alchemistry, used the word "kimiya". Kimiya moved from Egypt to Arabia and it flourished in whole world, in England it was named as "Alchemistry" (in India it was named as "*Ap-Rasayana*"). From this alchemistry our modern chemistry has been developed.

Medicines, drugs, paints etc. are all chemical products. Some processes such as distillation, fractionation, and purification, which we use widely now in modern chemistry are all have been given by the ancestors of chemistry, the Alchemistry.

The Modern Periodic Table developed by Dimitri Ivanovich Mendeleev is the modernized systematic formed of the elements (among them some were discovered by the Alchemists). The original table consists of only 107 elements but in this book all 119 elements have been discussed.

In this modern world of chemistry when I were engaged in my chemical lab, I were thinking that there should be something from which a student of chemistry will easily get all the information of chemistry.

Also a man who is anxious of chemistry, will start loving this subject, so to make the entrance of this chemical world easy and nice, I thought to write a book on chemistry, providing all the necessary information about chemistry [though the inorganic part].

While writing this book, I must first thank to the Supra-cosmic force for giving me the power to write this book, after this I must write the name of that person (My Lord) BABA

(Swami Arupanandaji), without whom the book would have been remained as an ordinary work.

And my respected Fairy Godmother my beloved Mummy (Mrs. Dianna Robinson) who gave me different ideas and thoughts.

While writing this book, some informations was needed and that is obvious so here comes the name of Dr. Paul Anderson, (my friend) Director Research Coordinator IUPAC, without whom the book would not have been finished.

I got an immense help from Dr. Palash Gangopadhyay School of Chemistry Central University, Hyderabad for the drawings and different applications such as CHEMSketch, Origin 6.0 and MathLab.

Also must thank to Mr. Dipak Dey and Mr Dipankar Dey of D.K. Scientific, (C-51 College Street Kol-7) for supplying the instruments and chemicals to me.

I am greatful to my three students Miss Ria Banerjee, Miss Moumita Banerjee and Miss Subhomita Banerjee for taking this colossal effort to publish this book.

And last but not the least my father, mother (for the computer) grand-mother and all other members of my family to finish this entire project.

At the end I must only appeal to my audience that, if you are interested in this book that will be my actual reward, I think and also do believe that, if any person is interested in the experimental chemistry, then from the guidance of A chemical Analyser's guide he or she will be able to set up his or her own chemistry lab). If you find any thing interesting at all then I will find that golden opportunity of sharing the secret of divine happiness and joy.

SANANDA CHATTERJEE

Contents

CHAPTER

21

Chemical Equilibrium and Kinetics

Introduction

In cold winter the work and the activities of a human being slows down specially its muscular activities in order to speed up, he must worm up himself. That is why the sports men are set for a worm up schedule before starting the game. This worm up enhances the chemical reactions which on the other hand stimulates the muscular activity they by increasing the ability of the individual. The speed of the chemical reaction also plays an important role in many other situations. The instructions for a test strip that medical laboratories use to test the presence of glucose (sugar) in the blood and the other factors in the urine state. "Proper read time is critical for optimal results". The glucose test is read 30 seconds after dipping the test strip in a urine sample and the blood test at 60 seconds.

The test for urine glucose is based upon two sequential reactions that are catalysed by enzymes. In the first reaction, glucose is converted into gluconic acid (a carboxylic acid) and hydrogen peroxide. In the second reaction the hydrogen peroxide reacts with potassium iodide to give colours ranging from green to brown on the concentration of glucose present. The blood testis based on a reaction of peroxide with a coloured indicator, and the reaction is catalysed by the hæmoglobin in blood. The colour that revels many range from orange to green, depending on the concentration of blood present. The timing of these tests is critical because they are based on the speed of the reaction that produce the colour.

Whenever we plan to run a chemical reaction, we should ask at least two questions, one is will the reaction produce the desired products in useful quantities? We can answer this question by making equilibrium and thermodynamic calculations, as ever going to discuss in this chapter, or we can use calculations, as ever are going to discuss in this chapter. We can use quantitative considerations such as the fact that the reactions of an acid with a base, yields a salt. The second question that occurs is "How rapidly will the reaction occur?" A reaction that takes 50 years to produce a product is as useless as one that not produces a product at all.

This chapter also discusses the answer of the question the rate at which it produces chemical compounds from respective reactions (chemical kinetics). We will examine factors that influences the rates of chemical reactions, mechanisms by which reactants produced and the quantitative techniques used to determine and to describe the rates of chemical reactions, mechanisms by which reactants produced. The quantitative techniques used to determine and to describe the rate at which reactions occur.

Rate of Reaction

The rate of reaction means that the time taken by the reaction to occur. We can determine the rate of reaction by measuring the change in the concentration of a reactant or a product per unit time. We could also measure a change that is proportional to the concentration such as a change in the pressure of a gas. In some cases the colours of the chemical reaction changes with the time and this too can be related to the concentration changes. However, we must be sure that changes we measure is due only to the reaction and does not reflect a change in volume of the system (which would also change the concentration) or the presence of one or more comparing reactions. When we measure the concentrations of hydrogen peroxide decomposes according to the reaction below :

$$\text{Hydrogen Peroxide} \longrightarrow \text{Water} + \text{Oxygen}$$

$$2H_2O_2 \longrightarrow 2H_2O + O_2$$

The change in concentration with time for such solution is shown in the Fig. 21.1 and in Table 21.1 below :

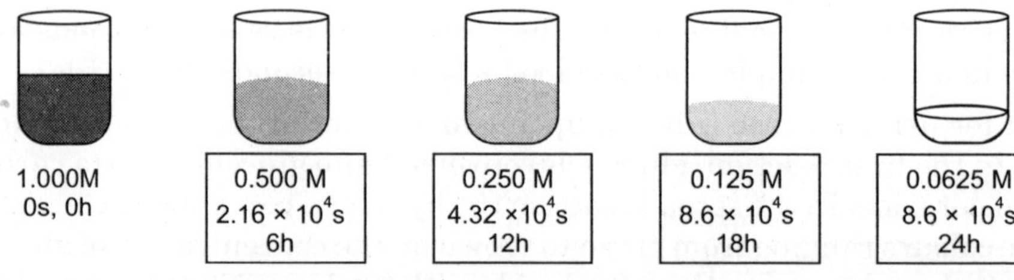

Fig. 21.1. The decomposition of Hydrogen Peroxide

Times, sec	H_2O_2 mol^{-1}	H_2O_2 mol L^{-1}	$\Delta^t S$	Rate of Deomposition
0	1000	0.500	2.16×10^4	2.3×10^5
2.16×10^4	0.500		2.16×10^4	1.16×10^5
		0.250		
4.32×10^4	0.250	–0.125	2.16×10^4	5.79×10^{-6}
6.48×10^4	0.125		2.16×10^4	2.9×10^{-6}
8.64×10^4	0.0625	–0.0625		

$$\text{Rate of Decomposition of } H_2O_2 = \frac{\text{Change in Concentration of Reactant}}{\text{Time Interval}}$$

$$= \frac{[H_2O]_{t_2} - H_2O_{t_1}}{t_2 - t_1} = \frac{H_2O_2}{\Delta^t}$$

We have used brackets to indicate molar concentration and the symbol Δ (delta) to indicate the rate of change. This $[H_2O_2]_{t_1}$ represents the molar concentration of hydrogen peroxide at some time $t_1[H_2O_2]_{t_2}$ represents the change in molar concentration of hydrogen peroxide during the interval Δt (that is $t_1 - t_2$). The negative sign is used to convert the change concentration of hydrogen peroxide to a positive rate.

A change in the number of moles of reactant or product in a litre of volume during some unit of time has units of moles per litre per second; which may be written mol L^{-1} S^{-1} because L^{-1} is identical to molarity, sometimes we see MS^{-1} (molar per sec).

Factors that Effect Reaction Rate

The rate at which the reactants are consumed and products are formed during chemical reactions vary, gently. We can identify four factors that affect the state of subdivision of reagents, the temperature of the concentration of the reactants and the presence of a catalyst.

The Nature of the Reacting Substance : The rate of a reaction depends on the nature of the participating substance. Any simple combination of two simple ions of appropriate charge usually occur very rapidly; the reaction of an acid with base is an example :

$$H_3O^+ + OH^- \longrightarrow 2H_2O$$

Water form as rapidly as we can mix solutions of hydrochloric acid and sodium hydroxide, the acid base reaction is almost instantaneous.[1] Water forms when hydrogen peroxide decomposes.

$$\text{Hydrogen Peroxide} \longrightarrow \text{Water + Oxygen}$$

$$2H_2O_2 \longrightarrow 2H_2O + 2O_2$$

However, the reaction is much slower than the acid base reaction; in some cases it can take months to occur. Reactions that involve the breaking of chemical bonds or the re-arrangements of molecules commonly are slower than reaction that involve the simple combination of two ions of opposite charge.

Similar reaction may have different rates under the same conditions if different reactants are involved, for example, when small pieces of the metals iron and sodium are left in air, the sodium reacts immediately and completely in air, while the iron is barely affected. The active metal calcium and sodium both react with water forming hydrogen and a base, but the calcium react moderately but the reaction with sodium is so violent that it explodes.

---

[1]Without any break or immediately.

The State of Subdivision of the Reactants : Except for the substance in the gaseous state or in solution, reaction occurs at the binary or interface, between two phases. Hence, the rate of a reaction between two phases depends largely upon the surface contact between them. Finely divided solids reacts rapidly because of their large available surface area than that of the same solid found in a big block.

For example; If a block of lycopodium powder is tried to ignite, it will not catch fire, but the same block if converted to powder, and blown over a candle flame explodes immediately.

Temperature of the Reactants : Chemical reaction proceeds faster at higher temperature. Wood will react with air to scald at ordinary temperature but at higher temperature it reacts fast enough to burn. Foods cook faster at higher temperature than at lower temperature. We use a burner or a spirit lamp to increase the speed of the reaction by heating the reaction mixture.

In many cases, the rate of a reaction in a homogeneous system is approximately doubled by an increase in temperature of only 10°C. This rule is a rough approximation; however and apply to those reactions which last only for a second.

Concentration of Reactants : The rates of reactions depend upon concentration of the reagents. Rates usually increase when the concentration of one or more than one of the reactants increase, for example the rate at which the lime stone $CaCO_3$ decomposes as a result of reaction with the pollutant sulphur dioxide in the air. Similarly the phosphorous burns much more rapidly in an atmosphere of pure oxygen than air, which is only about 20% of oxygen.

Presence of Catalyst : We have seen that hydrogen peroxide foams when it is applied to a cut, because an enzyme in a body catalyses its decomposition. A catalyst as we know that it does not take part in a chemical reaction but changes the rate of the chemical reaction.

If we run a reaction in a closed system so that the products cannot escape we often find that the reaction does not give a 100% yield of products. Instead some reagents remains after the concentrations of the reactants and products have stopped changing. At this point when there is no further change in concentrations of both the reactant and the product, we say that the reaction is at equilibrium.

A mixture of reactants and products is present in equilibrium. For example, when we place a sample of nitrogen dioxide (NO_2 a brown gas) in a flask, it forms a colourless gas called dinitrogen tetraoxide (N_2O_4)

$$2NO_2(g) \longrightarrow N_2O_4$$

The colour becomes lighter as NO_2 is converted to colourless N_2O_4 however, even after the concentrations of reactants and product stop changing, the brown colour of the mixture of the gases does contain some nitrogen dioxide remains.

The formation of N_2O_4 is a reversible reaction, in which reaction combine to form reactants. Not only can NO_2 react and form N_2O_4 but the N_2O_4 produced can also decompose to form NO_2. As soon as the forward reaction produces any N_2O_4, the reverse reaction begins and produces NO_2 by decomposition of N_2O_4. At equilibrium, the concentrations of N_2O_4 and NO_2 no longer change, because the rate of formation of NO_2 is exactly equal to the rate of formation of N_2O_4. Careful investigations have shown that equilibrium is a dynamic process; equilibrium is established in a system when reactants combine to form products at the same time product combine to form reactants.

Reaction Quotients and Equilibrium Constants

A general equation for a reversible reaction format can be written as

$$...mA + nB \longleftrightarrow xC + yD + zE$$

$$\therefore \text{ The reaction quotient } Q = \frac{[C]^x\,[D]^y\,[E]^z}{[A]^m[B]^n\,[C]^o}$$

When we use brackets to define molar concentrations "of". The reaction quotient is a ratio of molar concentration of the products of the chemical equation (multiplied together) and of the reactants (also multiplied together) each raised to a power equal to the co-efficient proceeding that substance in the balanced equation. The reaction quotient for reversible reaction.

$$2NO_2(g) \longleftrightarrow N_2O_4(g)$$

$$\text{The reaction quotient of this reaction is } Q = \frac{N_2O_4}{NO_2}$$

The numerical value of Q for a given reaction varies. It depends on the concentration of products and reactants present at the point. As the reaction proceeds, the value of Q increases as the concentration of the reactants simultaneously increase (see Fig. 21.2). The example selected here is the conversion of sulphur trioxide from sulphur dioxide and oxygen.

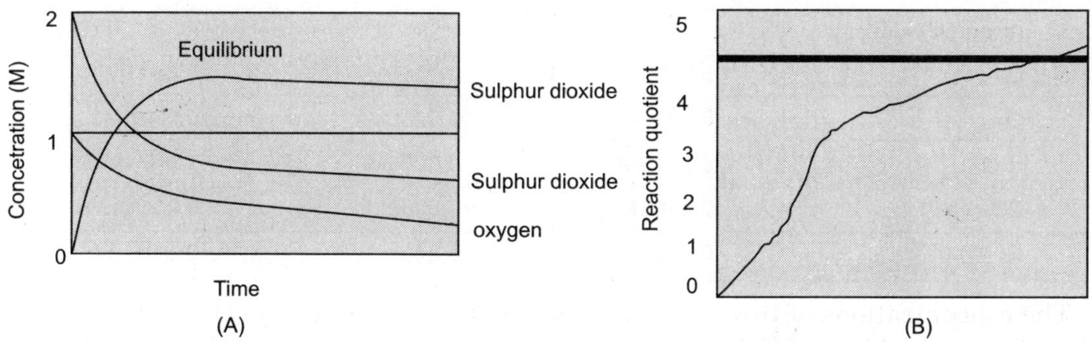

Fig. 21.2.

When the reaction reaches equilibrium the value of Q (reaction quotient no longer changes).

When a mixture of reactants and products of a reaction reaches equilibrium at a given temperature its reaction quotients always has the same value. This value is called the equilibrium constant, K of reaction at that temperature. When a reaction is an equilibrium constant K, for the reaction at that temperature.

The mathematical equation is

$$Q = K - \frac{[C]^x [D]^y}{[A]^m [B]^n}$$

Which is a statement of the law of mass action when a reaction has attained equilibrium at a given temperature the reaction quotient for the reaction always has the same value.

It is important to remember that an equilibrium can be established either starting from reactants or starting from products, for example equilibrium was established from mixture 2 in the Table 21.2 shown below when the products of the reaction were heated in a closed container. In fact, as one technique that is used to determine whether a reaction is truly at equilibrium is to approach equilibrium starting with reactants in one experiment and starting with products in another. If the same value of the reaction quotient is observed when the concentrations stop changing in both experiments, then we are certain that the system has reached equilibrium.

Fig. 21.2. The decomposition of Hydrogen Peroxide

Before Reaction	Mixture 1	Mixture 2	Mixture 3
CO	0.0243M	OM	0.0094M
H_2O	0.0243M	OM	0.0055M
CO_2	0.M	0.0468M	0.0005M
H_2	OM	0.0468M	0.0046M
At equilibrium			
O	0.013M	0.0260M	0.0074M
O	0.0315M	0.2260M	0.0260M
H_2O	0.0315M	0.0260M	0.0074M
CO_2	0.0108M	0.0260M	0.006M
Q(=K)	0.640	0.640	0.640

The concentrations of three mixtures before and after reaching equilibrium and the value of Q at equilibrium at 100°C

$$CO + H_2O \longleftrightarrow CO_2(g) + H_2(g)$$

Homogeneous and Heterogeneous Equilibria

We can divide equilibria into homogeneous and heterogeneous equilibria. However each of these classes can also be further subdivided according to the types of reactions involved. It should be noted that all the equilibria have some basic common features.

1. The reaction quotient for all reversible reactions

$$...mA + nB........ \longleftrightarrow xC + yD + ... \text{ is written as}$$

$$Q = \frac{[C]^x + [D]^y.............}{[A]^m + [B]^n.............}$$

2. No matter how a reaction gets equilibrium, at a given temperature. Its value Q is always is equal to the equilibrium constant for the reaction at the room temperature.

Homogeneous Equilibria: A homogeneous equilibria with a single phase, the reaction shown is at homogeneous equilibrium.

$$CO + H_2O \longleftrightarrow CO_2 + H_2$$

Homogeneous Equilibrium: Reaction between dissolved reactants that produce dissolved product gives one type of homogenous equilibrium. The chemical species involved can be molecules ions or a mixture of both, several examples and the expressions for their equilibrium constant follow :

$$C_2H_2 + 2Br_2 \longleftrightarrow C_2H_2Br_4$$

$$K = \frac{[C_2H_2Br_4]}{[C_2H_2][Br_2]^2}$$

(*a*) $I_2\,(eq) + I^- \longleftrightarrow I_3\,(aq)\ K = \dfrac{[I_3^-]}{[I_2][I^-]}$

(*b*) $Hg^{2+}\,(aq) + NO_3^- + 3H_3O^+\,(aq) \longleftrightarrow 2Hg^{2+}\,(aq) + HNO_2$

(*c*) $HF(aq) + H_2O\,(l) \longleftrightarrow H_3O^+ + F^-(aq)\ K = \dfrac{[H_3O]}{[HF]}$

(*d*) $NH_3(aq) + H_2O(l) \longleftrightarrow NH_4^+\,(aq) + OH^-\,(aq)\ K = \dfrac{[NH_4^+][OH]}{[NH_3]}$

In the last example, water is the solvent, so we do not write it in the expression for the reaction quotient, even through it is a product or a reactant.

Some of the homogeneous equilibria are so common that they have industrial names and their equilibrium constants have special symbols.

Homogeneous Equilibria among Gases: Gaseous reactants that produce gaseous products also give equilibria, we use molar concentrations in the expressions for the equilibrium constants in the following examples. But we will see shortly that partial pressure of the gases could be used as well.

$$C_2H_6(g) \longleftrightarrow C_2H_4(g) + H_2(g) \qquad K = \frac{C_2H_4[H_2]}{C_2H_6}$$

$$3O_2 \longleftrightarrow 2O_3(g) \qquad K = \frac{[O_3]^2}{[O_2]^3}$$

$$N_2(g) \longleftrightarrow 3H_2(g) \longleftrightarrow 2NH_3(g) \qquad K = \frac{[NH_3]}{[O_2]^3}$$

$$C_3H_8(g) \longleftrightarrow 3H_2(g) \longrightarrow 2NH_3(g) \qquad K = \frac{[NH_3]^2}{[N_2][H_2]^3}$$

$$C_3H_8(g) \longleftrightarrow 5O_2(g) \longrightarrow 3CO_2 + 4H_2O \qquad K = \frac{[CO_2]^3\,[H_2O]^4}{C_3H_8 + [O_2]^5}$$

It must be noted that the connection of water has been included in the last example because water is not a solvent in gas-phase reduction.

When ever the gases evolve in a reaction the pressure of each gas can be used (instead of concentration) in the equation for the reaction quotient or equilibrium constant because the pressure of a gas is directly proportional to its concentration at constant temperature. This relationship can be derived from the ideal gas reaction.

$$PV = nRT$$

The molar concentration of the gas, designed C, is the number of moles, n, is exactly 1 litre of the gas. Substituting 1L for V and rearranging the equation we get.

$$P(L) = nRT$$

$$P = (n/1L)RT$$

$$P = CRT \text{ (Since } n/1L=C)$$

Because R is constant, the quantity RT is constant when the temperature does not change. Thus, at constant temperature the pressure of a gas is directly proportional to its concentration.

Or
$$P = \text{Constant} \times C$$

For the equation

$$C_2H_6 \longrightarrow C_2H_4 + H_2$$

We can write the reaction quotient, using the gas partial pressure,

$$Q_p = \frac{(PC_2H_4)(PH_2)}{PC_2H_6}$$

In this equation we use Q_p to indicate a reaction quotient written with pressure PC_2H_6 is the partial pressure of C_2H_6, PH_2 is the partial pressure of C_2H_4

At equilibrium

$$K_p = Q_p = \frac{(PC_2H_4)(PH_2)}{PC_2H_6}$$

The subscript p in the symbol K_p designates an equilibrium constant found for the same reaction by using concentrations.

Conversion between a value for K and equilibrium constant expressed in terms of concentrations and a value K_p an equilibrium constant expressed in terms for pressure is straightforward. The relation between K and K_p.

$K_p = K (RT)\Delta_g$ where Δ_g is the difference between the sum of the co-efficients of the gaseous products and the sum of the co-efficients of the gaseous reactants of the reaction.

For the gas-phase reaction $mA + nB \longrightarrow xC + yD$

We have $\Delta_g = (x + y) - (m - n)$

The equation reacting K and K_p is therefore.

$$K_p = \frac{(P_c)^x (P_n)^y}{(P_A)^m (P_B)^n}$$

$$= \frac{(C_c \times RT)^x (D_c \times RT)^y}{(C_A \times RT)^m (C_B \times RT)n}$$

$$= \frac{(C_C)^x (C_D)^y \times (RT)^y}{(C_A)^m (C_B)^n \times (RT)^x}$$

$$= \frac{(C_C)^x (C_D)^y \times (RT)^{x+y}}{(C_A)^m (C_B)^n \times (RT)^{m+n}}$$

$$= K (RT)^{(x+y)=(m+n)} \text{ or } K(RT)\Delta_g$$

Heterogeneous Equilibria : A heterogeneous equilibrium is an equilibrium in which reactants and products are found in two or more phases. The phases may be any combination of solid phases, gas phases and solutions. When dealing with equilibria, it must be kept in mind that the activities of pure solids, pure liquids and solvents, so there species do not appear in reaction quotients or equilibrium constants.

Some heterogeneous equilibria involve chemical changes. Examples including:

$PbCl_2(s) \longleftrightarrow Pb^{2+} (aq) + 2Cl^-(aq)$ $\quad K = [Pb^{2+}] [Cl^{-2}]$

$CaO(s) + CO_2(g) \longleftrightarrow CaCO_3(s)$ $\quad K = 1/[CO_2]$

$C(s) + 2S(g) \longleftrightarrow CaCO_3(s)$ $\quad K = [CS_2]/[S]^{2-}$

Changes in Concentration at Equilibrium Le Chatelier's Principle

When a system is at equilibrium we can change its position of equilibrium (that is compounds of reactants and products present) by changing the presence in certain gaseous system, or the temperature of the system. The result of each change can be predicted from Le Chartelier's principle. *When a stress (such as a change in concentration temperature and pressure) the equilibrium shift both the ways to minimise the stress.*

Effect of Change in Concentration and Equilibrium : Sometimes we can change the position of the equilibrium by changing the pressure on a system. However the changes in pressure have a measurable effect only in systems where gases are involved and then only when the chemical reaction produces a change in the total numbers of gas molecules in the system.

As we increase the pressure of a gaseous system at equilibrium, either by decreasing the volume of the system or by adding more of the equilibrium mixture, we introduce Le Chartelier's principle chemical reaction that reduces the total number of molecules per unit of volume. According to Le Chartelier's principle chemical reaction that reduces the total number of molecules per unit volume will be favoured by a decrease in pressure.

Consider what happens when we increase the pressure on a system. On which NO, NO_2 and O_2 are in equilibrium.

$$2NO(g) + O_2(g) \longleftrightarrow 2NO_2(g)$$

The formation of additional amounts of NO_2 decreases the total number of molecules in the system, because each time two molecules of NO_2 form, a total of three molecules of NO and O_2 react. This reduces the total pressure exerted by the system and reduces but does not completely relieve, the stress of the increased pressure. On the other hand, decrease in the pressure on the system favours decomposition of NO_2. Let us now consider the reaction below :

$$2NO(g) + O_2(g) \longleftrightarrow 2NO\ (g)$$

Because these is no change in the total number of molecule in the system at equilibrium has a different effect. A change in temperature changes the value of the equilibrium constant. However, we can predict the effect of the temperature changes by treating it as stress in the system an applying Le Chartelier's principle :

When hydrogen reacts with iodine, energy is released and heat is evolved.

$$H_2(g) + I_2(g) \longleftrightarrow 2HI(g) \qquad \Delta H = -9.4kJ \text{ (exothermic)}$$

Because this reaction is exothermic, we can write it with heat as a product

$$H_2(g) + I_2(g) \longleftrightarrow 2HI(g) + 9.4kJ$$

Increasing the temperature of the reaction increases the amount of energy present. Thus increasing the temperature has the effect of increasing the amount of one of the products of this reaction. The reaction shifts to the left to relieve the stress of the reaction and there is an increase in concentration of H_2 and I_2 and a reduction in concentration of HI. Lowering the temperature of the HI system increases the equilibrium constant from 67.5 at 357°C to 50.0 at 400°C.

At equilibrium at the lower temperature; the concentration of HI has increased and the concentration of H_2 and I_2 have decreased. Raising the temperature decreases the value. When we change the temperature of this system, it reduces the amount of energy present the formation of hydrogen iodide. When we change the temperature of a system at equilibrium, equilibrium constant for the reaction changes. Lowering the temperature in the HI system increases the equilibrium, at the lower temperature, the concentration of hydrogen iodide has increased and the concentration of H_2 and I_2 has decreases. Raising the temperature decreases the value of the equilibrium constant from 67.5 at 357°C to 50.0 at 400°C

$$2NO_2(g) = N_2O_4(g) \qquad \Delta H = -57.20kJ$$

The negative value of ΔH tells that the reaction is exothermic and could be written

$$2NO_2(g) = N_2O_4(g) + 57.20kJ$$

At higher temperature the gas mixture has a deep brown colour, which indicates a significant amount of brown NO_2 molecules. If, however we put a stress on the system by cooling the mixture (with-drawing energy), the equilibrium shifts to the right to supply more energy lost be cooling. The concentration of colourless N_2, decreases, causing the brown colour to fade.

Effect of Catalyst on Equilibrium : A catalyst has no effect on the values of an equilibrium constant or on equilibrium concentration. The catalyst mearly increases the rates of both the forward and the reverse reactions to the some extent so that equilibrium is reached more rapidly.

All of these effects change in concentration or pressure, change in temperature and the effect of catalyst on a chemical equilibrium - play an important role in the industrial synthesis of ammonia from nitrogen and hydrogen according to the equation.

$$N_2 + 3H_2 \longleftrightarrow 2NH_3$$

Chemical Equilibrium and Thermodynamics

Thermodynamics means changing due to temperature (therm = heat, dynamics = movement). In chemistry there is a field which studies the different properties of chemical substance, which exhibits some astonishing properties. When temperature is applied, some reactions absorb heat while some produce heat, and some explode with the application of temperature. Here we are going to discuss the different aspects of thermodynamics, laws governing thermodynamics energy entropy, enthalpy to work.

Definition of Energy : Energy is defined as any property, which can be produced from or converted to work.

Heat and Energy

Heat is energy in Transit.

In view of the definition energy given above, heat must be regarded as a form of energy, since most of work can be readily converted into heat and heat can be at least partially converted into work. Heat is produced from mechanical work for instance, by means of friction and electrical work is transferred into heat by the passage of electricity through a resistance. On the other hand by means of suitable machine such as heat engine, a certain amount of heat can be converted into work. From the standpoint of thermodynamics heat is energy in transit. It is the form in which energy is transferred from one body to another, either by direct contact or by means of radiation, as the result of a difference of temperature. In evaluating the quantity of heat energy that has passed to a given body, the intensity factor is the change of temperature and the capacity factor is the heat capacity, the product of these quantities which can be stated in calories. If required is calorie is a measure of the heat energy transfer.

Heat Capacity

As a general rule, heat is expressed in terms of a unit known as *calorie,* which is defined as the quantity of heat required to raise temperature of 1 gm of water by 1°C in the vicinity of 15°C. The octal temperature is specified because the quantity of heat that must be supplied to raise the temperature of a given amount of water, or of any other substance for that matter, by 1°C depends to some extent in the temperature itself.

The heat capacity of a capacity of a body is the property which multiplied by temperature change gives the quantity of energy which has entered or left the body as heat when it is brought into contact with another body having a different temperature. Thus, if the temperature, of the body is raised fro T_1 to T_2 by the passage to it of an amount of heat capacity C of the body is given by

$$Q = C(T_1 - T_2)$$

According to this equation,

$$C = \frac{Q}{T_2 - T_1}$$

So, that the heat capacity is sometimes defined as the quantity of heat required to raise the temperature of the body by 1°. From the definition of the calorie given above, it is obvious that heat capacity of 1 gm of water in the vicinity of 15° is equal to 15°C calorie.

Heat capacity is the property proportional to the quantity of matter present and this must consequently be stated. Two particular quantities are commonly employed. The heat capacity is frequently referred to 1 gm of material is then called the specific heat. Thus, the specific heat of water is exactly 1 calorie per degree per gram in the region of 15°C. From the standpoint of the chemist a more useful form of the heat capacity is that referred to 1 mole, that is, the molecular (or formula) weight in grams. The quantity of heat is then known as the *molar heat capacity.* As defined in the above equation is thus the mean value for the temperature range from T_1 to T_2. In order to define the heat capacity at a given temperature it is necessary to make the temperature internal as small as possible, thus if the temperature increases $T_2 - T_1$ is referred by Δt, the true heat capacity is given by the equation.

$$C = \lim \frac{Q}{T_1 - T_1^{\Delta t}}$$

Since, the quantity of heat Q is usually stated in calories and Δt in degrees it is evident that the dimensions of heat capacity are calories/degree for that given amount of material, e.g., cal deg^{-1} or deg^{-1} mol^{-1}, that is calories per degree per gram. (or per mol).

It will be seen later that the heat capacity of a substance can have definite value only under certain specified conditions. The two most important are constant pressure and constant volume. It is necessary to indicate therefore the particular conditions under which heat is transferred to or from the given substance.

Equivalence of Work and Heat

The relationship between mechanical work and heat was first clearly seen be Count Rutherford in 1709. As a result of his observations made as the heat developed during the boring of cannon, he calculated that heat produced was related to the mechanical work done in the boring process. Some experiments were carried out H Davy (1799) appeared to indicate the connection between work and heat, but the most important result was concluded by *J.P. Joule* in an extended series of observations, commanded at 1840 and lasting for nearly 40 years. In a number of experiments Joule converted known number of work into heat and measured the amount of heat thus produced by determining the rise in temperature of a calorimeter of known heat capacity. Among the methods used for converting heat into work, mention may be made for the following. Stirring water or mercury by paddle wheels, compression or expansion of air falling water through capillary tubes, passage of an electric current through known resistance and passage of induced current through a coil of wire rotated in a magnetic field. As a result of his studies joule came to the highly significant conclusion that the expenditure of a given amount of work, neglecting the origin, always produced the same quantity of heat.

This fact is the basis of concept of a definite and constant ratio between the number of ergs, of mechanical work done and the number of calories produced by the conservation of this work into heat. According to the most resent experiments, one standard calorie (at 15°C) is equivalent to 4,1858 absolute joule of work, irrespective of the source of nature of the work. However, because heat quantities are decimal at the present time by comparison with the heat produced by electrical work, the value of, which is known in the internal joule it, has become the practice by chemists to adopt the relationship between work and heat. Thus a defined calorie is equal to 4.1850 internal joules.

The Conservation of Energy the First Law of Thermodynamics

The belief that perpetual theory of motion" is the production of energy of a particular type without the disappearance of an equivalent energy of another form. This was not possible and has long been accepted by scientists. No success had attended to attempts for construct a machine, which would produce mechanical work continuously without itself undergoing any change. The fundamental significance of this accepted view was not widely realised until 1847 when *H. Van Helmooltz* showed that the failure of active perpetual motion and the equivalence of work and heat described above were aspects of a wide generalisation. This has become known as the Law of Conservation of Energy. Which states that although energy can be transferred from one form of energy to the other form of energy, that is when one form of energy is produced the same form of energy must be consumed from other source.

Isolated System and the First Law

It was stated earlier that a combination of a system and its surroundings might be referred as an isolated system. The first law of thermodynamics requires that the total energy of an isolated system must remain constant, although there may be changes form one form to another. This means that any loss or gain of energy by a system must be exactly equivalent to the gain or loss respectively of energy by the surroundings. The forms of energy are not necessarily the same, but if there is to be a net conversion of energy the amounts must be equivalent. The conclusion desired from the first law that that of its surroundings must exactly compensate the energy change in a system is of great importance for the future chemists.

Energy and Heat Changes

The Energy Content : The energy possessed by a system, that is to say, its energy content, may be regarded as falling into main categories itself, this may include the translational energy of the moving molecules, as well as the energy of the electrons and the nuclei. On the other hand there is the energy which is determined by the position of the system in a force field, e.g., magnetic field or electrical or gravitational and by the motion of the system as whole, e.g., through space. In thermodynamics the energy in this second category is usually ignored and it is only that in the first which is taken in consideration. The latter is sometimes called internal energy, to separate from the it form the energy due to external factors but for present purpose it may be referred to simply as the energy of the energy content of the system. The term internal energy associated with the motions, e.g., vibration rotation etc. within the molecule, as distinct from the translational energy resulting from the motion of the molecule as a whole.

The motion of the molecule is of course not to be confused with the motion of the system as a whole, as stated is not included in thermodynamics energy content.

The energy content of the system must be depended upon its thermodynamic state. An increase of temperature at constant volume, for example brought about by the transfer for heat to the system from the surroundings must result in an increase of its energy. It will now be shown by utilising the first law of thermodynamics that the energy content is a property of the system of the kind that is determined only by the state of the system and not by the manner in which it reached that state. In other words it will be proved that the energy is a single-valence function of the thermodynamic variable of the system.

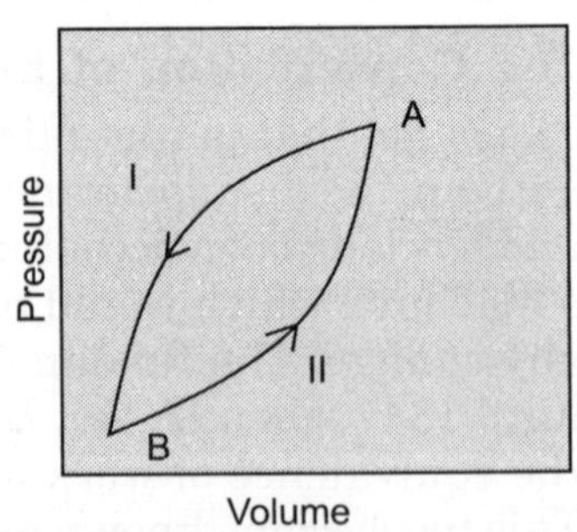

Fig. 21.3. The energy changes in direct and reverse paths

Consider any system represented by the state A in the Fig. 21.3, in which the co-ordinates are the observable properties, e.g. pressure, volume, that determine the energy content. Suppose conditions are now altered along the path I so that the state of the system is represented by the point B; the system is then returned to the exact point of origin by a series of changes indicated by the path II. It is then a direct consequence of the first law of thermodynamics that the total energy change of the system in part I must be identical in the magnitude, but opposite in sign to that in part II provided the surroundings remain unchanged. If this were not a perpetual motion the machine would be possible.

In thermodynamics the word change is frequently used to imply change in dynamic state, then the term path and process then refer to the means, whereby the given change in state is accomplished as well be seen below. A particular change may be achieved by following various different paths that is by different process.

Imagine for a system, that the increase of energy in the path I were greater than the decrease in returning by II, then by carrying out the change A → B by path I, must be numerically equal to that involved in the decrease in the reverse process B → A by path II.

It is possible to proceed from A to B by various path I, I, II etc. (See Fig. 21.4) that in II.

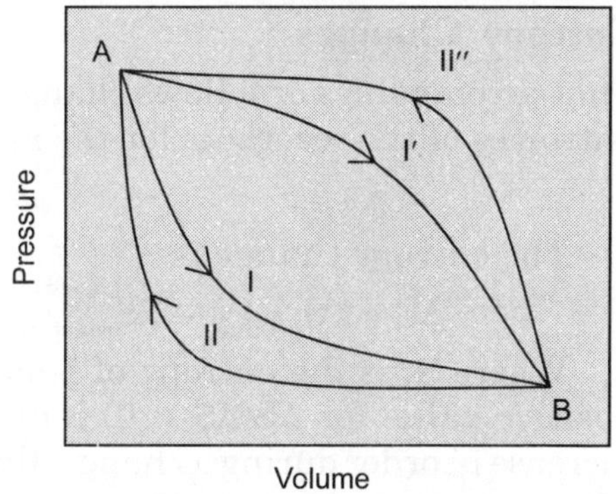

Fig. 21.4. The Energy change in the indipendant path

It is clear from the considerations that the energy of a system is a "property" of the system. It is, therefore, possible to format a definite value E to the energy or energy content of a particular system in a given state, this value will depend only on the state and not on the previous history, of the system. If E_A represents the energy in the thermodynamic state A and E_B that is the state B, then the increase of the energy ΔE in passing from A to B given by

$$\Delta E = E_B - E_A$$

And is independent of the path taken in the change of state A → B.

The system delta is applied to a change in a thermodynamics property accompanying a given process, represent the algebraic increase in the property. It is always equal to the value in the final state minus that in the initial state defined in this manner is positive. The property G has a larger value in the final than in the initial state. On the other hand, if ΔG is found negative, G is larger in the initial stage. The significance of –Δ is, of course, the reverse of ΔG.

Since the energy content of a system obviously depends on the quantity of material contained in the system it is apparent that the energy E is an extensive property. If the mass of the system is altered, the energy content will be affected in the same proportion. Similarly, the value of ΔE for any process depends upon the amount of material contained in the system, which undergoes the change.

Entropy

It is the amount of disorder of a substance (or a system) can be determined quantitatively is also called Standard Entropy (°S).

The standard entropy of a substance is the entropy content, perfectly equal to 1 mol of the substance at standard state condition (25.1°C and 1 atmospheric pressure). It must be noted that the solids have lower entropies and liquids have higher entropies, even gases have higher entropies than liquids. That in the same physical state the substance with different complicated molecules have more atoms per molecule to move about, they can exhibit greater disorder thus they have higher entropies. Hard crystalline substances like diamond, tend to be more ordered and have lower entropy than softer materials such as graphite. (This also happens in the case of three well known acids such as nitric acid, sulphuric acid and hydrochloric acid) among them the sulphuric acid has the lowest entropy, which does not break easily, so it is used for acidifying other chemical substances.

Entropy Changes

That accompanies or follows changes in a system in the difference between the sum of the entropies of the reactants for the reaction.

$$..mA + nB \longrightarrow xC + yD$$

The entropy change is

$$\Delta S = \Sigma S_{product} - \Sigma S_{reactant} = [(x \times S_C) + (y + S_D) - (M \times S_A) + (n \times S_B)]$$

Where S_A is the entropy of 1 mole of A, S_B is the entropy of 1 mole of B and so on. A positive value for $\Delta S(\Delta S < 0)$ indicates that there is a decrease in disorder that is an increase in order during a change; this is an negative value of ΔS which indicates a situation of non-spontaneous reaction.

Enthalpy Changes and Free Energy Changes

Generally the amount of heat q, that a system absorbs or evolves a it changes is not a state function, rather it varies with the way the process occurs.

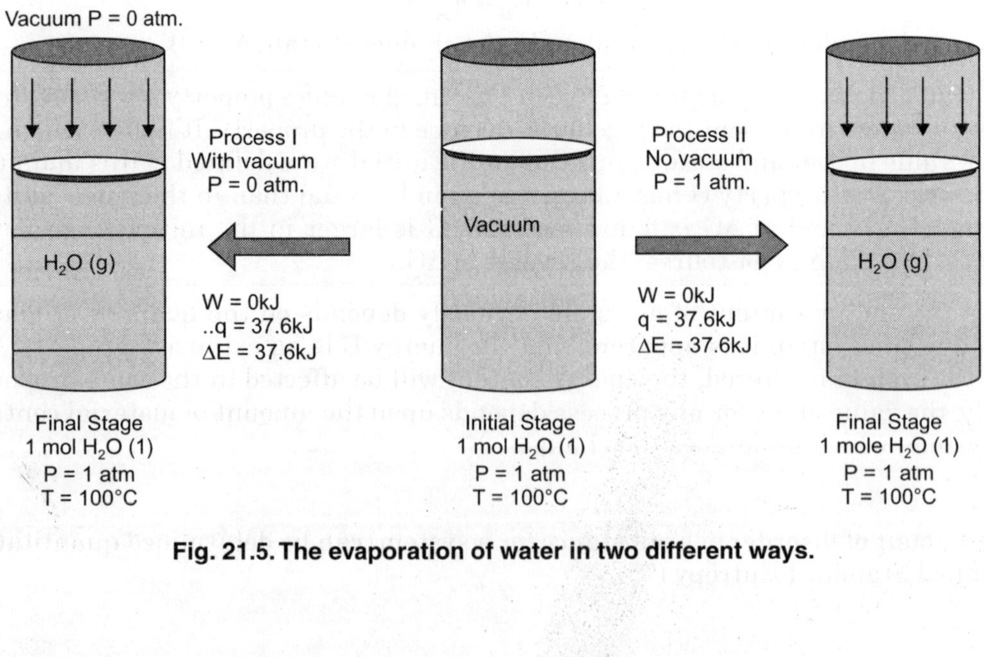

Fig. 21.5. The evaporation of water in two different ways.

However, if the change occurs in such a way that the only work done is to effect a change in volume of the system at constant pressure, them q reflects a change in a state function called enthalpy denoted as H. We cannot determine the enthalpy of a given system but we can measure the changes in enthalpy as the system changes. Having enthalpy as the state function is useful because it helps us to use the value of q to determine how the internal energy of a system changes and hence it is a guide to determine whether or not a reaction is spontaneous. The amount of heat, q exchanged during process that converts a system's reactants to products when the only work done by the system is expansion work at constant pressure, is called the enthalpy change, ΔH of the system ($\Delta H = Q$).

Most chemical reactions of interest occur under the essentially constant pressure, by measuring the heat transferred during the process. We can determine the internal energy for those reactions from the equation.

$$\Delta E = q + W = \Delta H + W$$

Because the reaction occurs at constant pressure and the only work done results from the change in volume of the system, the work, term in this case is $P(V_1 - V_2)$ or $-P\Delta V$.

We have previously seen that enthalpy changes include the standard enthalpy of formation $\Delta_f H°$ and enthalpy of combustion, vapourization and fusion.

Enthalpy of Combustion : Sometimes called heat of combustion, it is the enthalpy of change when generally 1 mole of a substance burns or combines with oxygen under standard state condition. For example, the enthalpy of combustion of carbon dioxide, –394 kJ mol^{-1}, is the amount of heat produced when 1 mole of carbon and 1 mole of oxygen gas both at 25°C and 1 atmosphere pressure, react and produce 1 mole of carbon dioxide gas at 1 atmospheric pressure and 25°C temperature.

$$C(s) + O_2(g) \longrightarrow CO_2(g) \text{ - - - - - - - - - } \Delta H = -394 \text{ kJ}$$

Enthalpy of Fusion : When we add heat to a crystalline solid its temprature increases until the solid reaches its molten state, at this point the temperature remains constant until all the solid has melted, after that the temperature again starts rising. *The amount of heat required to change exactly 1 mole of a substance from its solid state to the molten state is called enthalpy of fusion,* ΔH_{fus} of the substance. The enthalpy of fusion of ice is 6.0 kJ mol^{-1} at 0°C.

$$H_2O(s) \longrightarrow H_2O(L) \text{ - - - - - - - - - } \Delta H = -\Delta H_{fus} = 6.0 \text{ kJ mol}^{-1} \text{ at } 0°C$$

Fusion is an endothermic process.

Freezing is an endothermic process, in order for a substance to freeze, it must loose a quantity of heat equal to the amount it must gain to melt. Thus the enthalpy changes for freezing of water is –60 kJ mol^{-1} at 0°C.

As heat is removed from water at 0°C and the water changes to ice, its temperature does not change until all of the liquid gas frozen. Only at this point does the temperature fall as additional heat is removed.

Enthalpy of Vapourization : Heat is required for a liquid to evaporate. The cooling effect can be evident when you have a swimming pool or a shower. When the water on skin evaporates it removes the heat from your skin and causes you to feel cool.

For the temperature of a liquid to remain constant as a mole of liquid evaporates is called the enthalpy of vapourization of water which is 44.01 kJ mol^{-1} at 25°C.

$$H_2O\ (L) \longrightarrow H_2O(g) \quad \Delta H_{vap}\ (\text{at } 25°C) = 44.01\ \text{kJ mol}^{-1} \text{ at } 25°C$$

Equals that absorbed during evaporation.

At higher temperature, less heat is required per mole of liquid evaporated. At its boiling point (100°C) ΔH_{vap} of water vapour, but it does not increase the temperature of liquid water remains.

Enthalpy of Formation : A standard enthalpy of formation ΔH°f is an enthalpy change for a reaction is which with exactly 1 mole of pure substance is formed from free elements in their most stable states under state conditions. The standard enthalpy of formation of CO_2 (g) is –394 kJ mol^{-1}, which is the enthalpy for the exothermic reaction.

$$C(g) + O_2(g) \longrightarrow CO_2(g) \qquad \Delta°f = \Delta°_{298} = -394\ \text{kJ mol}^{-1}$$

Starting with the reactants at a pressure of 1 atmosphere and 25°C with the carbon present as graphie (the most stable form of carbon under this set of standard conditions) and ending with 1 mole of gaseous product CO_2 also at 1 atmosphere and a temperature of 25°C, for nitrogen dioxide, NO_2(g), ΔH°f is 33.2 which is ΔH for the endothermic reaction.

$$1/2N_2(g) + O_2(g) \longrightarrow NO_2(g) \qquad \Delta°f = \Delta°_{298} = 32.2\ \text{kJ mol}^{-1}$$

A reaction with 1/2 mole of N_2 and 1 mol of O_2 is correct in this case, because the standard enthalpy of formation refers to 1 mole of formation.

Hess's Law

There are three ways to determine the heat involved in chemical change. Measure it experimentally, look it up in a table of experimental values, or calculate it from other experiments determining enthalpy changes. Calculations generally involve the use of Hess's Law which states that *it is a process which can be written as the sum of several stepwise processes, the enthalpy change of the total process equal to the sum of the enthalpy changes of the various steps.* For example, carbon can react with oxygen to form carbon dioxide in a two step process. First carbon monoxide is formed, CO

$$C(s) + 1/2\ O_2(g) \longrightarrow CO(g)$$

Then carbon dioxide (CO_2) is formed from the carbon monoxide

$$CO(g) + 1/2\ O_2\ (g) \longrightarrow CO_2\ (g)$$

The equation determining the overall change of C to CO_2 is the sum of these two chemical changes.

Step 1 $\quad C\ (s) + 1/2\ CO_2\ (g) \longrightarrow CO\ (g)$

$\quad CO\ (g) + 1/2\ O_2\ (g) \longrightarrow CO_2\ (g)$

$$C(s) + 1/2\ O_2\ (g)\ \cancel{CO(g)} + 1/2O_2 \longrightarrow \cancel{CO(g)} + CO_2(g)$$

Because the compound formed in first step is consumed in the second step, the net change is

$$C(s) + O_2(g) \longrightarrow CO_2(g)$$

According to Hess's law, the enthalpy change of the reaction is equal to the sum of the enthalpy changes of the step 1. From the experimental enthalpy's of combustion we have,

$$C\ (s) + 1/2\ O_2\ (g) \longrightarrow CO\ (g) \qquad \Delta H°_{298} = -111\ kJ$$

$$CO\ (g) + 1/2\ O_2\ (g) \longrightarrow CO_2\ (g) \qquad \Delta H°_{298} = -293\ kJ$$

$$C(s) + O_2\ (g) \longrightarrow CO_2(g) \qquad \Delta H°_{298} = 399\ kJ$$

We can see the ΔH° value of the bottom reaction sum of the above two reactions.

Before we use Hess's law we must now the two important feature of ΔH.

1. The ΔH of the reaction in one direction is equal in magnitude and opposite in sign to ΔH for the reaction in the reverse direction. For example, ΔH for water is 6.0 kJ for 1 mole of ice.

$$H_2O(s) \longrightarrow H_2O(1) \qquad \Delta H = 60\ kJ$$

ΔH is –6.0 kJ for the reverse reaction 1 mole of water is 6.0 kJ.

$$H_2O(s) \longrightarrow H_2O(l) \qquad \Delta H = 6.0\ kJ$$

Then the enthalpy of fusion of 2 moles of water is twice as great.

Example: Determine the ΔH°f of $FeCl_3(s)$ from the enthalpy changes of the following two-step process that occurs under standard state conditions.

$$Fe(s) + Cl_2(g) \longrightarrow FeCl_2(s)$$

$$FeCl_2(s) + 1/2\ Cl_2(g) \longrightarrow FeCl_3(s)$$

The standard enthalpy of formation of $FeCl_2(s) = -341.8\ kJ\ mol^{-1}$

The enthalpy change for the second reaction is $-57.7\ kJ\ mol^{-1}$ of solid $FeCl_3$

We are trying to find the standard enthalpy of formation of $FeCl_3(s)$, which is equal to $\Delta H°_{298}$ for the following combustion reaction.

$$Fe(s) + 3/2\ Cl_2(g) \longrightarrow FeCl_3(s) \qquad \Delta H°_t = \Delta H°_{298}$$

We can write this reaction equation as the sum of two steps wth known ΔH values. The sum of the enthalpy changes of these two steps is equal to the ΔHf of the compound $FeCl_3$

$$Fe(s) + Cl_2(g) \longrightarrow FeCl_2(s) \qquad \Delta°_{298} = -341.8\ kJ$$

$$FeCl_2(s) + 1/2\ Cl_2(g) \longrightarrow FeCl_3(s) \qquad \Delta H°_{298} = -57.7\ kJ$$

$$Fe(s) + 3/2\ Cl_2(g) \longrightarrow FeCl_3(s) \qquad \Delta H°f = \Delta H°_{298} = (-341.8) + (-57.7)\ kJ = -399.5\ kJ$$

The enthalpy change for the formation of 1 mol of $FeCl_3(s)$ is –399.5 kJ so the enthalpy of formation, Δ°f, is $-399.5\ kJ\ mol^{-1}$.

Third Law of Thermodynamics

Although it is not possible to measure the value of some state functions, such as enthalpy H, and internal energy E, it is possible to measure the entropy, S of pure substance at any temperature. The reason for this is found in the third law of thermodynamics. The entropy of any pure perfect crystalline element or compound at absolute zero is equal to zero.

At absolute zero (0°K) all the molecular motion is at a minimum in an impure substance or a pure perfect crystalline substance has no disorder if entropy is zero. All molecular motion is also at a minimum in an impure substance or in a pure crystalline substance with defects at absolute zero, but the impurity or defects can be described in different ways giving rise to disorder (and a non-zero value of entropy).

That is why this law is called "zeroth law.

It will take a pure perfect crystalline substance at absolute zero (with entropy at zero) and measure the entropy change as its temperature is increased; the we have measured the absolute entropy of the substance at the higher temperature. Hence we can find entropy of pure substances. In contrast for internal energy and entropy, we can find only difference between two values, not the actual values. Standard entropies are enable to compare the relative amounts of disorder present in different pure substances and they can be used to determine entropy changes. However, we must maintain caution when using entropy because the entropy of a pure elemental substance at standard state of conditions is not equal to zero.

The Second Law of Thermodynamics

When a process occurs in an isolated system - a system in which no matter can get in or out $-\Delta H$ must be zero. When the process is spontaneous the free energy change ΔG, is negative, and because ΔH is zero, the entropy change ΔS must be positive. Thus the entropy increases in an isolated system, the universe includes everything, so nothing can get in or out. *Thus any spontaneous change that occur in the universe must be accompanied by an increase in the entropy of the universe.* This is the second law of thermodynamics. It is difficult to use the second law in its pure from to predict the spotaneity of chemical reactions. However, the second law does tell as that the disorder of the universe is increasing because spontaneous processes are occurring all the time.

We can understand how the order of a portion of the universe can increase (and its entropy decrease) if we divided the universe up into a system (the portion where the order increase) and surroundings. We can represent the change in the entropy of the system and its surrounding.

$$\Delta S_{univ} = \Delta S_{sys} + \Delta S_{surr}$$

Where ΔS_{surr} and ΔS_{sys} represent the entropy changes in the surroundings and systems respectively. The equation tells us that a spontaneous process with a decrease in disorder and a negative entropy tells us that a spontaneous process with a decrease in disorder and a negative entropy change can occur in the system if another process that leads to a greaer in crease in disorder occurs in the surroundings. In such a situation the over all entropy of the universe increases, even through the entropy in a small portion of universe (the system decreases).

CHAPTER

22

Electrochemical Process

Introduction

This chapter includes the study of chemical analysis where electricity rather a polarized electricity is involved, such as the flow of a direct current. There are many processes by which we can specifically analyze the different chemical properties of substance electrolysis is one of them. It is defined as the electrical *analysis by means of which some chemical changes are produced through reactions at electrodes in contact with electrolyte by the passage of a steady electrical curent.*

Electrolyte

It is a chemical compound which when fused or dissolved in certain solvents, usually water, will conduct electricity (e.g., common salt any acid potassium chloride, etc.) the passage of current is always accompanied by decomposition of electrolytes. All electrolytes in the fused state or in solution gives rise to ions, which conduct electric current. The phenomena of electrolysis in the fused state or in solution gives rise to ions, which conduct electric current. The Faraday's Law of electrolysis governs the phenomenon of electrolysis, which we are going to study later in this chapter.

Differentiation between electrolytes

Electrolytes are divided into two groups the strong and the weak electrolytes. Strong electrolytes are stable having a strong electrovalent bonding or ionic bonding, but weak electrolytes have a rather unstable and also co-valent bonding. The strong electrolytes get fully ionized in water but the weak electrolytes get partly ionized in water. A table following gives an idea of the names of the strong weak and non-electrolytes.

Non Electrolytes	Strong Electrolytes	Weak Electrolytes
Rubber	Potassium chloride	Carbon Tetra chloride
Silicon	Sodium chloride	Toluene
Distil Water	Any Inorganic Acid	Any Organic Acid

Electrodes

They have been named such because they convey electricity to the electrolytes and each electrode conveys a definite polarity of current. Therefore, they are named as *anode* and *cathode*. As anode catch the negative ions called *anions* and the cathode catches cations, the positive ions. (Anion catching electrode = anode, cation catching electrode = cathode).

Types of Electrodes

The electrodes are made up of different types and they are used in definite pairs, for definite electrolytes. They are showin in the Table 22.1.

Table 22.1

Types of electrodes	Electrolytes
Copper - Zinc	Sulphuric acid
Nickel - Cadmium	Cadmium Sulphate
Platinum	Any electrolyte

The Mechanism of Electrolysis

> Electrolysis is a process of producting metals or gases by passing electrical current through electrodes to electrolytes.

The passage of current through solutions of salts of such metals such as zinc, iron, nickel, cadmium, lead, copper, silver and mercury results in the liberation of these metals at cathode; from solutions of salts of the very base metals, e.g. alkali and the alkaline earth metals and from solutions of acids the substance set free is the hydrogen gas. If the anode consists of an attackable metal, such as one of those just discussed; the flow of the curent is accompanied by the passage of the metal into solutions. When the anode is made of an inert metal, e.g. platinum, an element is generally set free at this electrode; from solutions of nitrates, sulphates and phosphates, etc. Oxygen gas is liberated, whereas from halide solution, other than fluorides, the free halogen is produced. The decomposition of solutions by the electric current is resulting in the liberation of gases or metals, as described above, known as electrolysis.

The first definite proposal concerning the mechanism of electrolysis were made by Grotthuss (1806) he suggested that the dissolved substance consisted of particles with the positive and the negative ends, these particles being distributed throughout the solution. The potential was applied it was believed that the particles (molecues) became oriented in the form chains with the positive parts pointing in one direction (See Fig. 22.1) and the negative parts in the opposite direction (1).

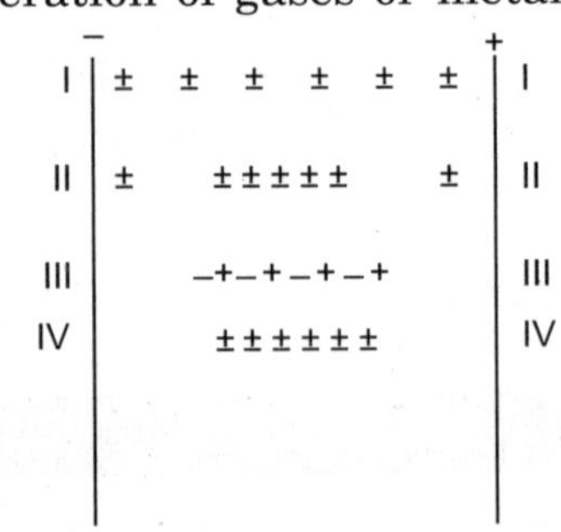

Fig. 22.1. The mechanism of electrolysis by Grotthuss Conduction.

It was supposed that the positive electrode attracts the negative part of one end particle in the chain, resulting in the liberation of the corresponding material, e.g. oxygen in the electrolysis of water. Similarly, the negative electrode attracts the positive portion of the particle e.g. the hydrogen of the water, at the other end of the chain, and sets, it free (II). The residual parts of the end units were then imagined to exchange partners with adjacent molecules, this interchange being carried on until a complete series of new particles is formed (III).

The current to give the correct orientation now rotates these particles (IV) followed by their splitting up, and so on. The chief objection to this theory of Grotthuss is that it would require a relatively high e.m.f., sufficient to break up the molecules, before any appreciable current was able flow, whereas many solutions can be electrolyzed by the application of quite small potentials all electrolytic conditions. Although the proposed mechanism has been described, as far as most electrolytic conduction is concerned, it will been seen later on in this chapter that a Grotthuss conduction occurs in solutions of acids and bases.

Objection against Grotthuss Theory

In order to account for the phenomena observed during the passage of electricity through solutions of electrolytes, Faraday (1833) assumed that the flow of electricity was associated with the movement of particles of matter carrying either positive or negative charges. These charged particles were called ions; the ions carrying charges positive in nature and moving in the direction of current, that is towards the cathode were referred as *cations,* and those carrying the negative charge and flowing against the current towards cathode are called *anions.* As it has been said earlier. The function of the applied e.m.f. is to direct the ions towards the appropriate electrodes where their charges are neutralized and they're set free as atoms or molecules.

The term ion has been derived from the Greek word *ion* meaning "wanderer" or "traveller" the prefixed *'ana'* and *'cata'* means "up" and "down" respectively; the anion is thus the ion moving up and the cation thus the ion moving down the potential gradient.

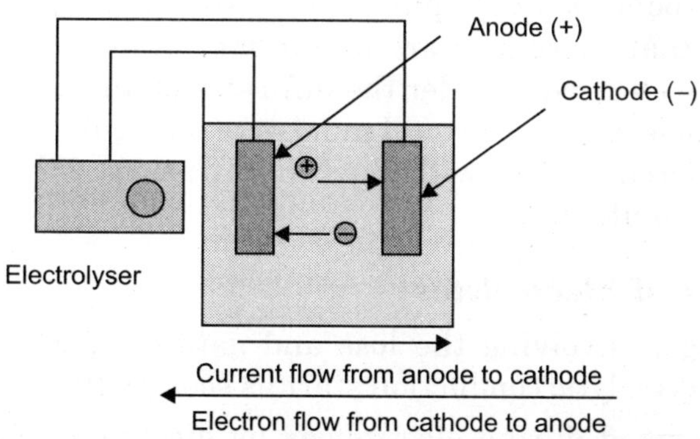

Fig. 22.2. The diagram showing the process of electrolysis and some electrochemical term.

Hydrogen is the only electro-positive non-metal.

The function of the applied e.m.f. is to direct the ions towards the appropriate electrodes where their charges are neutralized and they are set free as atoms or molecules. It may be noted that since the hydrogen and other metals only are discharged at the cathode (since they are electro-positive and all other non-metals are discharged at the anode being they are electro-negative).

It should be noted that the metallic portion of the electrolyte is produced in the cathode and other radical being electronegative is produced at the anode For example sodium chloride, sodium is produced at the cathode and the chloride radical is produced at the anode.

Although Faraday postulated the existence of charged material particles, or ions, in solution, he offered no explanation for their origin; it was suggested by *Clasious* (1851) that the positive and the negative parts of the solute molecules were not firmly connected, but were each in a state of vibration that often become vigorous enough to cause the portions to separate. These separated charged parts, or ions were believed to have relatively short period of free existence; while free they were supposed to carry the current. According to *Clasious* small fraction only of the total number of dissolved were split into ions at any instant, but sufficient ions were always available for carrying the current and hence for discharge at the electrodes. Since no electrical energy is required to break up the molecules, this theory is 31 arrangement with the fact that small e.m.fs are generally adequate to cause electrolysis to occur; the applied potential serves merely the to guide the ions to the electrodes where their charges are neutralised.

The Arrhenius Theory of Dissociation

From his studies of the conductance of aqueous solutions of acids and their chemical activity Arrhenius (1883) concluded that an electrolyte solution contained two kinds of solid molecules; these were supposed to be "active" molecules, responsible for electrical conducted and chemical action and inactive molecules, respectively. It was believed that when an acid base or salt was dissolved in water a *considerable portion,* consisting of so-called active molecules were spontaneously split up, or dissociated, into positive and negative ions; it was suggested that these ions are free to move independently and are directed towards the appropriate electrodes under the influence of an electric field. The portion of active, or dissociated, molecules to the total number of molecules, latter called II degree of dissociation was considered to vary with the concentration of the electrolyte, and to be equal to unity in dilute solutions.

Some Characteristics of Electrolytes

1. Chemical changes involving the loss and gain of electrons by the ions at the electrodes an electrolytic conduction start as soon as the current is switched off.
2. The products formed during electrolysis by discharge of ions are deposited or liberated only at the electrodes.

3. Electropositive elements (i.e. hydrogen and metals) are usually set free at the cathode while the electronegative elements or groups (oxygen and other non-metals) are deposited at anode.
4. All chemical reactions, which occur at the anode and cathode during electrolysis, are oxidation and reduction involving gain of electron or electrons by an ion.
5. In the process of electrolysis, electrons enter and leave the solution through chemical changes at the electrodes. The number of electrons entering in the solution is exactly same.

Practical Units of Electric Current and Quantity of Electricity

Ampere : It is the practical unit of electric current and is defined as the current which when passed for a second through a solution of silver nitrate deposits 0.001118g of silver at the cathode. This unite was named after the famous French physicist Andre Mare Ampere.

Coulomb : This is the unit of electricity named after the famous scientist C.A. Coulomb. It is the practical unit of quantity of electricity, it is defined as the quantity of electricity (Q) flowing through a conductor in time t, the quantity of electricity (Q) flowing is given by $Q = c \times t$. A coulomb is also defined as the quantity of electricity that liberated 0.00 1118g of silver form a solution of silver nitrate 0.0000104gms of hydrogen from an appropriate solution at the cathode. A coulomb is regarded as the smaller unit of quantity of electricity.

Faraday : Named after the famous English scientist Sir Michel Faraday. It is taken as the bigger unit of quantity of electricity and is approximately equals to 96,500 coulombs.

Now, 0.001118g of silver is liberated by 1 coulomb

:. 107.88g of silver are liberated by 107.88/0.001118 or 96,496 coulombs ≈ 96500 coulombs. Obviously 1 faraday is the quantity of electricity required to produce 107.88 g or 1 g-equivalent of silver. It is seen further that this quantity of electricity produces 1 gram-equivalent of any substance during electrolysis i.e., 1.00g of hydrogen, 8.00 g of oxygen, 31.77 g of copper and 35.46 g of chlorine are set free at the suitable electrodes. When 1 faraday amount of electricity passes through the appropriate electrolytic solution. So, the faraday is symbolised by F is the quantity of electricity requires depositing, evolving or dissolving one gram-equivalent of any substance at an anode.

Faraday's Law of Electrolysis

During his studies on electrolysis, Michel Faraday deduced the following laws first stated in 1834.

First Law : The weight of an ion liberated during electrolysis is proportional to the quantity of the electricity passed i.e., if the quantity of the electricity increases the weight of the ion will also increase.

WocQ oc **Ct**

w = CtZ

Where W = Weight of the ions liberated;

Q = Quantity of electricity in coulombs;

C = Strength of current in amperes; and t~Time in seconds; while

Z = is a constant of proportionality or electrochemical constant.

When c = 1 ampere and t = 1 second then W = C,

That is *electrochemical equivalent of an ion is the weight of an ion deposited by 1 ampere current in 1 second.*

Second Law : When the same quantity of electricity pass through different electrolytes, the weights of different ions liberated at the electrodes are in their ratio of their chemical equivalents.

For example, if the same current were passed through a number of voltmeters arranged in series as shown in Fig. 22.3 then the weights of hydrogen, copper silver and tin deposited on respective cathodes will be in the ratio of their chemical equivalents.

$$\frac{\text{Weight of Copper}}{\text{Weight of Silver}} = \frac{\text{Equivalent weight of Copper}}{\text{Equivalent weight of Silver}} \text{ and so on}$$

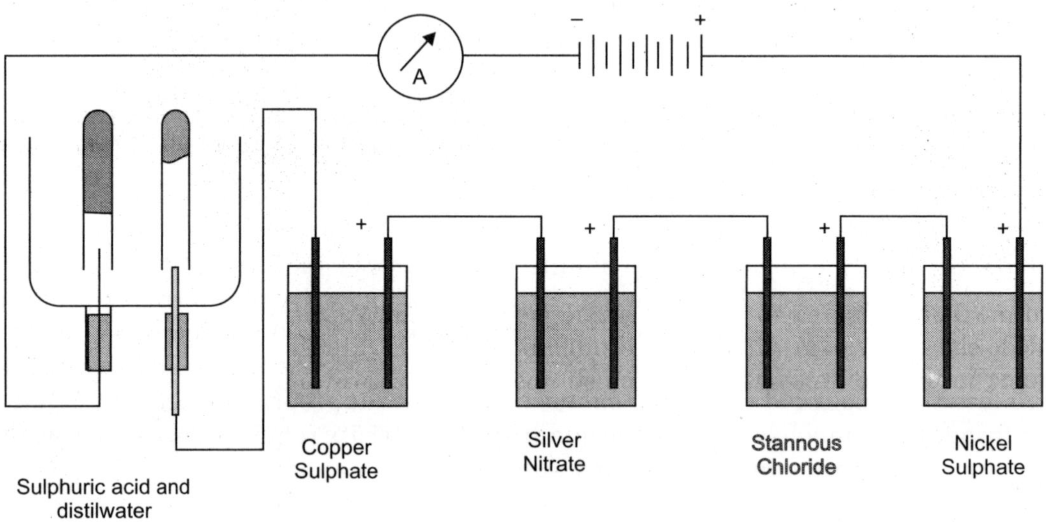

Fig. 22.3. The Experimental set-up for the arrangement of Voltameters in series.

If the passage of electric current is continued in the above voltameters until .008g of hydrogen (one gram equivalent weight) is evolved from the sulphuric acid solution, the weights of copper, silver and tin deposited will be found to be one gram equivalent weight in each case. It is found that 96,500 coulombs of electricity are required to liberate one-gram equivalent weight of any element. (This much quantity of electricity is called 1 Faraday). This fact can be used to determine the electrochemical equivalents of many other elements. For example

Electrochemical equivalent of Silver $= \dfrac{\text{Gram Equivalent Weight}}{965{,}500}$

$= \dfrac{107.88}{96{,}500} = 0.0011179 \text{ g}$

Electrochemical equivalent of Copper $= \dfrac{\text{Gram Equivalent Weight}}{965{,}500}$

$= \dfrac{31.78}{96{,}500} = 0.000329 \text{ g}$

Electrochemical equivalent of Hydrogen $= \dfrac{\text{Gram Equivalent Weight}}{965{,}500}$

$= \dfrac{1.008}{96{,}500} = 0.0000103 \text{ g}$

Combination of Faraday's Law of Electrolysis : If W be the weight in gram of a substance (of equivalent E) deposited or dissolved at an electrode by Q coulombs of electricity, then

From the first law $W \propto Q \propto Ct$ (when E remains unchanged)

And from the second law $W \propto E$ (when Q remains unchanged).

$\therefore$ $W \propto QE$ (when both E and Q vary) or $W \propto Ct. E$ or, $W = ct\ E/F$ (where I/F is the proportionality constant) ...(22.1)

When the quantity *Q* or *ct* is equal to *F* then *W* = *E*. That is *F*, represents a definite quantity of electricity which is required to deposit or dissolve 1 g equivalent of any substance in electrolysis. The value of F, called the faraday, has been experimentally found to be 96496 usually rounded up to 96,500 coulombs i.e. *F* = 96,500 coulombs.

Inserting the value of Faraday, Faraday's laws may be expressed by the following important form W = ct E/96,500.

The above relation is obeyed so rigidly that it may be applied to measure the quantity of electricity or current strength. If the amount of W of a given substance formed at an electrode is determined and the equivalent weight of the substance is known, the quantity of electricity that is $c \times t$ can be derived from the above relation. When t is measured, the current strength *c* assumed to be constant can also be found out.

Factors Influencing Electrolysis

When two or more species of icons of similar charge are present under similar conditions in a solution say, H^+ and Na^+ ; OH^- and SO_4^{2-}, one is preferential discharge of the ion depends on various factors:

(i) Position of the element (from which the particular ion is produced) *in the electro-chemical series :* Lower the position of the element in the electro-chemical series, greater is

the tendency of its ion to be discharged during electrolysis. Generally the selective discharge of ions occur according to the order shown in Fig. 22.4. Any ion will be discharged from solution in preference to those placed above it. In the figure, the arrangement of ions is the same as that of the corresponding elements in the electro-chemical series.

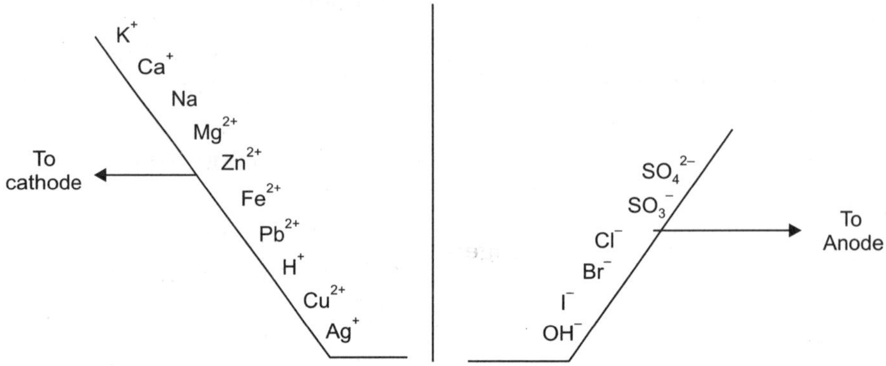

Fig. 22.4. The discharge of Ions in order

For example, when a solution of caustic soda containing W (produced by the dissociation of water) and Na^+ as positive ions is electrolyzed, W discharges in performance to Na^+; similarly during electrolysis of copper sulphate solution containing OH^- (from water) and sol' Selective discharge of ion may also be considered in the light of energy causing the discharge. If two or more different but similarly charged ions are attracted to given electrode, the ion to be discharged is that one which has the lowest discharge potential that is the ion which requires least energy for its discharge as neutral atom or radical.

(ii) Concentration of the electrolyte (i.e., the ions) in solution : Increase of concentration of anion tends to promote its discharge at the electrode.

(iii) Nature of the electrodes used : This factor may sometimes influence the selection of the ion for discharge. For example, if an aqueous solution of sodium chloride is electrolyzed between platinum electrodes, W ions will get discharged at the cathode in preference to Na^+ ion. If mercury cathode is used in place of platinum Na^+ ions are deposited at the cathode; sodium amalgam is the result. However, the concentration effect and the effect due to the nature of the electrode will be clear from the products of electrolysis of some important and typical solutions of electrolytes described below:

In certain cases, the primary products of electrolysis take part in secondary reactions to give secondary products.

A primary product is one, which is formed directly by the electric current rather than by subsequent chemical reaction.

Some Examples of Electrolysis

1 Electrolysis of water : Pure water is a very weak electrolyte and consequently a very poor conductor of electricity. It dissociates very slightly into W ions and OH^- ions.

In fact H^+ ions does not remain as such but forms Hydroxonium ions (H_3O^+) by combining with a molecule of water $H^+ + H_2O \rightleftharpoons (H_3O^+)$

If however, a little sulphuric acid (or any other strong electrolyte) is added to it, its conductivity i.e. ionization is greatly increased. The acidified water now contents W ions, Off and SO_4 Ions

$$H_2O \rightleftharpoons H^+ + OH^-; H_2SO_4 \rightleftharpoons 2H^+ + SO_4^-$$

During electrolysis with platinum electrodes, the positively charged W ions are attracted to the cathode and are discharged thereby gaining one electron each to give a hydrogen atom. The atoms then combine in pairs, which are evolved as gas. On the other hand, the negatively charged particles OH^- and SO_4^- ions are discharged to produce OH radicals. These radicals having no independent existence react with one another forming water and oxygen. Gaseous oxygen is thus evolved at the anode.

At the cathode : $H^+ + e \rightarrow H; H + H \rightarrow H_2\uparrow$

At the anode : $OH^- - e \rightarrow OH; 4OH \rightarrow 2H_2O + O_2\uparrow$

The amount of sulphuric acid added to water remains unchanged after electrolysis and only water is thus decomposed to produce hydrogen and oxygen.

The ratio of the volume of hydrogen and oxygen liberated on electrolysis of water measured under same conditions of temperature and pressure is always 2: 1. The invariable volume ratio for the gases is explained by considering the flow of four electrons through the electrolyte.

At the cathode : $4H^+ + 4e \rightarrow 2H_2$;

At the anode : $4OH^- — 4e \rightarrow 2H_2O + O_2\uparrow$

2. Electrolysis of Sulphuric Acid Solution : Electrolysis of dilute sulphuric acid is exactly equivalent to the electrolysis of water.

If 50% of the acid is electrolyzed using platinum electrodes, different products are formed at the anode. Sulphuric acid of this strength dissociates into W and HSO_4^- ions. The ions produced on dissociation of water are negligibly small. During electrolysis, hydrogen is liberated at the cathode as usual. HSO_4^- ions due to their presence in greater concentration are discharged at the anode giving HSO_4 radicals. Two such radicals then unite to yield a molecule of persulphuric acid ($H_2S_2O_5$), which reacts with water with the formation of hydrogen peroxide and regeneration of sulphuric acid.

At cathode : $H^+ + e \rightarrow H; H + H \rightarrow H_2\uparrow$

At anode : $HSO_4^- — e \rightarrow HSO_4; 2HSO_4 \rightarrow H_2S_2O_8$

$H_2S_2O_8 + 2H_2O \rightarrow 2H_2SO_4 + H_2O_2$

3. Electrolysis of Sodium Chloride

Electrolysis of Fused Sodium Chloride : Anhydrous sodium chloride on fusion produces only Na^+ Cl^- ions. On passage of electricity trough the molten salt using platinum cathode and graphite anode, the positively charged Na^+ ions migrate to the cathode and pick up one electron and become a neutral sodium atom. These deposited on the electrode. On the other hand, the negatively charged cr ions are attracted to the anode where they give up one electron each to produce chlorine atoms. The neutral chlorine atoms then unite in pairs to form chlorine molecules, which are liberated at the anode as chlorine gas. The electrolysis of fuse sodium chloride has been shown in Fig. 22.5.

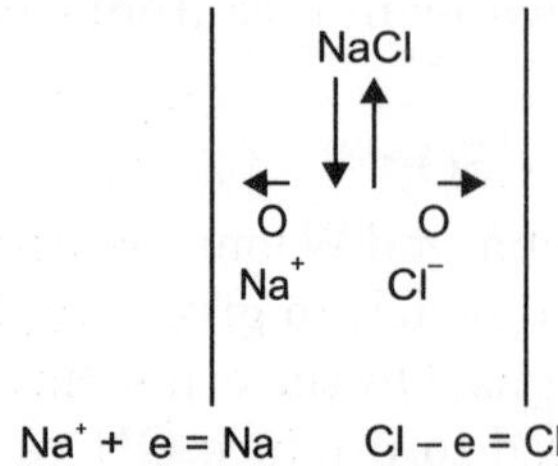

$Na^+ + e = Na \qquad Cl - e = Cl;\ Cl + Cl = Cl_2$

The net reaction involved are:

At cathode $Na^+ + e \rightarrow Na$

At the cathode: $Cl^- - e \rightarrow Cl; \quad Cl + Cl \rightarrow Cl_2$

The net reaction:

$$\begin{array}{l} 2Na + 2e \rightarrow 2Na \\ 2Cl^- - 2e \rightarrow Cl_2 \\ \hline 2Na^+ - 2Cl \rightarrow 2Na + Cl_2 \end{array}$$

Fig. 22.5. The pictorial representation of electrolysis of fused sodium chloride

Electrolysis of dilute aqueous solution of Sodium Chloride : Sodium chloride being a strong electrolyte dissociates completely in dilute aqueous solution yielding Na^+ Cl^- ions. Water also forms traces of H^+ and OH^- ions due to its very slight ionization.

$$NaCl \rightleftharpoons Na^+\ Cl^-;\ H_2O \rightleftharpoons H^+ + OH^-$$

When electricity is passed using platinum electrodes, both Wand Na^+ ions (cations) are attracted towards the cathode where W ions are discharged in preference of Na^+ ions producing hydrogen atoms. The hydrogen atoms thus formed then unite in pairs to form hydrogen molecules, which are released at the electrode as hydrogen gas. On the other hand, both OR and cr ions move to the anode where OR ions are deliberately discharged to produce OR-radicals. These radicals then react with one another forming oxygen, which escapes as gas at anode. The process is equivalent to the electrolysis of water.

At the cathode: $H^+ + e \rightarrow H;\ H + H \rightarrow H_2\uparrow$

At the cathode: $OH - e \rightarrow OH;\ 4(OH) \rightarrow 2H_2O + O_2\uparrow$

Electrolysis of Concentrated solution of Sodium Chloride : In the concentrated solution sodium chloride, there are Na^+ and W ions and OH^- due to the very slight ionization of water.

$$NaCl \rightleftharpoons Na^+\ Cl^-;\ H_2O \rightleftharpoons H^+ + OH^-$$

When electricity is passed through the solution using two electrodes made up of graphite, Na^+ and W ions (cations) are attracted towards the cathode and the chlorine Cl^- ions and OH^-, the hydroxyl ions towards the anode. At the cathode the *W* ions are discharged in the presence of Na^+ ions. Each H^+ ions gains one electron and is converted into *monatomic hydrogen.* Two hydrogen atom then combine to give a hydrogen molecule and hydrogen is also evolved as gas. Due to discharge of H^+ ions, the excess OH^- ions left in the solution forms NaOH with the neighbouring Na^+ ions immediately adjacent to the cathode. At the anode, Cl^- ions are discharged because these are present in much greater concentration than OH^- ions. Chloride ions (Cl^-) on giving up electrons from chlorine atoms. Chlorine atoms then combine in pairs to give molecules, which escape as the chlorine gas.

At the cathode : $H^+ + e \rightarrow H;\ H + H \rightarrow H_2\uparrow$

At the cathode : $Cl^- - e \rightarrow Cl;\ Cl + Cl \rightarrow Cl_2$

It is clear from the above changes that two electrons enter the solutions or are removed from the cathode to form a molecule of hydrogen and simultaneously two electrons leave the solution or are given to the anode by chloride ions to give a molecule of chlorine. The liberated electrons flow through external circuit from the electrode to the source of potential. The net result is the transfer of two electrons from the cathode side of the circuit to the anode side.

The above process is an example in which the order of discharge of an ion is reversed due to the concentration of the cr ions fall, discharge of Off ions occurs and a mixture of chlorine and oxygen comes off at the anode with the proportion of oxygen gradually increasing.

$$OH^- - e \rightarrow OH;\ 4OH \rightarrow 2H_2O + O_2\uparrow$$

However, should the chlorine happen to react with NaOH formed near the cathode, sodium chloride, hypochlorite may also be obtained.

$$2NaOH + Cl_2 = NaCl + NaOCl + H_2O$$

During electrolysis of a hot solution of sodium chloride in water with iron electrodes, H^+ ions and Cl^- ions are discharged and liberated as gaseous hydrogen and chlorine at the cathode and anode reacts with liberated chlorine forming sodium chlorate.

$$3Cl_2 + 6NaOH = NaClO_3 + 5NaCl + 3H_2O$$

If Mercury cathode is used (anode is of graphite), Na^+ ions are deliberately discharged at the cathode but not the W. Under such conditions, the discharge of Na^+ ions requires less energy than the discharge of H^+ ions to hydrogen gas. The liberated sodium dissolves in mercury yielding mercury amalgam and gaseous chlorine escapes at the anode.

At the cathode : $Na^+ + e \rightarrow Na;\ Na + Hg = Na/Hg$ (sodium amalgam)

At the anode : $Cl^- - e \rightarrow Cl;\ 2Cl \rightarrow Cl_2\uparrow$

This is an example where the nature of the electrode influences the usual order of discharge of an Ion.

(4) Electrolysis of Aqueous Copper Sulphate Solution : The product of electrolysis of an aqueous solution of copper sulphate depends on the nature of the electrodes used.

(A) Electrolysis of Aqueous Copper Sulphate Solution using Platinum Electrode. Copper sulphate in an aqueous solution yields Cu^{2+} and SO_4^{2-} ions upon dissociation and water being a weak electrolyte dissociates to a very small extent traces of H^+ ions and OH^- ions. Thus, the ions present in the aqueous solution of copper sulphate solution are Cu^{2+}, H^+, SO_4^{2-} and on, according to the following dissociation reactions.

$$CuSO_4 \rightleftharpoons Cu^{++} + SO_4^-,\ H_2O \rightleftharpoons H^+ + OH^-$$

When electricity is passed through solution, both Cu^+ and H^+ ions (cations) are attracted towards cathode where Cue ions are deliberately discharged. Each of the Cu^{++} ions takes up two electrons and forms the neutral copper atom and the copper atoms thus formed are deposited as a brown layer on the cathode. W ions remain in the solution.

Again, the negative SO_4^- and the OH^- ions (anions) migrate towards the anode where the On″ less electronegative than $SO_4''^-$ ions are discharged by giving electrons to the anode and arrange the OH group the OH group then interact to produce water and oxygen which evolve as a gas.

It is be noted that with platinum electrodes the blue colour of the electrolyte gradually fades with the progress of the electrolysis, as the Cu^{++} ions are responsible for the colour are discharged and deposited.

The SO_4^{--} ions are left in the solution. The residual solution is acidic as it contains H^+ and SO_4^- ions, *i.e.* H_2SO_4, the concentration of which increases progressively as the electrolysis continues.

At the Cathode : $Cu^{++} + 2e \rightarrow Cu$

At the Anode : $OH^- — e \rightarrow OH$; $4(OH) \rightarrow 2H_2O + O_2$

Thus, the electrolysis of an aqueous solution of copper sulphate using platinum electrodes or carbon electrodes results in the deposition of metallic copper on the cathode and the liberation of oxygen gas at the anode.

(B) Electrolysis of Copper Sulphate Solution with Copper Electrode : When electrolysis is carried out with copper electrodes cupric ions (Cu^{++}) are discharged as usual and deposited on the cathode as metallic copper. As a result, the weight of the copper cathode continually increases as the electrolysis proceeds. On the other hand, neither SO_4^- ions nor OH^- ions are discharged at the copper anode: but copper atoms from the anode pass into solution as cupric ions (Cu^{++}) since the electrons are removed more easily from a copper atom than from a SO_4^- or a OH^- ions to affect its discharge. The electrolysis continues until the anode dissolves completely so long the process of electrolysis the process of electrolysis goes on, the concentration of the copper sulphate as well as its colour will not change.

Cathode reaction : $Cu^{++} + 2e \rightarrow Cu\uparrow$

Anode reaction : $Cu - 2e \rightarrow Cu^{++}$

(5) Electrolysis of Aqueous Silver Nitrate Solution : Electrolysis of an aqueous silver nitrate with platinum electrodes results in the deposition of metallic silver on the cathode and Evolution of oxygen at the anode.

Dissociation : $AgNO_3 \rightleftharpoons Ag^+ + NO_3^-$; $H_2O \rightleftharpoons H^+ + OH^-$

At the Cathode : $Ag^+ + e \rightarrow Ag\downarrow$

At anode : $OH^- - e \rightarrow OH$; $4(OH) \rightarrow 2H_2O + O_2\uparrow$

The Solution contains H^+ and NO_3 – *i.e.*, HNO_3 and hence becomes acidic. During electrolysis with silver electrodes, silver from the anode dissolves and deposits on the cathode.

Determination of the E.C.E. of Silver

A clean platinum basin, carefully dried and weighted is taken. An aqueous solution of silver nitrate (approximately 15% pure) is taken in the platinum basin, which serves as both the cathode and the electrolytic cell. A thin anode of pure silver foil is partly immersed with the basin. The basin and the silver foil are then connected with the negative and the positive terminals of the of a battery through an ammeter (see Fig. 22.6) and the current is passed for some times after recording the current strength [c] from the ammeter and the time (t) from the stop watch, the current is switched off. The solution is then decanted and the basin containing the silver deposit is washed with distill water and then with alcohol dried, and carefully weighted from the increase in weight of the basin the E.C.E. of the silver can be calculated.

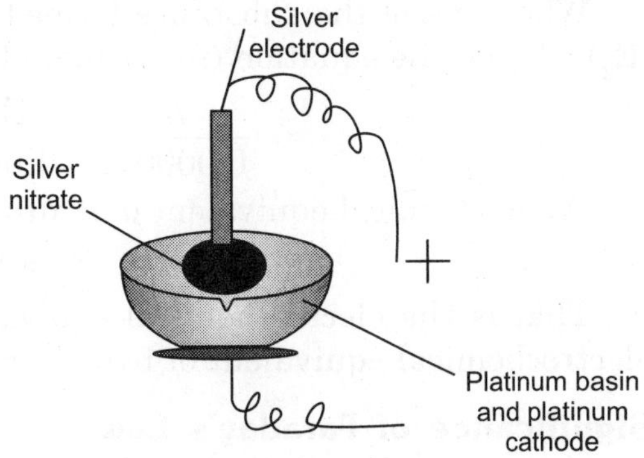

Fig. 22.6. The Experimental set up for the determination of ECE of Silver.

Let the weight of the basin before electrolysis is W_1 g and the weight after the electrolysis is W_2 grams.

$\therefore$ The weight of the silver deposited = $(W_2 - W_1)$ g. If a current c flows for t seconds, the quantity of electricity passed = $c \times t$ coulombs.

Now, $c \times t$ coulombs deposit $(W_2 - W_1)$ g of silver

$\therefore$ columb deposits $\dfrac{W_1 - W_2}{c \times t}$ g of silver

$\therefore$ the E.C.E. of silver (Z) = $\dfrac{W_1 - W_2}{c \times t}$.

Alternative Statement of Faraday's Second Law of Electrolysis

When the same quantity of electricity is passed through solutions of different electrolytes, the relative numbers of moles of the elements deposited or liberated are inversely proportional to the charges on the ions of each element respectively.

The least quantity of electricity required to liberate one mole of hydrogen ions, one mole of sodium ions, one mole of potassium ions.

> This is known as Faraday's Number.

Then the cathode reaction will be $M^+ + e \rightarrow M$ (M = the hydrogen or equivalent metal). So to discharge one mole of uni positive ion the charge carried by one mole of electrons is required.

$\therefore$ 1 faraday = the charge carried out by 1 mole or 6.023×10^{23} electrons.

Relation between Chemical and Electrochemical Equivalent

Let us, consider that the W_1 and the W_2 be the weights of two different substances formed at the electrodes by the passage of the same quantity (Q) of electricity and the equivalent weights of the substances be E] and respectively. Then, from the first Law, we get

$W_1 = Z_1 Q$ and $W_2 = Z_2\ Q \ \therefore\ W_1/W_2 = Z_1/R_2$ *(i)*

Again from the second law, we get $W_1/W_2 = E_1/E_2$ *(ii)*

Combining the two-equation *(i)* and *(ii)*, we get $Z_1/Z_2 = E_1/E_2$ *(iii)*

When one of the substance formed is hydrogen for E.C.E (Z_2) = 0.0000104 and C.E (E_2) =1.008, the equation (*iii*) reduces to

$$\frac{Z_1}{0.000014} = \frac{E_1}{1.008} \text{ or } Z_1 = E_1 \times \frac{0.000104}{1.008}$$

When chemical equivalent of hydrogen is taken to be 1, the above relation becomes

$$Z_1 = E_1 \times 0.0000104.$$

That is the electrochemical equivalent of a substance = Its chemical equivalent × electrochemical equivalent of hydrogen (approx.)

Significance of Faraday's Law

***e* = FIN; Relation between the Faraday** *Avogadro number and the change on the electron* : Besides the practical applications so far discussed, Faraday's laws have an important significance of theoretical interest. The laws have introduced the concept of atomic nature of electricity.

We know that during electrolysis, the transfer of electrons takes place between the electrodes through the ions approaching electrons from the cathode while the negatively charged anions are discharged by giving up electrons to the anode. From Faraday's laws, it is known then the discharge of gram equivalent of any ion (either positive or negative) requires the passage of 1 faraday of electricity. It means that 1 gram- equivalent of an ion carries 1 faraday (or 96,500 coulomb) of electricity.

Now, 1 gram-ion of any ion is numerically equal to 1 gram. 1tom of the element from which the ion is numerically equal to 1 gram-atom of the element from which the ion is formed by gain or loss of electrons and is therefore equal to the product of the 1 gram-equivalent of the ion and its valency.

That is 1 gram-ion = 1 gram-equivalent of the ion *x* valency.

If the valency of the ion were is *n*, it may reasonably be stated that 1 mole or 1 gram-ion will contain *n* gram equivalents and consequently will carry *n*F coulombs of electricity. Again the number N being equal to 6.023×10^{23}. So the charge carried by a single ion is nF /N.

Since *n* is an integer, 1 for a equivalent ion, 2 for a bivalent ion 3 for a trivalent and so on, it is true that a univalent ion carries the charge FIN, a bivalent carries 2FIN and the trivalent 3FIN and so on.

It follows that the charge carried by a single ion is the same (where $n = 1$) or a multiple of a fundamental unit charge FIN. This result obtained from Faraday's laws leads to the conclusion that the electricity like matter is discrete or atomic in nature and F/N is regarded as the atom or the unit of electrical charge.

This is $$e = \frac{F}{N}$$

By putting the values of F and N we get

$$e = \frac{96{,}500}{6.023 \times 10^{23}} = 1.602 \times 10^{-19} \text{ coulomb.}$$

Or minimum quantity of electricity and 1 faraday 96,500 coulombs of electricity = the charged carried by 6.023×10^{23} or 1 mole electron or the charge of a single electron is 1.602×10^{-19} coulombs. So $1.602 \times 10^{-19} \times 6.023 \times 10^{23} = 96{,}500$ coulombs. Thus the number of electrons passing through any point of a conductor in one second is

$$\frac{\text{The charge carried by an electron}}{\text{The Quantity of electricity passing}}$$

The relation e = F/n that is N=F/e states an excellent Method of calculating the accurate value of Avogadro's Number from the values of Faraday and experimentally determined electronic charge

Since a faraday is equal to the product of the Avogadro number and the electronic charge (F = N.e). It follows that a faraday contains the same number of unit electric charges as the number of molecules in a mole. Hence a faraday (F) may be taken as a mole of electricity. Finally, it may be mentioned that Faraday's Law has explained the phenomenon of electrolysis in a simple manner.

According to these Laws, the equivalent amounts cations and anions are discharged at the cathode and anode respectively for a given quantity of electricity. This means that when an electrolyte in solution or in fused state is electrolyzed, the number of electrons taken up by the cations at the cathode must be equal to the number of electrons given up simultaneously at the anode. These released electrons pass through the external wire from the anode to the cathode.

Now, a hydrogen ion is produced by the loss of the single electron of the hydrogen atom. The charge carried by a hydrogen ion is taken to be the unit of positive electricity. On the basis of Faraday's Laws, the charge carried by 1 gram atom of hydrogen ion equals to 96,500 coulomb.

$$\therefore \text{ The charge carried by 1H + } = \frac{96{,}500}{6.023} = 1.602 \times 10^{-19} \text{ coulombs.}$$

Thus the charge of an electron is equal but opposite to that carried by a hydrogen ion. The positive or the negative charges by some familiar ions are shown in the table :

Ion	Valency	Magnitude of carried electricity
Na^+, K^+, H^+, Ag^+	1	1.602×10^{-19} coulombs (+ve electricity)
Cl^-, OH^-, NO_3^-	1	1.602×10^{-19} coulombs (–ve electricity)
Ca^{++}, Cu^{++}, Zn^{++}	2	$2 \times 1.602 \times 10^{-19}$ coulombs (+ve electricity)
SO_4^{--}, SO_3^{--}	2	$2 \times 1.602 \times 10^{-19}$ coulombs (–ve electricity)
Al^{+++}, Fe^{+++}	3	$3 \times 1.602 \times 10^{-19}$ coulombs (+ve electricity)
PO_4^-	3	$3 \times 1.602 \times 10^{-19}$ coulombs (–ve electricity)

Behaviour of Electrolysis

1. Natural Behaviour of Electrolysis : Electrolysis lowers the freezing point of a solvent and also increases the boiling point more than non-electrolytes. For example when one gram mole of sugar is dissolved in 1 kg of water, the melting point is lowered by 1.86°C. One gram molecule of sodium chloride or hydrochloric acid when dissolved in the same volume of water that is 1 kg, lowers the freezing point nearly twice as much as sugar. Some electrolytes lowers the freezing point by three times as much as sugar does. Similarly the rising in boiling point of a solvent produced by some electrolytes is twice as much as the case of non-electrolytes like sugar. Some other raises the boiling point by three times as much by sugar.

The Theory of Ionization : To explain the natural behaviour of electrolysis, the Swedish Chemist Svante Arrhenius proposed the theory *of ionization* which has been previously discussed, according to Arrhenius the degree of ionization depends upon the following factors :

1. The Dilution - The ionization increases with the dilution of the solution.
2. The temperature - The ionization increases with the increase of temperature.
3. The nature of solute - The ionization is fast with the strong electrolytic solutes.
4. The nature of the solvent - The ionization is maximum with water.
5. The nature of the purity the solvent - The ionization is maximum if the solvent is pure.

Ionization and Dissociation

In the case of an electrolyte, the ions all ready exist as solid crystalline form. There the force, which joins them, is the electrostatic force. On dissolution, water merely decreases the attraction between the ions and gives them freedom of movement in solution. The two ions fall apart and start leading an independent existence. The process is called dissociation. We say that when sodium chloride is dissolved in water it dissociates into sodium and chlorine ions.

Heating produces the same effect, on heating the fused mass makes the ions mobile and the electrolyte starts conducting electricity (see Fig. 22.7).

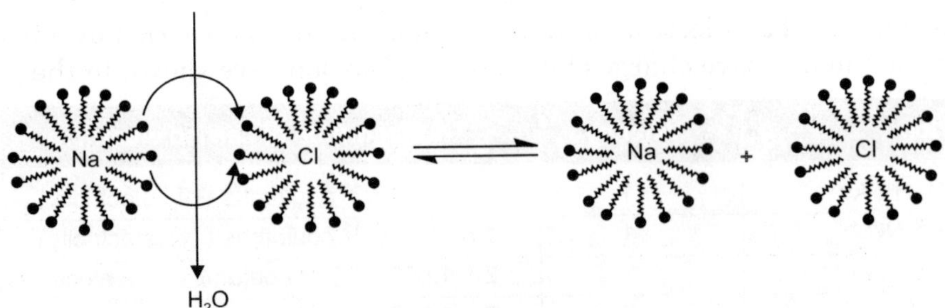

Fig. 22.7

Ionic Equations

These equations are used generally in the ionization reactions in electrolysis, the oxidation reduction reactions and in the explanation of chemical bonding (electrovalent bonding).

Most of the inorganic reactions are in fact ionic; for example when sodium and chlorine reacts to form sodium chloride we used to write.

Na + CI = NaCl, but according to the ionic theory it will be written as $Na^+ + Cl^-$ this is because Sodium ion is positive (Na 1 and having a single valency (only one positive sign) and chlorine ion is negative (Cl and having also a single valency (only one negative sign). According to the table of valency given we can learn the ionization potentials of the elements, we should keep in mind that the ionization potential increases or decreases with the valency. The electropositive elements gain one negatively charged electron and become a neutral element and the electronegative elements leave one electron and become a neutral element. We will now see some examples to learn these equations,

$$KCl \rightleftharpoons K^+ + Cl^-$$

$$BaCl_2 \rightleftharpoons Ba^{2+} \text{ or } Ba^{++} + Cl^-$$

$$BaSO_4 \rightleftharpoons Ba^{2+} + SO^{2+}{}_4 \text{ or } SO_4{}^{--}$$

Above reactions are all ionization reactions because the neutral compounds have broken up into respective ions, now the reverse reaction also takes place and this is called neutralization reaction of ions.

Applications of Electrolysis

Electrolysis is used in many practical and also industrial processes, without the application of the electrolysis some of the important elements like calcium carbide and aluminium would have reamed as a laboratory curiosity only. The prevention of the metals from corrosions, and a recent rescue action of the portions of the ship (named Titanic, the main engine was rescued and freed from corrosion by the process of electrolysis. The separation of metals from each other in the ionic state was impossible before electrolysis. The cells and batteries are also one of the main important applications of electrolysis, besides this there are many other applications such as electrotyping, electro-painting, Xerography (Xerox) also the printing process of printing in the computer printers (the ink-jet and laser) are the process of application of electrolysis. We shall discuss the following topics under this sub head.

(*i*) Electroplating
(*ii*) Electrotyping
(*iii*) Galvanic Cells
(*iv*) Corrosison Protection
(*v*) Photo-electric Reactions.

GALVANIC CELLS ELECTROLYTIC CELLS AND BATTERIES

Introduction

The word Galvanic comes from the name *Luigi Ganvani* (1737-1798). He was an Italian zoologist and he was explaining the anatomy of a toad. While doing this he noticed that when he was touching the leg of the toad, with a metal rod, the leg plunged up, as it was alive. Galvani investigated and found that, it was the flow of electrons, which did that stimulation. Galyani named this electricity as the zoological electricity.

All the other scientists did not accept this conception. Among them Alexandro Volta (1745-1827) denied and said that the cause behind the stimulation was electricity, is true, but the electricity didn't originated form the toad's body. It was formed by the chemical reactions due to the presence of common salt with the presence of two different metals caused the circulation of the electricity.

Alexandro Volta simplified this statement by making a cell called voltaic cell. He kept cooper and zinc discs, one above the other, between the plates he kept papers moistened in salt solution, this was also known as Voltaic Pile (see Fig. 22.8)

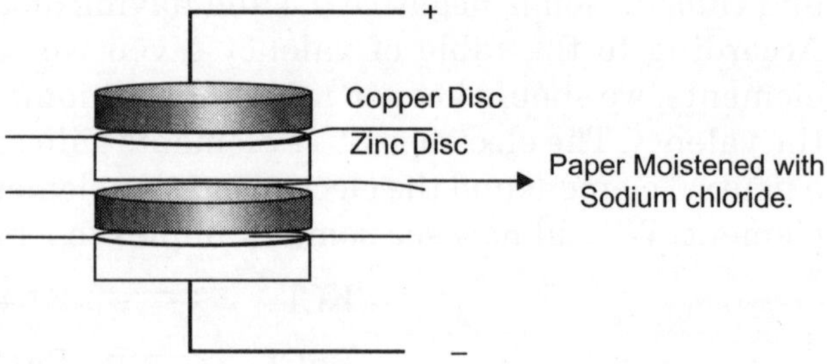

Fig. 22.8. The Voltaic Cell.

As Galvanic noticed the electricity first, Galvanic Cells were named after him.

Galvanic Cell

When we place magnesium metal in a solution of hydrochloric acid, the metal and the hydronium ion from the acid reacts spontaneously. Hydrogen gas and a solution containing magnesium ion are produced. An oxidation reduction reaction takes place.

$$\text{Magnesium(solid)} + \text{Hydronium Ion(aq)} \longrightarrow H_2(a) + \text{Magnesium Ion} + \text{Water}$$

An oxidation reduction reaction has two halves. In one half reaction the metallic magnesium is oxidised by loosing two electrons and becoming Mg^{2+} ions.

$$Mg(s) \longrightarrow Mg^{2+} + 2e^-$$

In other half of the reaction, two hydrogenium ions are reduced, they combine with two hydronium ions are reduced, they combine with two electrons giving rise to hydrogen gas and water. The net ionic reaction that describes the reduction half of the reaction is

$$2H_2O^+ \text{ (aq)} + 2e^- \longrightarrow H_2(g) + 2H_2O + (l)$$

The above reaction can be demonstrated by an experiment. In two beakers one containing magnesium chloride and the other containing the hydrochloric acid, they are connected with a salt bridge a magnesium chloride and a non-reactive platinum coil is dipped in the beaker consisting of hydrochloric acid (see Fig. 22.9).

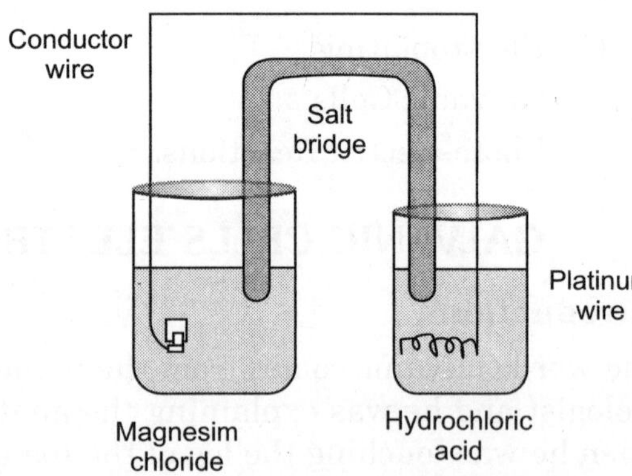

Fig. 22.9. The Galvanic cell reaction.

These reactions, the oxidation half and the reduction half of the overall reaction are called *half-reactions*. The net ionic equation that describes the overall reaction is the sum of the two half reactions. The co-efficients of the reactants and products of the half reactions are so adjusted that both the half reactions involve same number of electrons. This is necessary because the electrons produced by the oxidation half-reaction are consumed in the reduction half-reaction. In this example the co-efficient do not need any adjustment, because both of the half-reactions involve two reactions.

Oxidation half reaction : $Mg(s) \longrightarrow Mg^{2+}(aq) + 2e^-$

Reduction half reaction : $2H_3O^+ (aq) + 2e^- \longrightarrow H_2(g) + 2H_2O(l)$

Sum $Mg(s) + 2H_3O^- (aq) \longrightarrow Mg^{2+}(aq) + 2H_2O$

We can run the oxidation half and the reduction half of the reaction in two separate containers without magnesium coming in contact with hydrochloric acid.

A galvanic cell consists of two half cells with the oxidation half of the reaction occurring in the other half of the cell. The sum of the two half reactions is called the *cell reaction.* The half reaction that produces electrons is the oxidation half reaction and the electrode at which this reaction occurs is called anode of the cell. The half reaction that accepts the electrons is called reduction half reaction and the name of the electrode at which it occurs is called cathode of the cell.

Remember that anode and oxidation both start with a vowel, cathode and reduction both starts with a consonant.

To get energy from the galvanic cell we must connect the two half cells by a wire or a salt bridge through which the electrons will pass. This salt bridge consists of a porous dish, which allows ions flow but does not allow the mixing of solutions into the two half-cells. If the two solutions were allowed to mix, a direct reaction would take place, the half reactions will be shorted out and no current will flow through the wire.

The importance of the salt bridge or porous dish becomes even more apparent because when we consider what happens without one. When a wire connects two or more half-cells, not by a salt bridge a few electrons pass from the magnesium strip through the wire into the acid solution, but then the current stops. The right-hand half-cells become negatively charged because of the transfer of electrons into it and the left-hand half-cell reactions become positively charged, because the electrons leave it. The attraction of electrons by the positive charge in the right-hand half-cell prevent any further transfer of electrons, adding a salt bridge allows the two half-cells to loose excess charge and permits more electrons to flow.

Cell Potentials

A galvanic cell produces a driving force that pushes electrons through an external circuit (for example, the wire in Fig. 22.10). This driving force is called the cell potential E_{cell} of the cell. We measure a cell potential of a Danielle cell, described in the example.

Identify the anode, the cathode and the net ionic half-reactions in a Danielle cell, a battery first constructed by an English chemist *John Fedrich Danielle* and used as the source of electricity in early telegraph systems. A Danielle cell (Fig. 22.10) consists

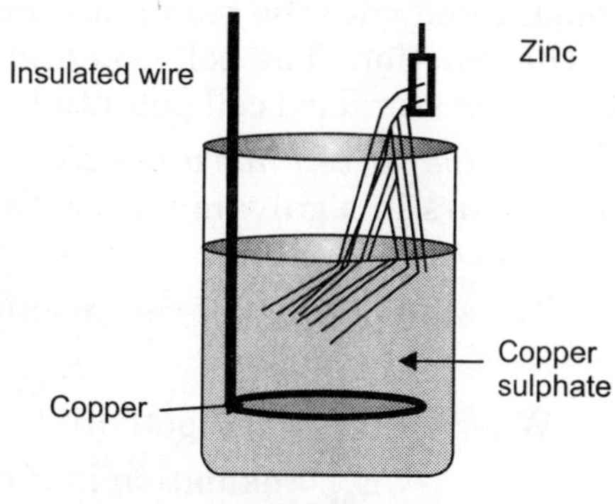

Fig. 22.10. The diagram of A Danielle Cell

of a piece of copper metal (Cu) with crystals of copper sulphate in a saturated solution of copper sulphate below a piece of metal (Zn) suspended over the top of the cell in a dilute solution of zinc sulphate. The crystals of copper sulphate simply keep the copper sulphate solution saturated. The zinc sulphate solution floats on the top of the denser copper sulphate solution, they mix very slowly and the boundary between them acts like a porous dish. The cell reaction is

$$Zn(s) + Cu^{2+}(aq) \longrightarrow Cu(s) + Zn^{2}(aq)$$

The anode is the half-cell where oxidation occurs. Zinc metal is oxidised so the anode consists of the zinc metal and the solution of zinc sulphate. The net oxidation half reaction at the anode is

$$Zn(s) \longrightarrow Zn^{2+}(aq) + 2e^-$$

The cathode is the half-cell where reduction occurs; copper (ii) ion is reduced to copper metal during the reaction, so the cathode consists of the copper sulphate solution. The net ionic reduction reaction at the cathode is

$$Cu^{2+}(aq) + 2e^- \longrightarrow Cu(s)$$

When the two metals are connected by a wire the zinc metal is oxidised and electrons from the zinc flow through the wire to the copper (ii) ions in the solution are reduced to metallic copper.

Let us now consider the half reaction of the equation

$$Zn(s) \longrightarrow Zn^{2+}(aq) + 2e^-$$

The cell consisting of two half cells, one having 1.0M zinc sulphate and one zinc strip and the other half of the cell consisting 1.0M copper sulphate with one copper strip. A salt bridge connects the two half-cells.

The sulphate ion is unchanged during the reaction, so it does not appear in the met ionic equations for the cell reaction

$$Zn(s) + Cu^{2+}(aq) \longrightarrow Cu(s) + Zn^{2+}(aq).$$

We measure the cell potential with a potentiometer placed in the circuit of the cell. The potential of the cell in (Fig. 22.8) determined by a potentiometer. This is because the cell is constructed with the reactants and products at standard state conditions at the beginning of the reaction. The cell potential is measured under standard state conditions and is known as standard cell potential, $E°_{cell}$.

A galvanic cell has a positive voltage when the cell reaction is spontaneous. A reaction that occurs in a galvanic cell with a positive potential also will be spontaneous when the reactants are mixed.

The sum of the half-cell reactions can be stated by the following reaction:-

$$E_{cell} \longrightarrow E_{ox} + E_{red}$$

Where E_{cell} = cell potential

E_{ox} = oxidation half reaction

E_{red} = half reduction reaction.

Calculation of the Cell Potential of a Standard Cell

Experiment 1. *The rechargeable cells used in the calculations and battery operated tools are based upon cadmium used nickel electrodes. What are cell reactions and the voltage of a rechargeable nickel cadmium battery (NICAD) if it is operated under standard state conditions. The standard half-cells potential are as follows :*

Anode: $Cd(s) + 2OH^-(aq) \longrightarrow Cd(OH)_2 + 2e^-$ $\quad E^\circ_{ox} = 0.40V$

Cathode: $NiO_2(s) + 2H_2O(l) + 2e^- \longrightarrow Ni(OH)_2(s) + 2OH^-$ $\quad E^\circ_{red} = 0.49V$

Solution : A battery is a galvanic cell; under standard state conditions a voltage equals the standard cell potential. The cell reaction is the sum of the anode and the cathode half reactions. The potential is the sum of the half potentials.

Anode : $Cd(s) + 2OH^-(aq) \longrightarrow Cd(OH)_2 + 2e^-$ $\quad E^\circ_{ox} = 0.40V$

Cathode : $NiO_2(s) + 2H_2O(l) + 2e^- \longrightarrow Ni(OH)_2(s) + 2OH^-$ $\quad E^\circ_{red} = 0.49V$

Sum : $Cd(s) + NiO_2(s) + 2H_2O \longrightarrow Cd(OH)_2\ (s) + Ni(OH)_2$ $\quad E_{cell} = E^\circ_{ox} + E^\circ_{red}$

$= 0.40 + 0.49 = 0.89V$

The standard state cell potential is .89V.

[It is to be noted that hydroxyl ion does not appear in the sum because due to the presence in both cathode and anode half reactions].

Standard Electrode Potential

We need to be careful with the term *electrode*. It is used in two ways (1) to refer to a wire or refer to a sum or other conductor that delivers electricity into a half-cell and (2) to refer to a complete half-cell. For example a solution of a zinc sulphate together with a zinc plate is a complete half-cell. The zinc plate is both the conductor and the reduced species; the metal ions are the oxidised species. In this section we shall discuss about those electrodes which are half-cells.

Although we can measure the potential of a galvanic cell, there is no satisfactory method for measuring the actual potential of an individual electrode. We can only measure the sum of the potentials of two electrodes. However, if one electrode is assigned to standard potential, potentials of other electrodes can be reported by comparison to the assigned value of the standard electrode. The electrode, which is most commonly used as the standard electrode, is called the *Standard Hydrogen Electrode,* which is having the potential of zero volts. The potentials of cell other electrodes are reported to the standard hydrogen electrode (This approach is similar to arbitrarily establishing sea level as zero elevation and reporting all the elevation as terms of how much higher or lower than this level they are).

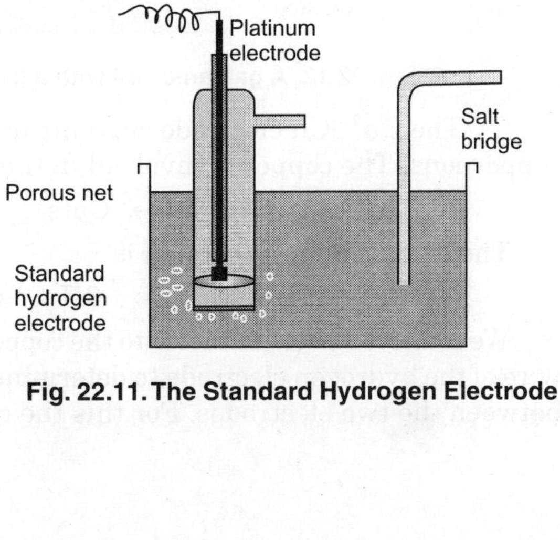

Fig. 22.11. The Standard Hydrogen Electrode

The standard hydrogen electrode is a gas electrode, a gas electrode, a half-cell with a gas one of the reagent. We can make a gas electrode by bubbling a reactive gas around an inert conductor a connector that conducts or carries electrons but does not react (see Fig. 22.11).

In a hydrogen electrode, hydrogen gas bubbles through a solution of hydrogen ions and around platinum foil or wire covered with very finely dried platinum is immersed in the solution. The net ionic equation is

$$2H_2O^+(aq) + 2e^- \longrightarrow H_2(g) + 2H_2O(l)$$

The potential of the gas electrodes changes with the changing of pressure of the gas and with changing concentration of the other components, so we must be careful to keep the pressure and concentrations constant at standard state conditions. When we measure standard potentials, a standard hydrogen electrode operates at a temperature of 25°C in a solution continuing 1M concentration of hydrogen ions.

To measure the potential of an electrode relative to that of the standard hydrogen electrode, we can use a galvanic cell consisting of the electrode being measured and a standard hydrogen electrode as reference. For example, we can measure the potential of a copper electrode by using a cell with the following cell reaction.

$$Cu^{2+}(aq) + H_2(g) = 2H_2O(l) \longrightarrow Cu(s) + 2H_3O^+(aq)$$

The cell is shown in the Fig. 22.12.

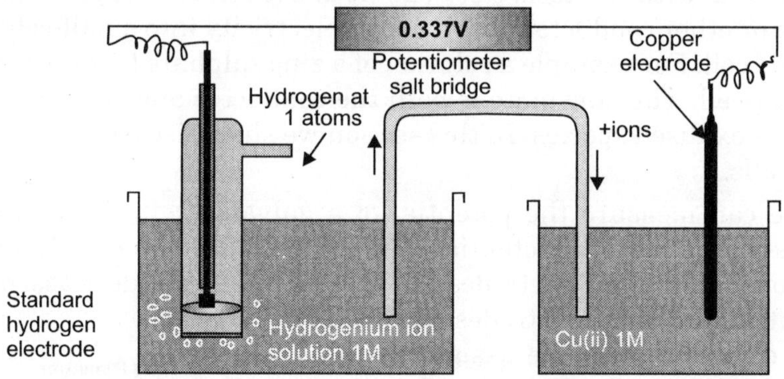

Fig. 22.12. A galvanic cell with a hydrogen electrode and a copper electrode.

The Cu^{2+}/Cu electrode currents of a copper stop in a contact with a 1M solution of copper ions. The copper is involved in the reduction half reaction.

$$Cu^{2+}(aq) + 2e^- \longrightarrow Cu(s) \qquad E^\circ_{red} = ?V$$

The oxidation half reaction is

$$H_2(g) + 2H_2O\,(l) \longrightarrow 2H_3O^+\,(aq) + 2e^- \qquad E^\circ_{ox} = 0.0V$$

We connect a potentiometer to the copper electrode in the Cu^{2+} solution and the platinum wire of the hydrogen electrode to determine the potential difference, $-E_{cell}$, the cell potential between the two electrodes. For this the cell will observe a potential of 0.337 volt.

$\therefore \quad E_{cell} = E^\circ_{ox}$ (for the hydrogen electrode) + E°_{red} (for the copper electrode)

$$0.337V = 0V + E_{red}$$

$$E^\circ_{red} = 0.337V.$$

By internal agreement, the values of electrode potentials are given for the reverse reaction in oxidation potential, is equal to the negative of the reduction potential. The reduction potential and the oxidation potential of an electrode have the same absolute value but are of opposite sign.

Once a standard potential has been established relative to the standard hydrogen electrode, it can be used to determine other potentials.

Determination of Electrode Potential

A galvanic cell at 25°C has an anode consisting of an iron strip immersed in a 1M solution of Ferrous perchlorate (Fe^{2+}/Fe electrode) E°_{ox} and a cathode consisting of a copper strip immersed in a 1M solution of cuprous perchlorate (Cu^{2+}/Cu electrode) $E^\circ_{red} = 0.337V$. A potentiometer shows the cell potential [E°_{cell}] to be 0.777.

The cell reaction is

$$Fe(s) = Cu^{2+}(aq) \longrightarrow Fe^{2+}(aq) + Cu(s).$$

What is the standard reaction potential of the Fe^{2+}/Fe electrode?

Solution : The standard cell potential (0.777V) is the sum of standard electrode potential of the anode and of the cathode. Iron metal is oxidised in the anode half reaction, copper ion is reduced in the cathode half reaction.

Anode $Fe(s) \longrightarrow$	$Fe^{2+}(aq) + 2e^-$	$E = E^\circ_{ox} = ?$
Cathode $Cu^{2+}(aq) + 2e^- \longrightarrow$	$Cu(s)$	$E = E^\circ_{red} = 0.337$
Sum $Fe(s) + Cu^{2+}(aq) \longrightarrow$	$Fe^{2+}(aq) + Cu(s)$	$E^\circ_{cell} = E^\circ_{ox} + E^\circ_{red}$

$$0.777V = E^\circ_{ox} + 0.337V$$

$$E^\circ_{ox} = E^\circ_{cell} - 0.337V = 0.440V$$

Thus iron electrode is the anode of the cell and is an oxidation half reaction. In order to find the standard reduction potential for the electrode we must reverse the sign of the oxidation potential.

$$E^\circ_{red} = -E^\circ_{ox} = 0.\ 440V$$

The standard reduction potential of the Fe^{2+}/Fe electrode is -0.440V

$$Fe(s) \longrightarrow Fe^{2+}(aq) + 2e^- \qquad E^\circ_{red} = -0.440V$$

The table below is the list of the standard half reaction potentials:

Half-Reaction	E°, V
$K^+ + e^- \longrightarrow K$	– 2.925
$Ba^{2+} + 2e^- \longrightarrow Ba$	– 2.90
$Ca^{2+} + 2e^- \longrightarrow Ca$	– 2.87
$Na^+ + e^- \longrightarrow Na$	– 2.714
$Mg^{2+} + 2e^- \longrightarrow Mg$	– 2.37
$Al^{3+} + 3e^- \longrightarrow Al$	– 1.66
$Zn(OH)_2 + 2e^- \longrightarrow Zn + 2OH^-$	– 1.245
$Mn^{2+} + 2e^- \longrightarrow Mn$	– 1.18
$Fe(OH)_2 + 2e^- \longrightarrow Fe + 2OH^-$	– 0.877
$Zn^{2+} + 2e^- \longrightarrow Zn$	– 0.763
$Cr^{3+} + 3e^- \longrightarrow Cr$	– 0.74
$Fe^{2+} + 2e^- \longrightarrow Fe$	– 0.440
$Cd^{2+} + 2e^- \longrightarrow Cd$	– 0.403
$PbSO_4 + 2e^- \longrightarrow Pb + SO_4$	– 0.356
$CO^{2+} + 2e^- \longrightarrow Co$	– 0.277
$Ni^{2+} + 2e^- \longrightarrow Ni$	– 0.257
$Sn^{2+} + 2e^- \longrightarrow Sn$	– 0.136
$Pb^{2+} + 2e^- \longrightarrow Pb$	– 0.126
$2H_3O^+ + 2e^- \longrightarrow H_2 + 2H_2O$	0.00
$Sn^{4+} + 2e^- \longrightarrow Sn^{2+}$	+ 0.15
$AgCl + e^- \longrightarrow Ag + Cl^-$	+ 0.222
$Hg_2Cl_2 + 2e^- \longrightarrow 2Hg + 2Cl^-$	+ 0.27
$Cu^{2+} + 2e^- \longrightarrow Cu$	+ 0.337
$NiO_2 + 2H_2O + 3e^- \longrightarrow Ni(OH)_2 + 2OH^-$	+ 0.49
$I_2 + 2e^- \longrightarrow 2I^-$	+ 0.5355
$MnO_4^- + 2H_2O + 3e^- \longrightarrow MnO_2 + 4OH^-$	+ 0.588
$Fe^{3+} + e^- \longrightarrow Fe^{2+}$	+ 0.771
$Hg_2^{2+} + 2e^- \longrightarrow 2Hg$	+ 0.789
$Ag^+ + e^- \longrightarrow Ag$	+ 0.7991
$Br(l) + 2e^- \longrightarrow 2Br^-$	+ 1.0652
$Pt^{2+} + 2e^- \longrightarrow Pt$	~ +1.20
$O_2 + 4H_3O^+ + 4e^- \longrightarrow 6H_2O$	+ 1.23
$Cl_2 + 2e^- \longrightarrow 2Cl^-$	+ 1.3595
$Au^{3+} + 3e^- \longrightarrow Au$	+ 1.50
$MnO_4^- + 8H_3O^+ + 5e^- \longrightarrow Mn^{2+} + 12H_2O$	+ 1.51
$PbO_2 + SO_4^{2+} + 4H_3O^+ + 2e^- \longrightarrow PbSO_4 + 6H_2O$	+ 1.685
$F_2 + 2e^- \longrightarrow 2F^-$	+ 2.87

It is to be noted that when the concentration of the hydrogenium ion is 1×10^7 M in the distilled water, the electrode potential decreases to –0.41 volt.

The position of metal in the table of reduction potentials, when the electrodes are arranged from the most negative to the most positive, is the same as in the activity series.

Calculation of Cell Potential

The values in the table provide information that we can utilise in determining saturated state cell potentials. However, we must remember two points when using these values:

- The E° values are for reduction half reactions and the sign of a reduction potential must be reversed when it is used as a potential for an oxidation half reaction sign when the direction of a reaction is reversed.
- Changing the stochiometric co-efficient of a half cell reaction does not change the value of E°, because electrode potential are intensive properties for e.g., from table 1.

$$Ag^+(aq) + e^- \longrightarrow Ag(s) \qquad E^\circ_{red} = 0.799V$$

E°_{red} does not change when the quantities are changed.

$$3Ag^+(aq) + 3e^- \longrightarrow 3Ag(s) \qquad E^\circ_{red} = 0.799V$$

This might be unrealistic but we see in batteries all the time. The voltage (potential) of a battery does not change with the size. A larger D-size alkaline battery and a small AA size alkaline battery both have the same voltage.

Conductivity and Conductance

This term is most frequently used in the electrochemistry. The term conductance means the case with which a current flows through the conductors. It is usually written as C= 1/R. This conductance is the reciprocal of resistance.

As resistance (R) is measured in ohms conductance is expressed in mhos.

Specific Conductance (k)

Specific conductance is the reciprocal of specific resistance. *It is the resistance offered by the conductor to the passage of electricity through it and is proportional to the length of the conductor (l) and inversely proportional to the area of the cross section(s)* to the specific resistance R.

$$\frac{R = \rho \times 1}{S} \quad \text{where } \rho \text{ (rho) is a constant}$$

The value of which depends upon the material of the conductor when *l* and s are equal then, R = ρ, from which it can be stated *that the specific resistance of and electrolyte is the resistance in ohms which a centimetre cube of it offers to the passage of electricity.*

Its is expressed in $ohm^{-1}\ cm^{-1}$ here *l*/s = cell constant.

Thus, specific conductance = cell constant × conductance.

$$\therefore \text{Cell constant} = \frac{\text{Specific conductance}}{\text{Conductance}}.$$

In the table below there are some of the specific conductance of electrolytes.

Substance	Specific Conductance
Water (pure)	6.0×10^{-5}
0.1 M Hydrochloric Acid	3.5×10^{-2}
0.1 m Nitric Acid	4.7×10^{-4}
0.1 M Sodium Chloride	9.2×10^{-2}
Iron Metal	1.9×10^{-5}
Silver Metal.	5.0×10^{-5}

Equivalent Conductance

It is the property, which is suitable for characterising electrolytic conductors (e.g HCl, KCl etc). In such cases, the value of conductance also depends upon the concentration of the electrolyte or the number of ions present in solution. One mole of potassium chloride ionises to give one mole of K^+ ions and one mole of Cl^- ions. The total charge carried out by their ions is equal to two faradays. On the other hand, one mole of potassium sulphate ionises to give two modes of K^+ and one mole of SO_2^{-2} ions. The total charge carried by these ions is equal to 4 faradays. Clearly two solutions containing one mole of KCl and the other one consisting of half mole of potassium sulphate will give comprehensive results.

Hence, for measuring conductance of electrolytes, a more significant term is commonly used, which is called *equivalent conductance.*

Equivalent conductance is developed in the conducting power of all the ions produced by one gram equivalent of an electrolyte in given solution and it is denoted by λ_v.

Equivalent Conductance = Specific Conductance × V

Where V is the volume in cc containing 1 gram equivalent of electrolyte.

Thus $\lambda_v = k \times V$.

Equivalent conductance (λ_v) is expressed as $ohm^{-1}\ cm^2/gm$ eq. Relation between specific conductance and equivalent conductance of a solution is generally not found directly but is calculated from the specific conductance. The relation between the specific conductance and the equivalent conductance is as follows.

In the equivalent conductance the volume of the electrolytic is concerned which is related to specific conductance for e.g in a cell designed with 1 cm × 1 cm × 1 cm × 1 cm walls with two of the walls are conducting electrodes the cell having 1 cm^3 of the electrolyte. The cell thus having a conductance will be regarded as the specific conductance. Now if the volume of the electrolyte is increased (by 9 times $\therefore$ 1 $cm^3 \times = 9\ cm^3$) the conductance still remains as the equivalent conductance, but the value of the equivalent conductance will change. From this if the conductance of each say that if the conductance of each cm^3 of the solution is equal to its specific conductance so that the total conductance, that is equivalent conductance of the solution is nine times the specific conductance. Now if the solution is diluted by 1000 time there will be 1000 cm^3 of the solution and then the equivalent conductance will be:

Equivalent Conductance $\lambda_v = k \times 1000/C$

where k is the specific conductance and C is the molar concentration (that is number of moles of electrolyte dissolved per litre)

Or more clearly $\lambda_v = k \times V$ or, $ohm^{-1} \dfrac{cm^{-1} \times cm^3}{\text{Gram equivalent}}$ or, $ohm^{-1}\ cm^{-1}(g\text{-}eq)^{-1}$

Molar Conductance

The molar conductance of a solution is the conductivity power of all the ions furnished by one gram-mole of electrolyte solution. It is usually represented by μ. As in the case of equivalent conductance, the molar conductance is related to the specific conductance by the relation

$\mu = k \times V.$

where k is the specific conductance and V is the volume.

Measurement of Conductivity

Since the conductance is just the reciprocal of the resistance, determination of conductance of a solution amounts to the measurement of its resistance. Measurement of resistance is more frequently carried out by using a whetstone bridge. The principle of Whetstone Bridge may be explained with the help of Fig. 22.13.

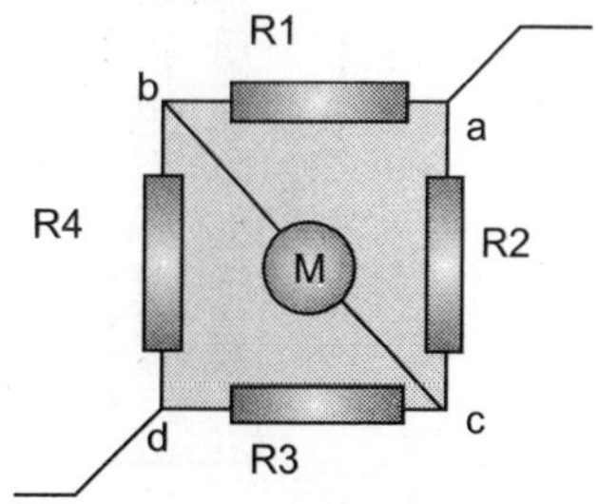

Fig. 22.13. The Whetstone Bridge.

The four arms ab, bd, ac and cd have resistance R1, R2, R3 and R4. A source of A C is connected between a and d. A meter is also connected between b and c.

Let E_1, E_2, E_3, E_4 be the fall of potentials across the four arms with resistance R_1, R_2, R_3 and R_4 respectively and also let the current of the four arms be I_1, I_2, I_3 and I_4. According to Ohm's law.

$$E = R \times I \quad \therefore \quad \left.\begin{aligned} E_1 &= R_1 \times I_1 \\ E_2 &= R_2 \times I_2 \\ E_3 &= R_3 \times I_3 \\ E_4 &= R_4 \times I_4 \end{aligned}\right\} \qquad \ldots(1)$$

If the resistance are adjusted in such a way that there is no flow of current through the meter M the bridge is said to be balanced. Then at this condition the following condition is applicable.

$$\frac{R_1}{R_2} = \frac{R_3}{R_4}$$

If the resistance of the three arms use known the resistance of the forth arm can be calculated from the above relation.

The solution whose conductance is to be determined is placed in a special type of cell called conducting cells. The conductivity cells which are commonly used has been shown in Fig. 22.14.

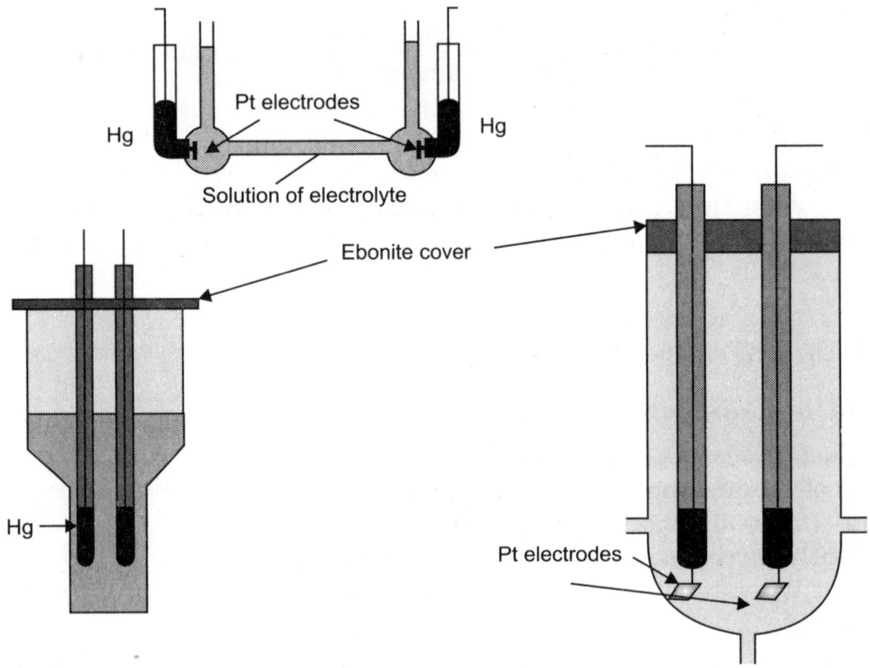

Fig. 22.14. Types of conductivity cells.

The electrodes consist of platinum discs coated with finely divided platinum black and wielded to platinum wires fixed in two glass tubes. The glass tubes contain mercury and are firmly fixed in an ebonite cover of cell. So that the distance between the electrodes may not alter during the experiment. Dipping copper wire in the mercury contained in the tubes makes contact with the platinum plates. The major difficulty in determining the resistance of the solution of an electrolyte is that electrolysis of the solution takes place simultaneously with the conduction of the current and produces two complications:

1. *Polarisation* soon sets in, which causes the resistance to alter.
2. *The concentration* of the solution alters.

To avoid these complications and alternating current is employed and instead of a detecting meter, a headphone is used. The whole apparatus is shown in Fig. 22.15.

C is a conductivity cell, containing the solution under examination. R is the resistance box. TH is the headphone to detect the current. XY is the uniform wire and a sliding contact point A moves over it.

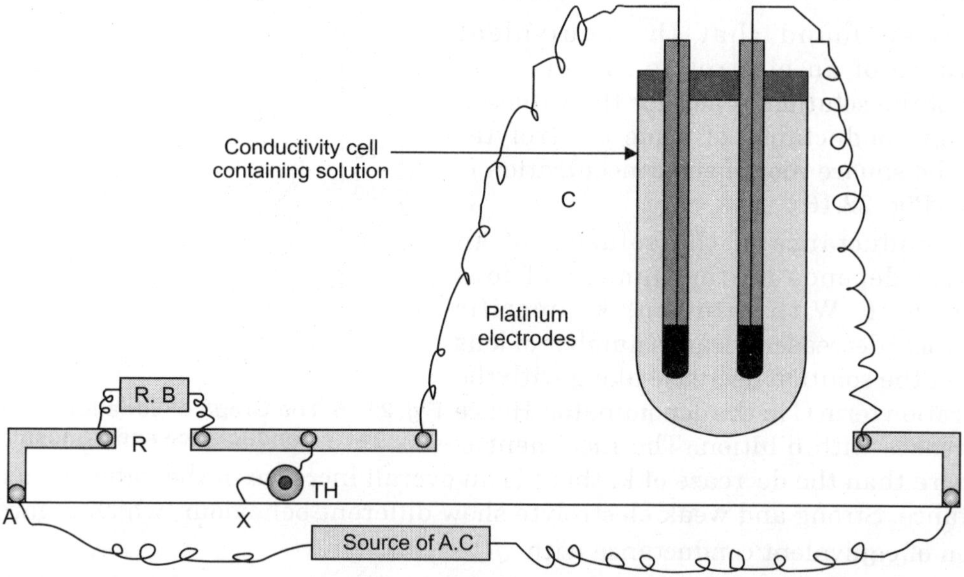

Fig. 22.15. The Experiment of Measuring the Resistance of an Electrolytic Cell by using AC

To start with a known resistance which is approximately of the same order (as that of solution) is introduced from the resistance box. Now the sliding contact A is moved as the uniform wire and a point of minimum sound is detected. Let this point be A. Thus

$$\frac{\text{Resistance of Solution}}{\text{Resistance R}} = \frac{\text{Length AY}}{\text{Length AX}}$$

The AX and AY can be measured on the scale. Hence, the resistance of the solution is determined. The reciprocal of these values give the conductance of the solution.

FACTORS DETERMINING THE CONDUCTANCE OF AN ELECTROLYTE

The conductance of an electrolyte is mainly dependent upon the following factors: -

(*i*) Nature of Electrolyte

(*ii*) Temperature of Electrolyte

(*iii*) Concentration of the Electrolyte

(i) Nature of the Electrolyte : The conductance of the solution depends upon the nature of the electrolyte. On the basis of conductance measurements of the solutions of different electrolytes these have been classified as strong and weak electrolytes they are as follows:

(a) Strong Electrolyte : Substances like mined acids and bases which do not dissociate completely in saturation and thus process low values of equivalent conductance are known as weak electrolytes.

Variation of Conductance with Concentration of Strong and Weak Electrolytes

It has been found that the equivalent conductance of an electrolyte increases on dilution of the solution. A plot for the values of equivalent conductance of some electrolytes against the square root of the concentration is shown in Fig. 22.16.

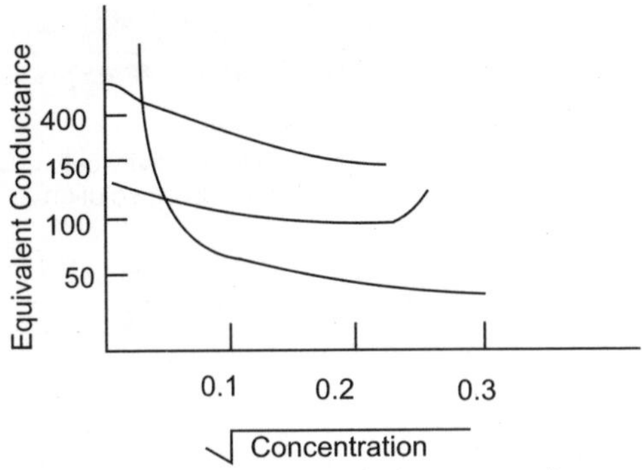

Fig. 22.16. The Graph of Variationof Equivalent Conductance and Concentration.

The conductance of the solution of an electrolyte depends on the number of ions present in it. With dilution, k (specific conductance) decreases because number of ions per cm^3 of the solution decrease along with the concentration term C in the denominator. Hence 1/C increases with dilution. The increment is much more than the decrease of k, there is an overall increase in the value of equivalent conductance. Strong and weak electrolyte show different behaviour, which regard to the variation of equivalent conductance with $\sqrt{C}$.

1. In strong electrolytes like HCl or KCl, there is a tendency for equivalent conductance to approach a certain limiting value. When the concentration approaches zero, the value of the equivalent conductance is called *equivalent conductance at zero concentration or at infinite dilution,* which is denoted as λ. This value can be obtained by replotting the line in the graph shown in Fig. 22.15 to zero concentration.
2. In weak electrolytes such as acetic acid, it can be observed that no limiting value can be attained even when the concentration approaches zero. Hence, equivalent conductance of weak electrolyte cannot be obtained by replotting. As discussed earlier, the value at zero concentration can be obtained individually by applying Kohlrausch's Law. We know that the conductance of solution depends upon the degree of ionization of the electrolyte. As the degree of ionization increases with dilution, we say that the value of equivalent conductance also increases with dilution.

   When the limiting value of equivalent conductance is reached, the degree of dissociation is unity, here the complete dissociation of the electrolyte is obtained. This happens in the case of strong electrolytes, but in the case of weak electrolytes the degree of dissociation ($\propto$) is not complete even at great dilution. Thus the degree of dissociation is expressed in the form of the following expression.

$$\propto = \frac{\lambda_0}{\lambda_k} = \frac{\text{Equivalent Conductance at given Concentration}}{\text{Equivalent Conductance at zero Concentration}}$$

   The conductance ratio provides a method for measuring the degree of dissociation and hence the degree of dissociation of the weak acid. λ_∞ is found by measuring the conductance of a solution of known concentration of the weak acid and λ_∞ can be found out in consideration of Kohlrausch's Law of migration of ions.

Kohlrausch's Law of Migration of Ions

Kohlrausch studied that equivalent conductance at infinite dilution (λ_∞) for a number of pairs of strong electrolytes having the same cation or the same anion. Thus, he calculated the difference of these values, λ_∞ for each pair. A few of the results are given in the table.

Electrolyte	λ_∞	Difference
KCl NaCl	149.86 126.45	23.41
KNO_3 $NaNO_3$	144.96 121.55	23.41
KBr NaBr	151.92 128.51	23.41
KBr KCl	151.92 149.86	2.06
NaBr NaCl	128.51 126.45	2.06
KBr NaBr	151.92 199.86	2.06
LiBr LiCl	117.09 115.03	2.06

It is evident that the replacement of potassium ion by sodium ion in any of the electrolyte having a common anion produces the same difference (23.41 $ohm^{-1}\ cm^2$) in, λ_∞, irrespective of the nature of the anion.

Similarly, the replacement of the bromide ion by chloride ion in any of the electrolytes having a common cation produces the same difference (2.) 6 $ohm^{-1}\ cm^2$) in λ_∞. In other words, the difference in conductance of any two cations or anions appear to be the same irrespective of the nature of the other ion with which they are associated. This led Kohlrausch to conclude that: At infinite dilution, when dissociation is complete, each ion makes a definite contribution towards equivalent conductance of the electrolyte. Irrespective of the nature of the other ion with which it is associated and that the value of the equivalent conductance at infinite dilution for any electrolyte is given by the sum of the contribution of two ions.

Thus $\lambda_\infty = \lambda^0_c + \lambda^0_a$ where λ^0_c = cation and λ^0_a = anion at infinite dilution.

It is also a fact that the ionic conductance of any ion is found at a given temperature what ever might be the nature of the other ion associated with it.

Conductimetric Titration

The titrations in which experiment or equivalent point is determined by means of conductance measurement, are called conductimetric titrations. They are based upon the principle that conductance of an electrolyte solution depends upon the number of ions, their charges and mobilises. As a reaction proceeds during a titration, number of the ions or concentrations of ions starts changing (see Fig. 22.17).

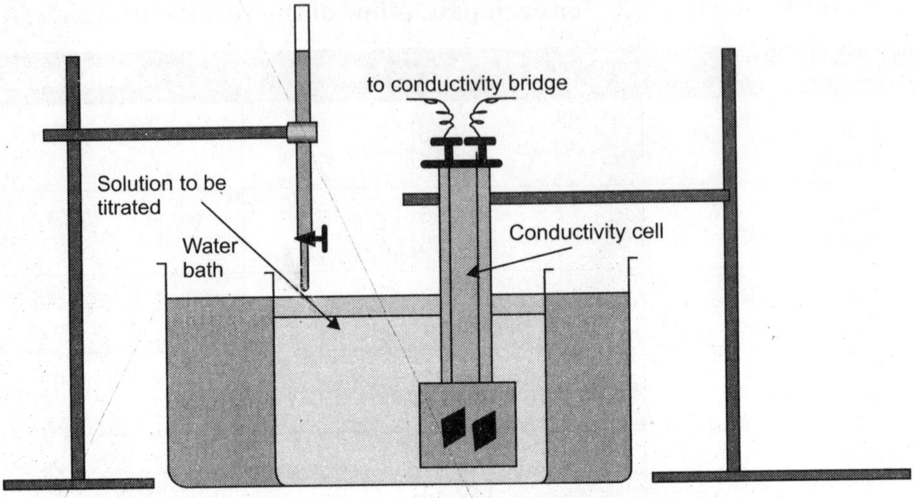

Fig. 22.17. The Apparatus set up for Conductimetric Titrations.

Also the ions of one kind are replaced by the other kind, differing in the mobility. As a result, the conductance of the solution changes with every addition of the titrant. At the end point a sharp change in conductance is observed. Thus, if the conductance of solution is recorded during the course of titration and plotted against the volume of the titrant added, the end point could be conveniently obtained by plotting the straight lines and finding out the point of their intersection. Titrations are usually carried out by taking a solution in a beaker, which is kept in the water bath at a constant temperature and in which a conductivity cell of the type shown in Fig. 22.16 is used. The cell is connected to a conductivity bridge circuit (Whetstone Bridge). The titrant is added from the burette, the solution is stirred and the conductance is measured after each addition. For an accurate result, there should be no appreciable change in the volume of the solution being titrated during conductimetric titrations. This is best achieved by using a solution of titrant 20 to 100 times concentrated than the solution to be titrated and using a micro-burette for measuring the solutions.

Experiment 1. Consider the solution of hydrochloric acid (a strong acid) with sodium hydroxide a strong alkali. A measured volume of standard acid (say N/10) is taken in the burette. We know the conductance of the hydrochloric acid is due to H^+ ions will go decreasing instead Na^+ ions appear due to the formation of barium chloride.

$$HCl + NaOH \longrightarrow NaCl + H_2O$$

Or $$H^+ + Cl^- + Na^+ + OH^- \longrightarrow Na^+ + Cl^- + H_2O$$

Compared to H^+ ions. Na^+ ions are slow moving. Thus as NaOH is being added the conductance of the solution will go on decreasing until the acid is completely neutralised.

Further addition of the sodium hydroxide results in increasing in conductance due to presence of fast moving hydroxyl ions. Hence, we say that in the acid-alkali titrations, the gradual addition of alkali shows a continuous decrease in conductance, reaches a minimum value and again starts increasing. Now if a graph is plotted from the conductance against the volume of alkali added, we get two lines AB and CD (see Fig. 22.18).

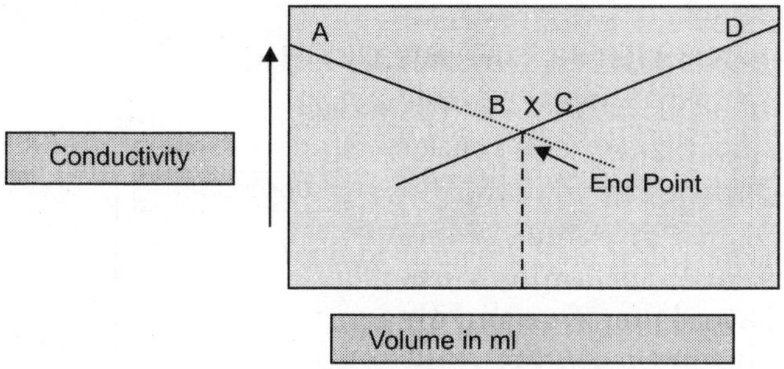

Fig. 22.18. The graph of titration

The two lines intersect at point X this point gives the volume of alkali required for the neutralisation of definite volume of the acid.

Experiment 2. Consider the titration of a weak acid (say oxalic acid) against a strong alkali (say sodium hydroxide) a measured volume of acid is taken in the conductivity vessel and strong alkali is taken in the burette. We know that weak acid is slightly ionised and, therefore, its conductance will be low. When strong alkali, say sodium hydroxide is gradually added hydrogen ion gets neutralised by hydroxyl ions (OH^-) and also sodium acetate is formed. Sodium acetate is highly ionised and therefore the conductivity of the cell increases again.

$$(COOH)_2 + NaOH \longrightarrow (COONa)_2 + H_2O$$

or $$(COOH)_2 + Na^+ + OH^- \longrightarrow COOH^- + Na^+ + H_2O$$

Now when the acid is completely neutralised further addition of sodium hydroxide solution shows a sharp increase in conductance than before, because of presence of fast moving hydroxyl ions. The graph for the conductance against the volume of the alkali added, shows two lines, which intersect at the end point (see Fig. 22.19).

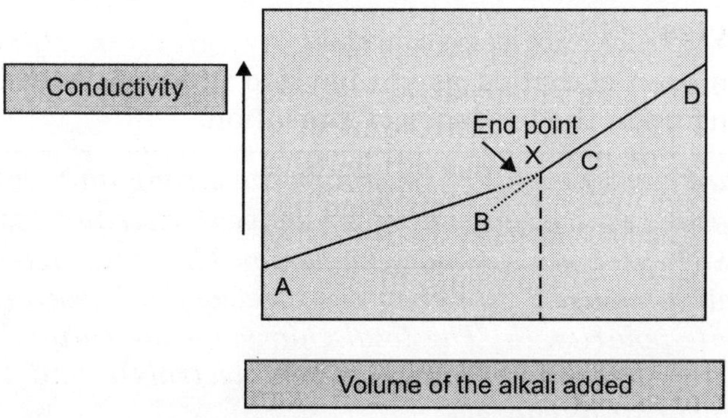

Fig. 22.19. The titration graph

Experiment 3. The titration of strong acid (say HCl) against a weak base (say NH_4OH).

A measured volume of acid is taken in a conductivity vessel and weak alkali is taken in the burette. We know that strong acid is highly ionised and its conductance will be high. As alkali is gradually added, the conductance will go on decreasing. The decrease in concentration is due to the fact that H^+ ions are continuously removed and instead slow moving NH_4^+ ions formed.

$$HCl + NH_4OH \longrightarrow NH_4Cl + H_2O$$

Or $$H^+ Cl^- + NH_4OH \longrightarrow Cl^- + H_2O$$

When the acid is completely neutralised and then further addition of ammonium hydroxide does not cause any appreciable change in conductance, as ammonium hydroxide is very weak and therefore slightly ionised. Thus, after the end point the value of the conductance remains almost constant (see Fig. 22.20)

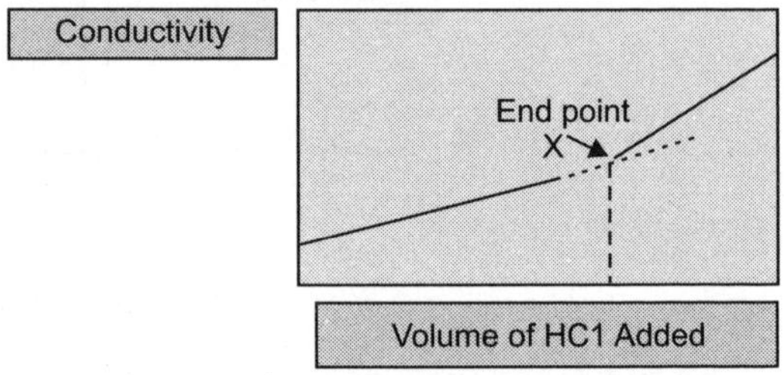

Fig. 22.20. The graph of titration

Experiment 4. The reaction between a weak acid (say acetic acid) and a weak alkali (say ammonium hydroxide). The conductivity in the beginning decreases due to the formation of a strong CH_3COONH_4, (ammonium acetate), which decreases the dissociation of the weak acid any further. With more addition of the ammonium hydroxide, the conductance increases due to increased concentration of the salt, which is slightly ionised. This continues until at end point considerable hydrolysis of the salt occurs. However, beyond this, 'the excessive base added superposes the salt hydrolysis. Thus the conductance remains mearly constant (see Fig. 22.21)

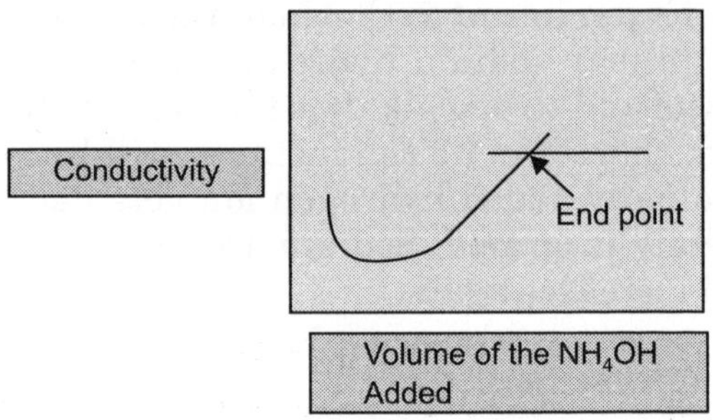

Fig. 22.21. The titration curve

Theory of Electrolytes

Electrolytes have been classified as earlier into two groups called strong and weak electrolytes depending upon the tendency of ionisation.

1. *Electrolytes are classified into two groups the strong and weak.*
2. *Strong electrolytes ionize completely, while weak does not easily.*
3. *Arrhenius (1887) proposed a theory known as Theory of electrolytic dissociation, the main postulates are : (a) When the electrolyte dissociates it produces ions having opposite polarity, (b) The total charge on the cation is just opposite and equal to that of the anion present in the electrolyte and therefore the whole electrolyte is neutral in charge.*

Strong electrolytes are almost completely ionised into solutions while weak electrolytes undergo very little in ionization. To study the behaviour of the solutions of electrolytes, *Arrhenius* (1887) proposed a theory known as *theory of electrolytic solutions.* Its main postulates are as follows :

- Molecules of an electrolyte when dissolved in water split up into ions.
- The cations and the anions have equal but opposite charges so the atom is neutral electrically.
- The ions in solution move freely but under the influence of electricity, they are direct towards opportunity charged electrodes. Electric current does not produce ions but it has a directive influence.
- The electrolyte in aqueous solution may not be completely ionised. The ratio of the dissociated molecules to the total number of molecules is termed as the *degree of dissociation.*

The ions present in the solution also reunite to form undissociated molecules; thus equilibrium is maintained between the ions and the undissociated molecules, such as electrolyte BA is dissociated in water.

$$BA \longleftrightarrow B^+ + A^-$$

The behaviour of weak electrolytes has been successfully explained on the basis of Arrhenius Theory. However, if fails to explain the behaviour of the strong electrolytes.

Demerits of Arrhenius Theory

1. The Ostwald's dilution Law based on dissociation theory is obeyed by weak electrolytes, but not all by strong electrolytes.
2. Degree of dissociation, ∞ is calculated from the conductance measurement, i.e. $\infty = \frac{\lambda_k}{\lambda_\infty}$ for concentrated solution of the strong electrolyte was found to be different from the one calculated by colligative property measurement.
3. It is now known that even in crystal of sodium chloride the ions exits separately, the Na^+ and Cl^- ions. It is difficult to explain how the undissociated molecules,
$$Na^+ + Cl^- \longleftrightarrow NaCl.$$
4. The value of equilibrium constants of strong electrolytes have been found to vary with the concentration of the electrolyte. The existence of the ionic equilibrium as explained by the Arrhenius is doubtful.
5. It is an observed fact that the equivalent conductance of both strong and weak electrolyte increases with dilution. This increase in case of weak electrolyte, is easily understood because dilution ionises more of the undissociated molecules of weak electrolyte. In case of strong electrolyte, which is completely ionised even in the solid state, the increase of equivalent conductance with dilution cannot be explained.

To explain the above difficulties, a theory for the strong electrolytes was put by *Debye* and *Huckel* in 1923. Their theory is based on the assumption that strong electrolytes are completely dissociated into ions. The observed difficulties or deviations from the ideal behaviour are associated to inter ionic attractions. Although Arrhenius theory assumed the existence of ions, it did not take into account the electrostatic attractions between the ions in the solutions.

The effect of interionic attractions in case of the solution is not large. Hence, the effect of interionic attractions would be noticeable. The effect will be still mere significant for the electrolytes having high valencies of ions.

Debye Huckel Theory

This theory is based on the assumption that strong electrolytes completely dissociate into ions.

According to Debye and Huckel, the increase in equivalent conductance with dilution (in strong electrolyte) is due to increase in the mobilities of ions; rather the increase in the degree of ionisation as suggested by Arrhenius. Degree of ionisation, according modern theory of strong electrolytes remains unity over a wide range. In the same way the decrease in equivalent conductance of strong electrolyte with increasing concentration is due to decrease of mobilities of the ions. Thus *Inter-ionic Effect* plays great role in understanding the behaviour of the solutions of the strong electrolytes. Inter-ionic effect is caused due to the increase in the concentration of the solution. As the concentration is increased, the ionic mobility falls due to mutual attraction of oppositely charged ions. Thus, their contribution towards electrical conductance decreases. On the other hand, if the electrolyte is present in a very dilute solution, inter ionic attractions become negligible, as the ion remains far apart. The ions contribute their full share towards total conductance, as they are independent of each other. Therefore, interionic attractions are the cause of decrease in conductance of strong electrolytes with increasing concentration.

Debye Huckel Equation

P. Debye and *E. Huckel* (1923) mathematically expressed the magnitudes of (1) asymmetric effect, (2) elctrophoretic effect in terms of such factor as ionic concentration, the valency of the ion and viscosity of the medium. The expression for uni-valent electrolyte :

$$\lambda = \lambda_\infty \left[\frac{\varepsilon_k}{(DT - 273)\eta} + \frac{\varepsilon^2 k}{(DT - 273)\lambda_\infty} \right] \times \bar{C}$$

This equation is also known as Onsagar equation because later on developed by L. Onsager

Where λ_∞ = Equivalent Conductance in infinite dilution.

$\varepsilon^2 k$ = Electrical Force field of the asymmetric effect(B)

D = Dielectric constant of the cell.

T -273 = Temperature in centigrade

ε_k = Electrical Force field of the electrophoretic effect (A)

C = Concentration in moles per litre

η = Co-efficient of viscosity.

This equation accounts for the difference between the equivalent conductance of the strong electrolyte at any concentration and equivalent conductance at infinite dilution (λ_∞). Specific conductance is always found less than λ_∞. This is due to asymmetry and electrophoretic effect. The first term in the bracket refers to the electrophoretic effect and the second term in the bracket refers to the asymmetry effect. The above equation for a given solvent and at a given temperature is expressed as

$$\lambda = V_0 - (A + BV_0)\sqrt{C} \text{ where A and B are constants.}$$

If a solution approaches infinite dilution, λ is almost zero. In such a case, λ_∞ is almost equal to λ.

Debye-Falkenhagen Effect : Conductance under High Frequency Current

Debye-Falkenhagen studied the conductance behaviour of a solution of strong electrolytes. They applied alternating current of different frequencies on the solution. According to them, if the frequency of the alternating current is high then the asymmetry effect will be absent. Under such conditions, the retarding effect due to asymmetry will be absent and the conductance will increase. Here we can say that the conductance will be high. Hence, we can say that the conductance of a solution changes with the frequency of the alternating current used. Higher the frequency of the current, higher the conductance of the cell. This can be proved by the experimental procedures.

Wien Effect : Conductance under High Potential Gradient

It has been found that the speed of the ions depends upon the applied potential gradient. An ion may have a speed of 100 cm sec^{-1} under potential gradient of 20,000 volts per cm. Under such conditions the ion will moving so that there will be no time for building up of concentrated atmosphere of any kind. Hence the asymmetric and the electrophoretic effects are not formed. The conductance of strong electrolyte in aqueous solution will increase with increase in potential gradient applied. This observation was experimentally verified by M. Wien (1927) and so named as *Wien effect.*

Electrolytic Cells

These cells behave opposite to the galvanic cells, the galvanic cells produces electricity from chemical reaction which was developed by *Alexandro Volta.* The electrolytic cells consume electricity and produces chemical reaction.

The difference between the galvanic cell and the electrolytic cell is shown in the table below :

Galvanic Cell	Electrolytic Cell
1. In galvanic cell, the oxidation reduction reactions takes place indirectly and the electrical energy is produced.	1. In electrolytic cell, the electrical energy is used and electrochemical reaction is carried out.
2. In galvanic cell reaction is spontaneous.	2. In electrolytic cell the reaction is non-spontaneous.
3. In galvanic cell the anode is negative cathode is positive.	3. In electrolytic cell anode is positive and the cathode is negative.
4. Two electrodes are set up in two different vessels.	4. In electrolytic cell both the electrodes are electrodes are suspended in the same vessel in the same electrolyte.
5. The electrolytes are kept in two half cells, which are separated.	5. Only one electrolyte is taken.
6. Salt bridge or porous pot acts as a connector.	6. There is no such thing required.

We shall now study the practical applications of galvanic cells commercially known as batteries.

The practical application of electrolytic cells that is the electroplating, which we are going to study after this discussion on batteries.

Batteries are galvanic cells constructed so that they can deliver large current and tolerate rough handling. Some instruments like torchlight require only one cell. There are other forms of cells also which can give high voltage and high current they are also required in the automobiles and other heavy applications. Now it is a problem to increase the e.m.f of the cell rather the potential difference of the cell. The cells are connected in series. Actually when a single galvanic unit is considered it called *cell*, these single galvanic units when connected in series or parallel then it is called *battery* (the word battery comes from the word battalion). Thus a battery can provide more potential than a cell.

These cells are subdivided into two groups they are (1) Primary Cells and (2) Secondary Cells.

Primary Cell

The galvanic cell produces power due to the internal chemical reactions, *those cells that can produce power without any secondary means rather no external sources is called primary cell*. In these types of cells the internal chemical reaction is not reversible, therefore it immediately stops the voltage output when the reaction comes to equilibrium.

Among the primary cells Daniel Cell, Leclanche Cell, Smee's Cell, Grove Cell and Dichromate Cell they are all well known. These cells are not rechargeable that means, when the internal chemical reaction is stopped it cannot be started again by applying an electrical current.

Most inexpensive flashlight cells are Leclanche Cells.

Leclanche Cell

It consists of a glass vessel containing a strong solution of ammonium chloride having a zinc rod, which serves as the negative electrode (see Fig. 22.22).

The positive consists of a carbon rod placed in the porous vessel containing manganese dioxide and gas carbon.

The hydrogen is oxidised to water and manganese dioxide is reduced to a lower oxide of manganese.

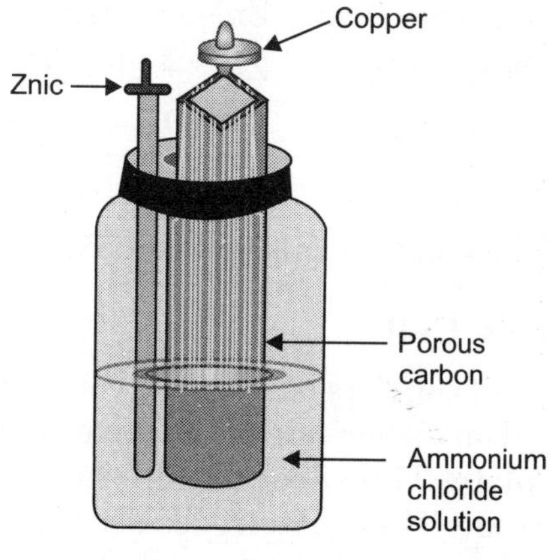

Fig. 22.22. The Lechlanche Cell.

$$2NH_4Cl + Zn \longrightarrow ZnCl_2 + 2NH_3 + H_2$$
$$H_2 + MnO_2 \longrightarrow Mn_2O_3 + H_2O$$

The solid oxidiser being comparatively slow in action, the E.M.F. of the cell falls of very quickely. If the cell is kept in rest for a while the oxygen in the air reacts with the lower oxide of the manganese and turns it to manganese dioxide and the cell starts producing E.M.F again. The E.M.F. of this cell is 1.4V.

Advantages of this Cell

1. It is not expensive.
2. It does not emit any injurious fumes.
3. It lasts for many months with a little attention.
4. There is particularly no action when not in use.

Disadvantages of this Cell

- It quickly stops so not suitable for continuous work.
- The cell mainly consists of liquid chemicals which therefore makes the cell immobile and the chemical reagents needs to be saved form leaking.

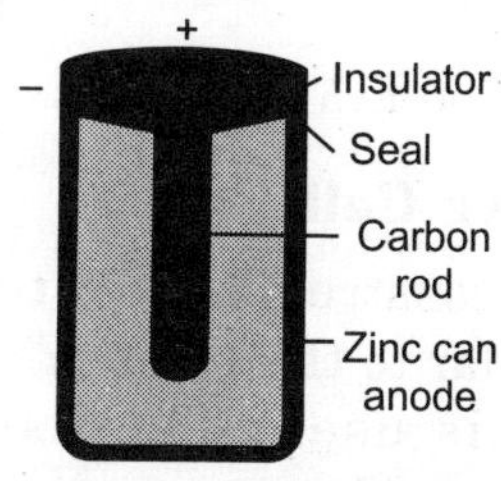

Fig. 22.23. The Dry Cell.

So for these disadvantages in modern days Leclanche cell has been modified into a new type of cell called dry cell, the chemical reagents are same but the structure of cell is quite different (see Fig. 22.23)

A dry cell consists of a zinc container, which serves as the negative electrode and the carbon rod as the positive electrode. The chemical reagents are ammonium chloride, manganese dioxide and saw-dust. This mixture is separated from the zinc containers by a porous paper liner, which acts as a salt bridge.

When the cell produces current, the zinc electrode is oxidised forming zinc ions and electrons, which flows through the external circuit to the carbon rod and reduces the manganese dioxide. The reaction is not completely known as because of complicity but one simplified reaction is stated below :

Negative Pole : $Zn \longrightarrow Zn^{2+} + 2e^-$

Electrolyte : $Zn^{2+} + 2NH_4Cl + 2OH^- \longrightarrow [ZnCl_2(NH_3)^+{}_2] + 2H_2O$

Positive Pole : $2MnO_2 + 2H_2O + 2e^- \longrightarrow 2MnO(OH) + 2OH^-$

Net reaction : $2Zn + 2NH_4Cl + 2MnO_2 \longrightarrow [ZnCl_2(NH_3)] + 2MnO(OH)$

The e.m.f. of this cell is 1.5V.

Smee's Cell

This cell consists of a thin silver plate on which platinum is deposited and two zinc plates, placed on either side of the silver plate. The two zinc plates are connected together to form one plate. The silver plate forms the positive electrode and the zinc forms the negative electrode. These electrodes are placed in the glass vessel containing dilute sulphuric acid. The silver plate being very thin is usually supported by a crosspiece. The following chemical reaction takes place. The following chemical action takes place inside the cell.

$$Zn + H_2SO_4 \longrightarrow ZnSO_4 + H_2$$

The disadvantage of this cell is that the current starts varying after sometimes.

Bunsen Cell

It consists of a glass porcelain vessel containing dilute sulphuric acid and a zinc vessel P containing strong nitric acid, in which a carbon rod is dipped, the whole thing is placed in sulphuric acid. The hydrogen liberated by the action of sulphuric acid and zinc passes through the porous vessel and is oxidised by strong nitric acid and no polarisation takes pace. The following equation supports the chemical action of the cell.

$$Zn + H_2SO_4 \longrightarrow ZnSO_4 + H_2$$

$$H_2 + 2HNO_3 \longrightarrow 2H_2O + 2NO_2$$

The E.M.F. of this cell fairly high about 1.9V.

Grove Cell

The construction and the chemical action of this cell is exactly similar to the Bunsen's Cell, instead of carbon rod a platinum foil is used, The usual shape of the sell is shown in Fig. 22.24.

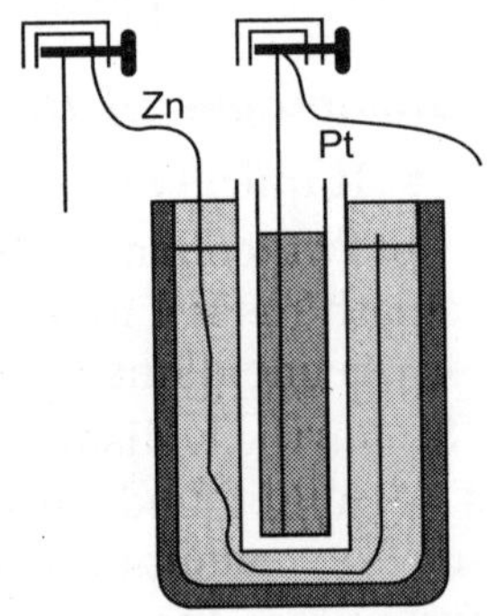

Fig. 22.24. The Grove's Cell

Both the above cells have a fairly high E.M.F. in order of 1.9 volts and are used where strong currents are required but the trouble is that they produce obnoxious fumes.

Dichromate Cell or Bichromate Cell

A strong solution of potassium dichromate or chromic acid is added to a dilute solution of sulphuric acid contained in a glass vessel. A carbon rod acting as positive and the zinc rod acting as negative electrode. As soon as the hydrogen is evolved it is oxidised by the dichromate or chromic acid see Fig. 22.25.

The zinc rod can be taken out as it is acted upon the acid due to the presence of impurities.

The chemical actions taking place inside the cell are a bit complicated. However, by the action dichromate, we get chromic acid. This acid oxidises the hydrogen liberated by the action of sulphuric acid in zinc. The following reaction equation can be deduced from the internal reaction taking place inside the cell.

(*a*) Action of sulphuric acid on potassium dichromate gives chromic acid and potassium sulphate.

$$H_2SO_4 + K_2Cr_2O_7 + H_2O \longrightarrow 2H_2CrO_7 + K_2SO_4$$

(*b*) Action sulphuric acid on zinc gives hydrogen and zinc sulphate.

$$Zn + H_2SO_4 \longrightarrow ZnSO_4 + H_2$$

(*c*) The hydrogen liberated in the action (b) above acts on the chromic acid

$$3H_2 + H_2SO_4 \longrightarrow Cu_2O_3 + 5H_2O$$

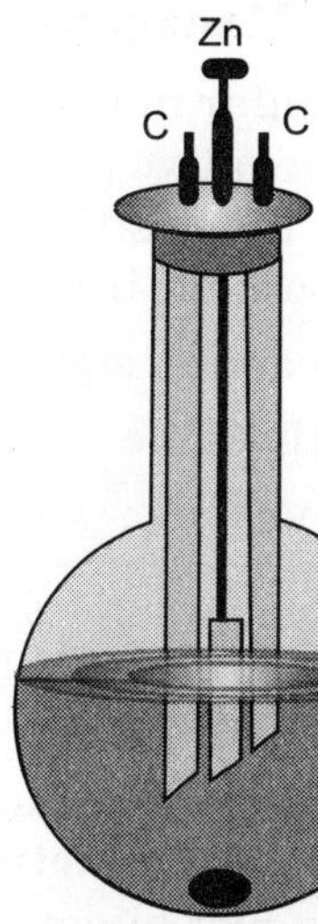

Fig. 22.25. The Dichromate Cell.

In the above reaction potassium sulphate and the chromium sulphate, which are, formed produce chrome alum. This crystallises on the plates and the cell is weakened. Therefore chromic acid should be used instead of potassium dichromate.

Daniel Cell

This cell consists of a copper vessel, which serves as a positive pole consisting a strong solution of copper sulphate. The zinc rod is placed in a porous pot containing dilute sulphuric acid or zinc sulphate. The hydrogen liberated in the porous pot, on its way to copper plate acts on copper sulphate and reduces it to copper, which is deposited on the vessel.

$$Zn + H_2SO_4 \longrightarrow ZnSO_4 + H_2$$
$$CuSO_4 + H_2 \longrightarrow H_2SO_4 + Cu$$

As the copper sulphate will be constantly reduced, hence to make up its strength crystals of copper sulphate are placed on a perforated self fixed to the inside of the cell near the top.

The E.M.F of this cell is very low being 1.07V but it is fairly constant.

Mercury Cell

This cell consists of negative pole made up of compressed amalgamated zinc powder and graphite. In between an electrolyte paste is used called potassium hydroxide. The reaction takes place under.

Negative Pole	$\longrightarrow$	$Zn + 2OH^- \longrightarrow ZnO + H_2O + 2e^-$
Positive Pole	$\longrightarrow$	$HgO + H_2O + 2e^- \longrightarrow Hg + 2OH$
Net reaction	$\longrightarrow$	$Zn + HgO \longrightarrow Hg + ZnO$

Alkaline Managenese Cell

This is similar in principal to the dry cell but is constructed in the manner of the mercury cell. The negative pole is made up of powdered zinc formed into a paste with the electrolyte KOH and the positive pole of compressed graphite and manganese dioxide are separated by an absorbent impregnated by an electrolyte See Fig. 22.26.

The reaction equations are as follows :

Negative Pole : $Zn + 2OH^- \longrightarrow ZnO + H_2O + 2e^-$

Positive Pole : $2MnO_2 + H_2O + 2e^- \longrightarrow Mn_2O_3 + 2OH^-$

Net Reaction : $Zn + MnO_2 \longrightarrow ZnO + Mn_2O_3$

Secondary Cells

These are also galvanic cells, but they involve such chemical reactions, which are easily reversible just by an external D.C. source of same potential that of the cell. These cells are therefore called secondary cells. They have also another name called accumulator, as because they can store the electrical charge in the form of chemical energy.

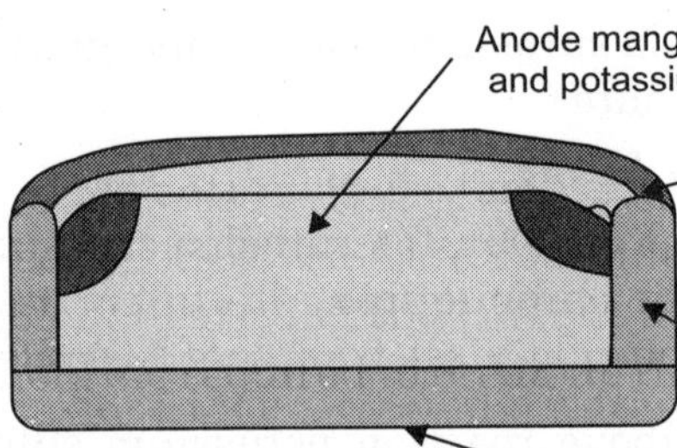

Fig. 22.26. The Alkline Cell.

Nickel Cadmium Cell

This cell consists of nickel as the positive electrode and the cadmium hydroxide as the negative electrode and the electrolyte is the cadmium hydroxide solution. The charging reactions are :

Negative Pole : $Cd(OH)_2 + 2e^- \longrightarrow Cd + 2OH^-$

Positive Pole : $Ni(OH)_2 + 2OH^- \longrightarrow 2H_2O + 2e^-$

Net Reaction : $Ni(OH)_2 + Cd(OH)_2 \longrightarrow NiO(OH) + Cd + 2H_2O$

During discharge these reaction is reversed. A crucial feature of the construction of cell is that oxygen produced at the positive pole during charging by the side reaction.

$$4OH^- \longrightarrow 2H_2O + O_2 + 4e^-$$

Can migrate readily to the negative pole to recombine with the reaction.

$$O_2 + H_2O + 2Cd \longrightarrow 2Cd(OH)_2$$

But for this rapid migration and recombination the cell cannot be sealed.

Lead Storage or Lead Acid Battery

The lead in an automobile is a lead acid battery See Fig. 22.27.

Its electrodes are lead battery grids. One set of grids is coated with lead peroxide and the other set of grids is coated with spongy lead metal. Dilute sulphuric acid is used here as the electrolyte. When the battery delivers a current during discharge the spongy lead is oxidised to lead ions, which combines with sulphate ion of the electrolyte and coat the lead electrode with insoluble lead sulphate. The reactions are as follows:

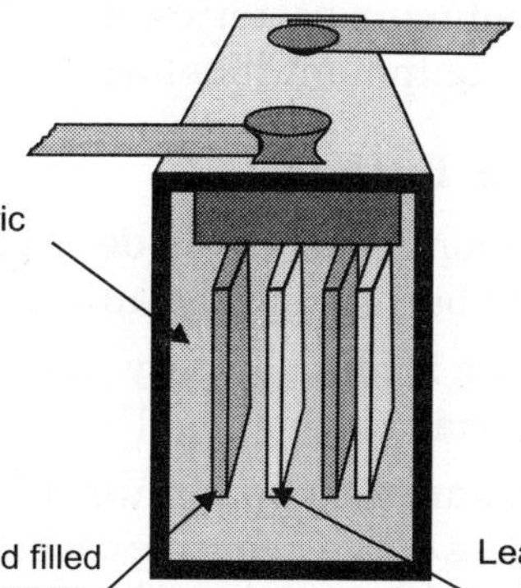

Fig. 22.27. The Lead Acetate Battery.

Negative Pole: $Pb + SO_4^{2-} \longrightarrow PbSO_4 + 2e^-$

Positive Pole: $PbO_2 + 4H_3O^+ + SO_4^{2-} + 2e \longrightarrow PbSO_4 + 6H_2O$

Net Reaction: $Pb + PbO_2 + 4H_3O + 2SO_4^{2-+} \longrightarrow 2PbSO_4 + 6H_2O$

During recharge and external voltage is applied that pushes the electrons in the reverse direction through the cell. The cell becomes electrolyte cell during charging. The above reacts are just reversed during charging.

Advanced Batteries

In recent years there have been a progressive change in the field of cells and batteries used in consumer and industrial applications.

Conventional batteries have many limited applications, so there is a need for developing more advanced batteries, which can serve longer and require zero maintenance.

Emergence of NiMH Batteries

The first usable NiMH batteries appeared 10 years ago uses a hydrogen absorption alloy for the negative electrode. Like the nickel cadmium battery, which stimulated the speed of portable equivalent in the 80s, the NiMH battery is a 1.2V alkaline battery having an energy density of 50% higher that of nickel cadmium battery. Using more environment friendly substances, it is capable of high current charging and discharging and lasts longer than other types of batteries. It is now widely used because of these features. The power generation elements of this battery are the positive nickel electrode which uses nickel hydroxide as an active substance, the negative electrode uses a hydrogen absorbing alloy; the separator which separates the negative and the positive electrode used here is an alkaline solution called potassium hydroxide. This battery is sealed air tight in a cylinder or prismatic metal containers. The positive electrode is a sponge metal electrode (SME) which is also used in high energy density Nickel Cadmium batteries. The high energy density of this battery comes from the high capacity high metal alloy used for the negative electrode, made up of a highly porous nickel substitute filled at high density with nickel hydroxide as an active material, also contributes to the improved capacity.

NiMH batteries after considerably better performance for their cost compared to nickel cadmium batteries. The electric capacity of the NiMH battery is 1,450 mAH (miliampere hour) maximum for UM-3 size 5% higher than that of standard nickel cadmium battery.

Nickel cadmium batteries have strong discharge characteristics, making them especially suitable for equipment like large cassette recorders and etc.

Lithium Ion Battery

The lithium ion battery was developed in response to the determined for a high capacity rechargeable battery similar and lighter than NiMH batteries.

Typically a lithium battery uses a lithium cobalt oxide $LiCoO_2$ for the positive electrode, carbon material for the negative electrode and an organic electrolyte.

During charging, the lithium of the $LiCoO_2$ material in the positive electrode is inserted between layers of the negative carbon material as ions pass through the organic electrolyte. Reverse process takes place during discharging. In the charging and discharging process, lithium travels not as metal but as lithium ions between the positive and the negative electrodes. This is why the battery is called lithium ion battery.

Lithium Polymer Batteries

Conventional batteries use a solution - an electrolyte - obtained by dissolving a salt-on-elute into a solvent. One draw back of this technology is electrolyte leakage. In polymer batteries the electrolyte is a polymer therefore leakage does not occur, nor do the polymer batteries require a container to hold the liquid electrolyte. Researches during 80s have been trying to develop a polymer battery that is thin and can take conventional forms. At present the most promising polymer battery exhibits an ion connectivity of 10g/cm or ion conductivity equal to that of liquid electrolyte.

A battery of this kind with a small thin design would be promising as a power source for mobile information terminals and various kinds of communication equipment.

Recent innovation in solid state polymeric electrodes have led to the developed of solid state batteries using a lithium anode a solid cathode and a lithium anode a solid cathode and a solid polymer electrolyte that both separates and provides the ionic pathway between the anode and the cathode. The absence of liquid eliminates leaks. The solid electrolytes function at temperatures as high as 140°C.

The battery consists of a lithium anode and a composite cathode consists of vanadium oxide, carbon and a small amount of electrolyte. Polythene oxide is a widely used electrolyte. During discharge, lithium is oxidised at the interface of the lithium anode and the polymer electrolyte. The positive ion is transferred across the electrolyte, lithium ions from the electrolyte pass into the positive electrode. These processes are reversed when the battery is charged.

The polymer electrolytes flat cells are being subjected to thermal management, cell balance, failure and protection tests prior to commercialisation.

Maintenance Free Lead Acid Battery

The lead acid battery, used in the automobile industry, consists of a spongy lead negative and a lead oxide positive with sulphuric acid as the electrolyte. It has a normal voltage of

2 and requires periodic maintenance, depending in its use. The convention lead acid battery is improved by developing VRSLAB (valve regulated sealed lead acid battery) with absorbent glassmat or gel technology. These batteries need no addition of water during their normal lifetime. There is no spillage of electrolyte or release of gas.

The batteries are charged at a constant potential mode. Calcium alloys are used as grid material due to their lower self-discharge and increased conductivity. Oxygen and hydrogen evolve from positive and negative respectively during recharge. Oxygen evolved at the anode reacts with hydrogen at the cathode to produce water. A valve releases the excess gas, preventing accident during over charging.

VRSLAB finds its use in electronic equipment, standby system like UPS and remote power supply where it is only solar energy to charge the batteries.

Zinc Air Batteries

The zinc air cell uses a metallic zinc anode 0.5 mm air (oxygen) cathode with an aqueous solution of an alkali metal hydroxide (KOH). The cathode catalytically promotes the reaction of atmospheric oxygen with the aqueous electrolyte and the resulting electrochemical activity follows the equation:

$$H_2O_2 + H_2O + 2e^+ \longrightarrow 2OH^- \ (0.4V)$$

Oxygen from air, water from the electrolyte and electrons are consumed to form hydroxyl ion, with the consumption of electrons leaving the cathode positively charged. The zinc anode reacts with the electrolyte following the reaction below:

$$Zn + 2OH^- \longrightarrow ZnO + H_2O + 2e^- \ (1.25V)$$

At the anode, the zinc oxide, water and electrons with the local formation of electrons give the anode a negative charge. The electrolyte conducts the hydroxyl ions, which are produced at the cathode and consumed at the anode. The over all zinc air system equation is

$$Zn + O_2 \longrightarrow ZnO \ (1.65V)$$

The rate of conversion of zinc to zinc oxide is 0.82 Ah per gram of zinc. The capacity of oxygen cathode is 4.79 Ah per litre of molecular oxygen. As the atmosphere provides limitless supply of oxygen the amount of zinc contained in the cell is the factor that determines its capacity.

Fuel Cells

Fuel cells are galvanic cells in which electrode materials in the form of gases are supplied continuously and are consumed to produce electricity. A typical fuel cell is based on the reaction of hydrogen and oxygen forming water (see Fig. 22.28). Hydrogen gas is diffused through the anode; a porous electrode with a catalyst such as finely divided platinum or palladium on its surface. Oxygen is diffused through the cathode, a porous electrode impregnated with a cobalt oxide, platinum, or silver as a catalyst. The two electrodes as separated by a concentrated solution of sodium hydroxide as an electrolyte. The anode half reaction is

$$2H_2 + 4OH^- \longrightarrow 4H_2O + 4e^-$$

The cathode half reaction is

$$O_2 + 2H_2O + 4e^- \longrightarrow 4OH^-$$

The overall cell reaction is the combination of hydrogen and oxygen to produce water.

Anode:	$2H_2 + 4OH^- \longrightarrow 4H_2O + 4e^-$
Cathode:	$O_2 + 2H_2O + 4e^- \longrightarrow 4OH^-$
Cell Reaction:	$2H_2 + O_2 \longrightarrow 2H_2O$

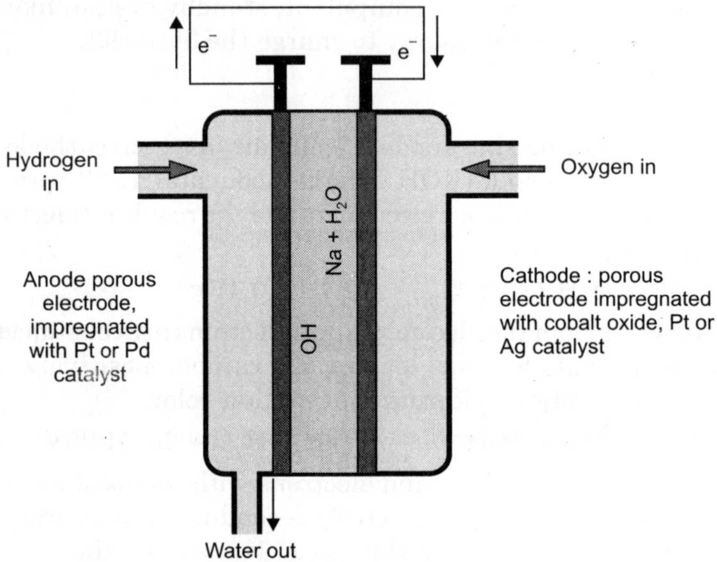

Fig. 22.28. The Fuel Cell.

Fuel cells have the potential for providing electricity from methane and other hydrocarbon, much more effectively than, steam-driven generators. Thus a great deal of effort is being spent investigating fuels such as methane and electrode system for these fuels.

Electroplating Invention and History

Electroplating is the electrodeposition of metals, on metals, alloys and non-metallic substances. Historically, the operation is over 100-year old, and was first applied by Sir Michel Faraday.

Electroplating has the express purpose of altering the surface properties of metals or non-metals. Its beneficial results may include improved appearance; improved resistance to corrosion, tarnish, chemical attack and wear, improved frictional and non-galling behaviour and increased hardness.

Electroplating of metals include the broader aspects of electroplating, that of cold casting one metal on the surface of another metal or on the same metal to replace the metal worn away.

The process of coating one metal by another metal to protect the base metal is called galvanisation.

Applications

Plating of Appearance (Decorative Plating) : There is a natural overlapping between decorative plating and protective plating, which is associated with prevention of corrosion and turnish.

Silver, gold, nickel, chromium, brass, and copper are the most commonly used metals for decorative coating. For special decorative effects, tin, zinc, cadmium, lead, palladium and platinum can be used. Metals and alloys surfaced by electroplating are steel, brass, copper, nickel silver, white-metal alloys of lead, zinc, or tin base, and aluminium alloys. The plate may involve only one single metal, it can also be an alloy of two metals or a coating of multiple desecrate layers of different metals placed on top of another as a composite plate.

Millions of parts daily are plated in connection with the manufacture of automobiles; air craft; refrigerators etc.

The well known "chrome plate", more properly called chromium plate, is the plate most widely used for appearance today. Other metal plate or plates are generally under chromium to provide the major portion of the protection. The chromium provides the lasting appearance because of its excellent resistance to tarnish abrasion and chemical attack.

Plating for Protection : Electroplating is used mainly for protection of the base metal against corrosion, sometimes for decoration also, but protection is the prime factor. Generally steel is protected from corrosion by plating it with zinc, cadmium and tin. Zinc is best in the rural or industrial area; cadmium is best near the seashore. Tin does not generally provide the protection offered by the zinc and cadmium, but under certain conditions it does, mostly notable in application to food containers and food handling equipment. Tin not only protects the metal but also does not contaminate the food. Copper, brass, nickel, chromium and lead plates are used to protect steel in special situations. Each depends upon an envelop effect, that is, upon completely enclosing the base metal, since none of these metals can generally be expected to provide galvanic protection.

When protection also involves appearance, steel is usually plated with successive layers of copper, nickel and chromium or nickel and chromium. The copper-nickel protects it from rusting and the chromium improves its appearance.

Copper and brass are generally chromium plated. Zinc, zinc die castings, aluminium and aluminium die castings are practically copper plated before nickel or chromium plating, while ornaments of silver are gold plated for decoration.

Electroplates are also applied for temporary use in metal treatments. A large number of steel parts are copper-plated before carburizing to prevent carburization at undesired locations.

Improved performance and service life of cast-iron or steel glass moulds is attained through the application of low-concentration chromium plate.

Plating on Non-metallics : Non-metals are plated for decoration, preservation, or to obtain conductance and also strength.

The base metals used here are synthetic resin, wood, paper, cloth, glass, porcelain and other ceramic materials and even leather, eggs, and dried leaves or other plant forms for special occasions.

The Process of Electroplating

Electroplating is carried out in a tank, the anode is the metal to be electrodeposited that is a rod of the pure metal to be deposited and the cathode is made up of articles to be electroplated, e.g., cup dish etc. (see Fig. 22.29).

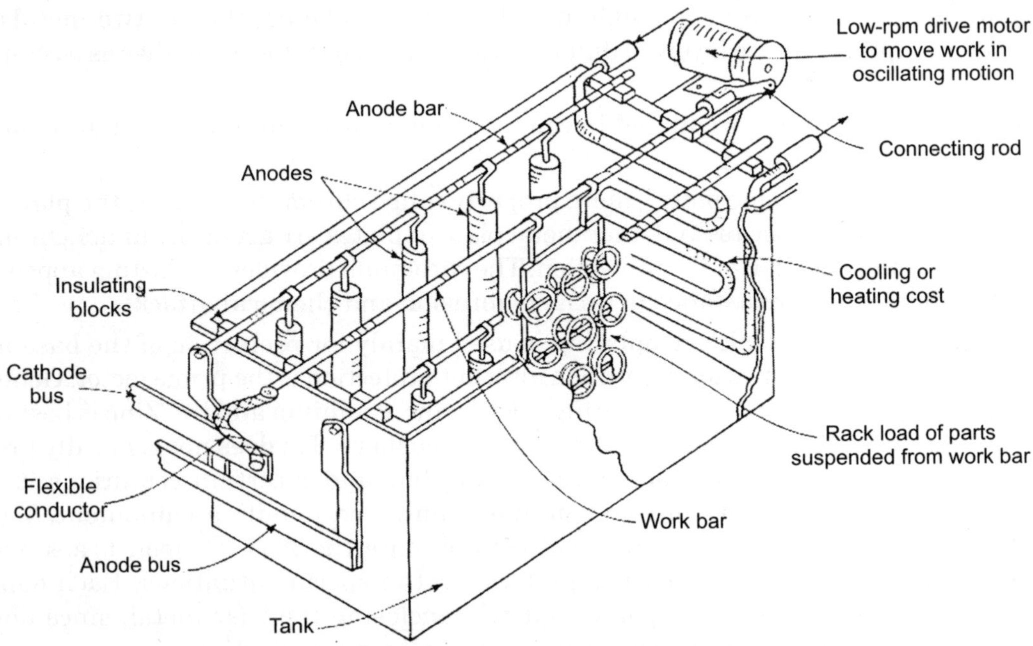

Fig. 22.29. The Electroplating Tank

Corrosion

Many metals, particularly iron, undergo corrosion when exposed to air and water. Iron will not rust in dry air or in water that is free of dissolved oxygen. Both air and water are involved in the corrosion process. The presence of an electrolyte in the water accelerates the corrosion, particularly when the solution is acidic.

Corrosion is an electrochemical process (see Fig. 22.30). When iron is in contact with water with even a microscopic drop the iron tends to oxidise and give up electrons.

Anode : $Fe \longrightarrow Fe^{2+} + 2e^-$

The electrons pass through the iron to the edge of the drop, where they reduce oxygen from the air to hydroxide ion.

Cathode: $O_2 + 2H_2O + 4e^- \longrightarrow 4OH^-$

The ferrous ions and hydroxide ions diffuse together and form insoluble ferrous oxide.

$$Fe^{2+}(aq) + 2OH^-(aq) \longrightarrow Fe(OH)_2(s)$$

This precipitation is rapidly oxidised by oxygen to rust, a ferric compound with the appropriate composition $Fe_2O_3\,H_2O$.

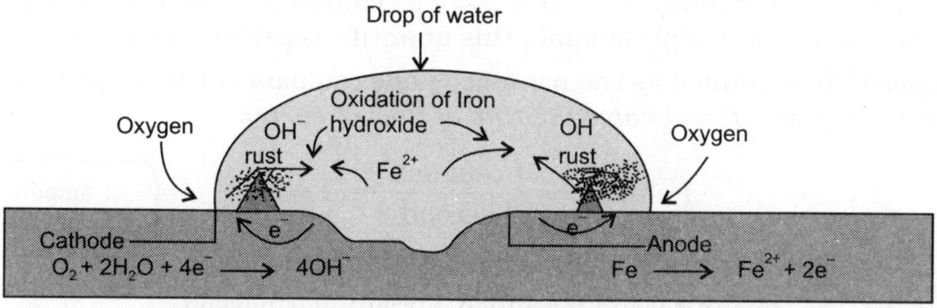

Fig. 22.30. The Electrochemical Process of Corrosion of Iron by a drop of water

Many methods and devices have been employed to prevent corrosion. We protect iron by coating it with paint, lacquer and greeze or asphalt. Galvanisation is also used as a successful process but it has also some draw back.

There is another method of preventing iron from resting is the method of cathodic protection. This is entirely an electrochemical process. Cathodic protection is used in the case of iron and steel. If the iron is connected by a wire to a more active metal, such as zinc, aluminium or magnesium, the iron becomes a cathode at which oxygen is reduced, rather than the anode where the iron is oxidised. The difference in activity of the two metals causes a current to flow between them producing corrosion of the more active metal and protecting the iron. With magnesium used as the more active metal, the following reactions occur.

Anode : $2Mg \longrightarrow 2Mg^{2+} + 4e^-$

Cathode : $O_2 + 2H_2O + 4e^- \longrightarrow 4OH^-$

Sum : $2Mg + O_2 + 2H_2O \longrightarrow 2Mg^{2+} + 4OH^-$

In this series of reaction no iron is oxidised. The active metal is slowly consumed and must be replaced periodically, but this is less expensive than replacing the iron.

Some of the more active metals such as cadmium and magnesium, which might be expected to corrode rapidly, are protected by a tightly adhering oxide coat of the metal oxide that forms when the metal is exposed to air. The metal is made passive by coating.

NUMERICAL CHEMISTRY
PROBLEMS OF ELECTROCHEMISTRY

Electrical Units

(i) ***Coulomb:*** It is the practical unit of electric charge. 1 Coulomb is equal to the quantity of charge transferred in one second by a constant current of 1 ampere.

The current passing through a point is expressed conveniently in the terms of the movement of electrons and is expressed as charge in coulomb. 1 mole of electrons carry a charge of 96500 coulomb, this quantity is called *Faraday Constant.*

(ii) ***Ampere :*** It is defined as the passage of one coulomb of charge per second passed at a given point. *It indicates the rate of flow of current.*

$$1 \text{ ampere} = \frac{1 \text{ coulomb}}{1 \text{ second}} \text{ or, } I = Q/t$$

1 ampere = 1000 milliampere.

Where Q = The quantity of electricity passed in coulomb.

T = the time in second for which current was passed.

I = the rate of change of current in ampere.

The charge of a single electron is 1.6022×10^{-19} coulomb

From the Faraday's Second Law of electrolysis

Charge carried by one mole of electron = Charge of an electron × number of electrons

$(1.6022 \times 10^{-19} C/e^-)$ $(6.022 \times 10^{23} e) = 96486$ C/mol e^-

or 96500 coulomb (round Figure)

$\therefore$ One faraday = 96500 coulomb

1 faraday = Movement of 1 mole of electrons (6.023×10^{23})

PROBLEMS

1. What is the number of electrons in one coulomb?
2. The charge on an electron is 1.602×10^{-19}, how many electrostatic unit is this?
3. Find the charge in coulombs on 1 gram ion of N^{3-}.
4. From the values of a faraday (96500 coulomb) an Avogadro's number (6.023×10^{23}) calculate the coulombic charge on an electron.

CHAPTER

23

Oxidation Reduction

Introduction

Oxidation (old concept) : Generally the oxidation is the term used to represent the reaction which is involved with the direct addition of oxygen to an element or a compound or increase in the proportion of oxygen in a compound. The substance with which oxygen unites is said to be oxidised. Magnesium, sulphur etc when reacts with oxygen forms oxides and thus are oxidised.

$$2Mg + O_2 \longrightarrow 2MgO \qquad S + O_2 \longrightarrow SO_2$$

Due to chemical nature, hydrogen is regarded as a kind of "chemical opposite" of oxygen; so the reaction in which hydrogen is removed (partly or wholly) from a compound is also called oxidation.

When heated, manganese dioxide and concentrated hydrochloric acid react with each other producing chlorine. Sulphur is precipitated when hydrogen sulphide gas is passed through bromine water. Here hydrochloric acid and hydrogen sulphide are said to be oxidised to chlorine and sulphur respectively by giving up their hydrogen.

$$MnO_2 + 4HCl \longrightarrow MnCl_2 + Cl_2 + 2H_2O; H_2S + Br_2 \longrightarrow 2HBr + S$$

The definition of oxidation has further been modified according to the mode of the chemical reactions. We know that oxygen is a typical electronegative element and hydrogen is electropositive element. Hence the concept of oxidation has been extended logically to include all reactions involving addition of any electronegative element and removal of any electropositive element.

So oxidation is defined as, *it is a process which introduces or increases the proportion of an electronegative element or oxygen to a compound or substance; or removal of hydrogen or any electropositive element from any compound or substance is called oxidation.*

Following reactions are examples of oxidation :

(*i*) $Fe + S \rightarrow FeS$ (*ii*) $2FeCl_2 + Cl_2 \rightarrow FeCl_3$

(*iii*) $2KI + H_2O_2 \rightarrow KOH + I_2$

Here in the reaction (*i*) between iron and sulphur, iron has been oxidised, because sulphur an oxygen like electronegative element has been added to it. In the reaction (*ii*), ferrous chloride has undergone oxidation because more chlorine which is electronegative oxygen like element has been added. In the equation (*iii*) the electropositive hydrogen element potassium has been removed from the potassium iodide by hydrogen peroxide and free iodine has been liberated, so the reaction is said to be oxidation.

It is evident that there are oxidation reactions where oxygen does not play any part.

Reduction : It is the reverse process of oxidation. In a restricted sense, the term reduction is used to describe the reaction, which involves removal or decrease in the proportion of oxygen from a compound. Moreover, addition or increasing the proportion of hydrogen to a substance (either an element or compound) is also called reduction. The substance, which gives up oxygen or takes up hydrogen, is said to be reduced.

Consider the reactions following :

$$CuO + H_2 \longrightarrow Cu + H_2O,$$
$$ZnO + H_2 \longrightarrow ZnO + C \longrightarrow Zn + CO$$
$$2Al + Fe_2O_3 \longrightarrow 2Fe + Al_2O_3$$

In the above reactions oxygen is removed from the metals.

Again hydrogen directly reduces the elements by reacting with them. The example are given below :

$$N_2 + 3H_2 \longrightarrow 2NH_3;$$
$$H_2 + Cl_2 \longrightarrow 2HCl;$$
$$H_2S + I_2 \longrightarrow 2HI + S$$

From the above explanations reduction can be defined as, *the process in which it introduces hydrogen or hydrogen like electropositive element to other compounds or substances; or removes oxygen like electronegative element from the substance or compound is called reduction.*

Consider the reactions below :

$$FeCl_3 + [H] \longrightarrow FeCl_2 + HCl;$$

It is to be noted that in the above reaction the nascent hydrogen has reacted with ferric chloride producing ferrous chloride, that is it has removed the oxygen like electronegative element chlorine from the compound thus chemically reducing it.

Valency Change in the Oxidation and Reduction

The change of valency in the Redox (reduction-oxidation) should be noted. Oxidation involves in the increase of positive ions and the removal of the positive valency (or decreasing of the negative valency). The following examples will clear the above statement.

$$4FeO + O_2 \longrightarrow 2Fe_2O_3; 2FeCl_2 + Cl_2 \longrightarrow 2FeCl_3$$
$$6FeSO_4 + 3H_2SO_4 + 2HNO_3 \longrightarrow 3Fe_2(SO_4)_3 + 4H_2O + 2NO$$

Decrease in the positive valency (or increase in the negative valency) of an element is associated with the process of reduction, e.g., (*i*) $2FeCl_3 + SnCl_2 \longrightarrow 2FeCl_2 + SnCl_4$

(*ii*) $2KMnO_4 + 5SO_2 + 2H_2O \longrightarrow K_2SO_4 + 2MnSO_4 + 2H_2SO_4$

In the reaction (*i*) mentioned above, ferric chloride has been reduced to ferrous chloride; trivalent ferric ion is changed to divalent ferrous ion.

In the reaction (*ii*), permanganate salt has been reduced to heptavalent manganese. It is converted into bivalent manganese i.e., change of valency of manganese is from + 7 to + 2 (higher to lower).

Oxidising and Reducing Agents

According to old concept, *the substance which in chemical reaction supplies oxygen or any other electro-negative element to a compound, or withdraws hydrogen or any other electro-positive elements from a substance is called oxidising agent or oxidant.* Examples of oxidizing agents are - oxygen, hydrogen peroxide, halogens, nitric acid, potassium permanganate, and potassium dichromate etc.

Again *the substance which in a chemical reaction, removes oxygen or any other electro-negative element partly or wholly from a compound or supplies hydrogen or any other electro-positive element to an element or a compound is called reducing agent or reductant.*

Some examples of reducing agents are hydrogen sulphide, hydrogen, sulphur dioxide, stannous chloride.

Simultaneous Occurrence of Oxidation and Reduction

The two processes, oxidation and reduction, are complementary and must occur simultaneously. Neither an oxidation nor a reduction can take place alone. An oxidising agent brings about the oxidation of a substance. In other words, an oxidising agent cannot exhibit its oxidising properties unless there is a reducing agent in the reaction system. This is the main feature of the oxidation-reduction reaction. Such reactions are referred to as reduction-oxidation reaction or Redox reactions in short.

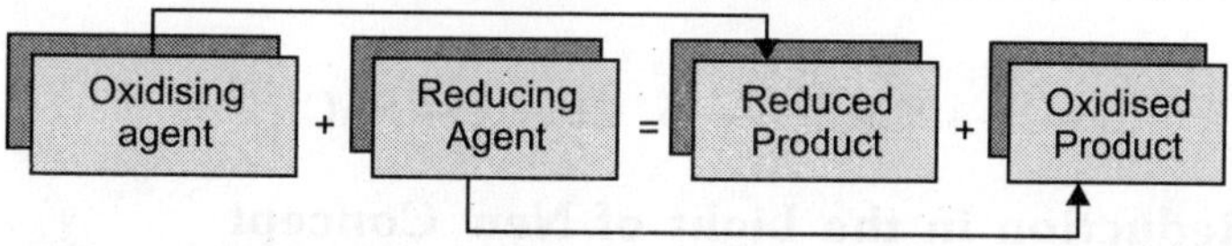

Examples : (*a*) When hydrogen is passed over heated cupric oxide, the latter is reduced to metallic copper by the removal of oxygen but the reducing agent hydrogen is itself oxidised to water (steam) by gaining oxygen.

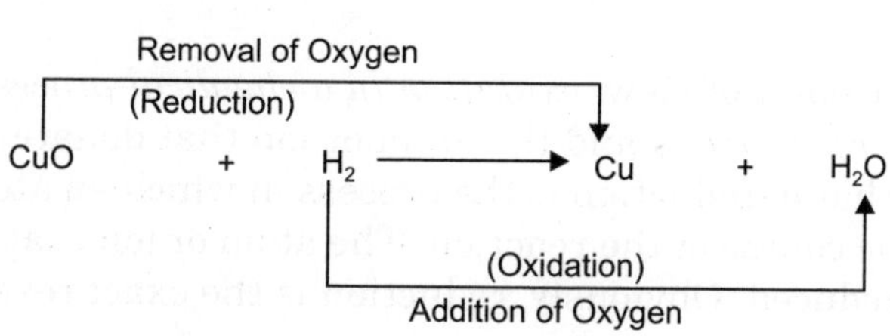

(*b*) In this reaction below bromine reacts with hydrogen sulphide producing hydrogen bromide, and here the bromine is connected with hydrogen an electropositive element, the reduction occurs. On the other hand, hydrogen sulphide looses hydrogen, an electropositive element, oxidation occurs.

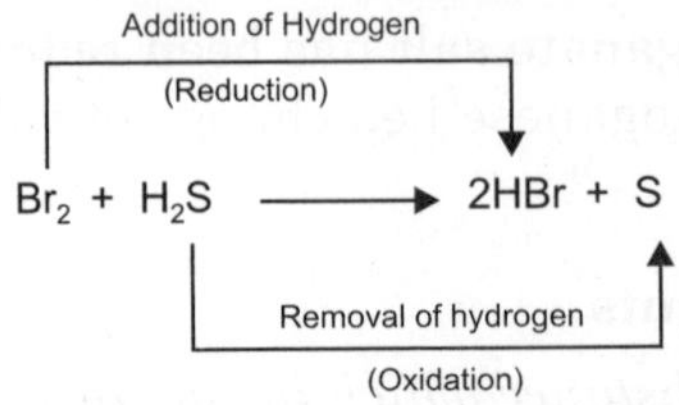

(*c*) Reaction below shows that the ferrous chloride reacts with chlorine producing ferric chloride, here the electronegative chlorine is added to ferrous chloride, hence oxidation. On the other hand the electropositive iron is added to the ferrous chloride, the reduction is the result.

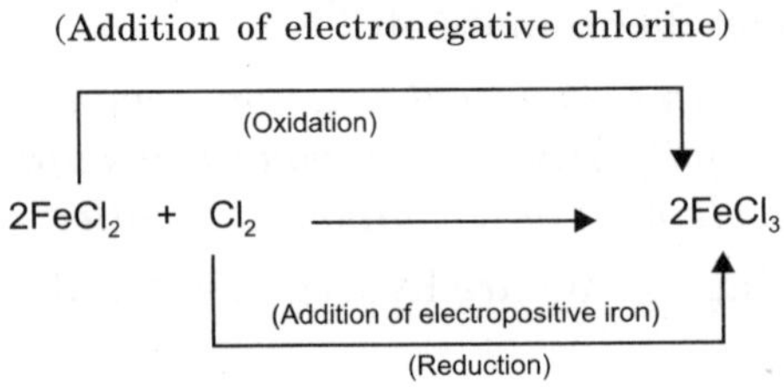

(*d*) When hydrogen peroxide oxidises lead sulphide into lead sulphate, it itself is reduced into water.

$$PbS + 4H_2O_2 \longrightarrow PbSO_4 + 4H_2O$$

(*e*) In the reaction between conc. sulphuric acid and hydroiodic acid, hydroiodic acid is oxidised to iodine and sulphuric acid gets reduced to sulphur dioxide.

(*f*) Stannous chloride reduces ferric chloride into ferrous chloride by decreasing the proportions of chlorine but at the same time it is oxidised into stannic chloride by taking up of chlorine.

$$2FeCl_3 + SnCl_2 \longrightarrow 2FeCl_2 + SnCl_4$$

Oxidation and Reduction in the Light of New Concept

The oxidation-reduction reaction is now-a-days explained in a generalised and understricked way in terms of electrons. The most recent definition of oxidation-reduction process expresses the changes taking place through electron transfer from element to element in the course of chemical reaction.

From the electronic point of view, *oxidation in a chemical process whereby an atom or an ion looses one or more electrons* and the atom or ion that donates electron is said to be oxidised. On the other hand, reduction is the process in which an atom or an ion gains one or more electrons in the course of the reaction. The atom or ion that accepts electrons in a reaction is said to be reduced. Obviously, reduction is the exact reverse of oxidation.

From the forgoing discussions, oxidation-reduction reaction may be looked upon as a chemical reaction resulting in a change in the electrical charges on the reacting particles.

Oxidation	Reduction
$Na - e \rightarrow Na$	$Na^+ + e \rightarrow Na$
$Zn - 2e \rightarrow Zn^{2+}$	$Zn^{++} + 2e \rightarrow Zn$
$Fe^{++} - e \rightarrow Fe^{2+}$	$Fe^{2+} + e \rightarrow Fe^{++}$
$2I^- - 2e \rightarrow I_2$	$I_2 + 2e \rightarrow 2I^-$
$Cl^- - e \rightarrow Cl$	$Cl + e \rightarrow Cl^-$
$S^{2-} - 2e \rightarrow S$	$S + 2e \rightarrow S^{2-}$

It is clear that oxidation results in the conversion of:

1. A neutral metallic atom into its ion
2. A metallic ion (cation) in the lower valency state (ous-ion)
3. A non-metallic ion (anion) into the non-metallic element.

In a reverse process, reduction results in the conversion of:

1. a non-metallic element into its ion
2. An 'ic' cation into an 'ous' cation
3. The metallic ion (cation) into the metal.

In sort, oxidation may be referred to as *de-electronation* and the reduction as *electronation.*

Modern Definitions of Oxidising and Reducing Agents

The substances which can accept or more electrons and undergo reduction easily are called oxidising agent. The reducing agents are the substances, which are readily oxidised by the donation of electrons.

Thus, oxidation and reduction are complementary process of electron loss and electron gain respectively. The electrons lost by the reducing agent must be accepted by an oxidising agent present. So, in a reaction, whenever one substance is oxidised, another substance is corresposnsdingly reduced. This leads to the conclusion that there cannot be a reaction in which only oxidation or only reduction takes place.

Examples : (*a*) When magnesium is burnt in air, it is converted into magnesium oxide. Here, magnesium is oxidised by losing two electrons per atom. These electrons are taken up by oxygen atoms, which are reduced as a result. Magnesium (donor of electrons) is the reducing agent.

$$Mg \longrightarrow Mg^{2+} + 2e \text{ (oxidation)} \quad ...(i)$$

$$O_2 + 2e \longrightarrow O^{2-} \text{ (reduction)} \quad ...(ii)$$

By adding (*i*) and (*ii*), $Mg + O_2 \longrightarrow Mg^{2+}\ O^{2-}$

Or $2Mg + O_2 \longrightarrow 2Mg^{2+}O^{2-}$ (or $2MgO$)

Combination of iron and sulphur to give ferrous sulphide is a similar process.

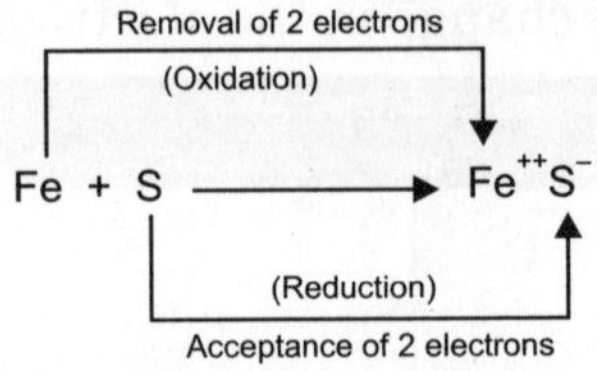

(*b*) When chlorine is passed through a solution of ferrous chloride, the latter is converted into ferric chloride. This reaction is stated as :

$$2FeCl_2 + Cl_2 \longrightarrow 2FeCl_3$$

According to the ionic theory, this reaction is primarily a reaction between Fe^{2+} ions and chlorine ion. Each ferrous ion Fe^{2+} (reducent) is oxidised to ferric ion, Fe^{3+} by loosing one electron and the neutral chlorine atom (oxidant) is reduced to chloride ion by gaining the electron.

$$Fe^{2+} \longrightarrow Fe^{3+} + e \text{ (oxidation)} \qquad (i)$$

$$Cl + e \longrightarrow Cl^- \text{ (reduction)} \qquad (ii)$$

Here, the number of electrons lost by each Fe^{2+} ion is equal to the number of electrons gained by chlorine atom. So, by adding equations (*i*) and (*ii*), we get

$$Fe^{2+} + Cl \longrightarrow Fe^{3+} + Cl^-$$

$$2Fe^{2+} \longrightarrow 2Fe^{3+} + 2e$$

$$Cl_2 + 2e \longrightarrow 2Cl^-$$

$$\overline{2Fe^{2+} + Cl_2 \longrightarrow 2Fe^{3+} + 2Cl^-}$$

(*d*) When zinc is added to a solution of copper sulphate, copper is precipitated and the zinc goes into the solution in the form of zinc sulphate.

$$Zn + CuSO_4 \longrightarrow ZnSO_4 + Cu$$

Ionically,

$$Zn \longrightarrow Zn^{2+} + 2e \text{ (oxidation)}$$

$$Cu^{2+} + 2e \longrightarrow Cu \text{ (reduction)}$$

$$\overline{Zn + Cu^{2+} \longrightarrow Zn^{2+} + Cu}$$

Here, copper gains two electrons lost by zinc. Hence, copper is reduced and zinc is oxidised.

(*e*) The reduction of hot cupric oxide by hydrogen is expressed as

$$Cu^{++}O^{--} + H_2 \longrightarrow H_2O$$

It is evident that cupric ion, Cu^{2+} is reduced by electron gain as

$$Cu^{2+} + 2e \longrightarrow Cu$$

The electrons are made available by the interaction between the oxide ion O^{2-} and a molecule of hydrogen consisting of two atoms.

$$O^{2-} + H_2 \longrightarrow H_2O + 2e$$

Thus, by combining with oxygen in this way and supplying electrons to the metallic ion hydrogen shows reducing properties. The oxide ion is oxidised by electron loss and the oxygen remains in combination with hydrogen as water.

(*d*) when a solution of ferric chloride is treated with a solution of stannous chloride, ferric chloride is reduced to ferrous chloride while stannous chloride is oxidised to stannic chloride.

$$2FeCl_3 + SnCl_2 \longrightarrow 2FeCl_2 + SnCl_4$$

In this reaction, Fe^{3+} and Sn^{2+} actually react: the chloride ions, Cl^- play no part.

Here, each ferric ion, Fe^{3+} (oxidant) gains one electron and is reduced to ferrous ion, Fe^{2+}

$$Fe^{3+} + e \longrightarrow Fe^{2+} \text{ (reduction)} \qquad (i)$$

Each stannous ion, Sn^{2+} (reductant) on the other hand is oxidised to stannic ion, Sn^{4+} by releasing two electrons.

$$Sn^{2+} \longrightarrow Sn^{4+} + 2e \text{ (oxidation)} \qquad (ii)$$

As the number of electrons lost by a reductant must be the same as the number of electrons captured by an oxidant, the partial equation (*i*) is multiplied by 2. Now

$$2Fe^{3+} + 2e \longrightarrow 2Fe^{2+}$$

and $$Sn^2 \longrightarrow Sn^{4+} + 2e$$

By adding $$2Fe^{3+} + Sn^{2+} \longrightarrow 2Fe^{2+} + Sn^{4+}$$

Similarly, the following oxidation-reduction reactions are expressed in terms of electrons.

(*i*) $$2KI + Cl_2 \longrightarrow 2KCl + I_2$$

Partial equation for the oxidising agent

$$Cl_2 + 2e \longrightarrow 2Cl^- \qquad (i)$$

$$2I^- \longrightarrow I_2 + 2e \qquad (ii)$$

By adding (*i*) and (*ii*) $$2I^- + Cl_2 \longrightarrow I_2 + 2Cl^-$$

(*ii*) $$Br_2 + H_2S \longrightarrow 2HBr + S$$

Partial equation for the oxidising agent

$$Br_2 + 2e \longrightarrow 2Br^- \qquad (i)$$

$$H_2S \longrightarrow 2H^+ + S^\circ + 2e \ldots \qquad (ii)$$

Adding (*i*) and (*ii*) $$Br_2 + H_2S \longrightarrow 2H^+ + 2Br^- + 50$$

So indicates the electrical neutrality of sulphur. An element in the uncombined state is considered to have zero valency.

Oxidising Agents : Substances, which are capable of giving oxygen to other substances or contain an atom or group of atoms which readily gains electrons, are called oxidising agents. For example, during oxidation of lead to lead oxide, oxygen is supplied by air. In this case air is the oxidising agent. Some important oxidising agents are Oxygen, ozone, hydrogen peroxide, chlorine, nitrogen dioxide, nitric acid, sulphuric acid, potassium permanganate and potassium dichromate.

Following experiments may be performed to test whether the given substance is an oxidising agent.

Exp 1 : Treat the given substance with hydrogen sulphide. A colloidal precipitate of sulphur is obtained, if the sample is oxidising agent.

Exp 2 : Add potassium iodide solution to the given substance when iodine is liberated which turns starch paper blue.

Exp 3 : Freshly prepared ferrous sulphate does not give any colour with ammonium sulphocyanide. Add ferrous sulphate solution to the given substance and then treat the mixture with ammonium sulphocyanide solution. It gives a blood red colour.

Reducing Agents : A substance, which brings about reduction by supplying electrons is called a reducing agent. Some important reducing agents are: Hydrogen, sulphur dioxide, hydrogen sulphide, carbon, nitrous acid, hydrogen bromide, hydrogen iodide and stannous chloride.

Following experiments may be performed to test whether the given substance is a reducing agent.

Exp 1 : Treat the given substance with acidified potassium permanganate solution (pink). It is decolourised if the substance is a reducing agent.

Exp 2 : Add acidified potassium dichromate solution (orange) to the given substance. It changes green.

Exp 3 : Pour some bromine water (red) into the substance and shake red colour disappears.

SOME MODERN CONCEPTS

Oxidation Numbers : Oxidation in any reaction in which an atom or ion looses electrons. On the other hand, reduction is defined as any reaction in which an atom or ion gains electrons.

Change of ferrous (Fe^{2+}) to ferric (Fe^{3+}), stannous (Sn^{2+}) to stannic (Sn^{4+}), manganate (MnO_4^{2-}) to permanganate (MnO_4^-) are all cases of oxidation as each one of then involves a loss of electron.

$$Fe^{2+} \longrightarrow Fe^{3+} + 1 \text{ electron}$$

$$Sn^{2+} \longrightarrow Sn^{4+} + 2 \text{ electrons}$$

$$MnO_4^{2-} \longrightarrow MnO_4^- + 1 \text{ electron}$$

Similarly, when mercuric ion is converted to mercurous ion, and chloride ion is formed from chlorine atom also are the examples of oxidation where loss of electrons takes place.

In order to keep track of electron shifts in oxidation-reduction reactions, it is convenient to use the concept of *oxidation number or the oxidation sate of various atoms involved in oxidation-reduction* reactions. *The oxidation number is defined as the charge which an atom appear to have when electrons are counted* in accordance with the following rather arbitrary rules:

(*a*) Electrons shared between two like atoms are counted with more electronegative atom for example:

The electron pair shared between H and Cl in H^{+1} : Cl^{-1}: is counted with more electronegative Cl. As a result of its hydrogen having lost share in the electron pair appears to have +1 and chlorine appears to have –1 charge. Hence oxidation numbers of H and Cl are +1 and –1 respectively.

(*b*) Electrons shared between two like atoms are divided equally between the two sharing atoms. For example, in hydrogen molecule H : H, the electron pair is equally shared between the two atoms. Thus both the atoms appear to have no charge, i.e, oxidation number of hydrogen is 0 in hydrogen molecule.

In the molecule of water given in the margin, the two electron pairs shared between oxygen and the two hydrogen atoms are counted with the more electronegative oxygen atom. Hence in water oxidation number of each H is +1 and that of the O-atom is –2.

$$H^{+1} : O^{-2} : H^{+1}$$

Counting of electrons like this is very laborious. The following operational rules derived from the above will be found very convenient:

1. In the element or uncombined state, the atoms are assigned an oxidation number is zero.
2. In the compounds, the oxidation number of fluorine is always –1.
3. In compounds, the group I elements (Li, Na, K, Rb, Cs and Fr) have an oxidation number +1 and the group II elements (Be, Mg, Ca, Sr, Ba and Ra) have an oxidation number +2.
4. Oxidation number of hydrogen in compounds is generally +1 except in metallic hydrides where the oxidation number is –1.
5. In compounds, the oxidation number of oxygen is generally –2 except in Fe_2O where the oxidation number of fluorine is -1 and that of oxygen is +2. In hydrogen peroxide molecule, the electron pair shared between O and H is counted with O but the other electron pair shared between two O-atoms is equally shared. The number of electrons counted with each o is, therefore, seven (i.e., one more than its own electrons). The oxygen atom, therefore, appears to have -1 charge or its oxidation number in H_2O_2 is –1.
6. In neutral molecules, the sum of the oxidation number of all the atoms is zero.
7. For complex ions (charged species), the oxidation number of all the atoms is equal to the net charge on the ion.

With the help of these operational rules, we can calculate the oxidation number of an atom present in molecule or complex ion as illustrated in the following examples:

Examples : What is the oxidation number of S in (a) H_2SO_4 (b) $H_2S_2O_7$ and (c) $Na_2S_2O_3$

(*a*) Let oxidation number of S is x.

Sum of the oxidation number of various atoms in H_2SO_4

$= 2 \times (+1) + x + 4(-2)$
$= 2 + x - 8$
$= x - 6$

this must be zero according to (rule 6) hence $x - 6 = 0$ where $x = 6$ or the oxidation number of sulphur in H_2SO_4 is 6.

(*b*) Sum of the oxidation numbers of various atoms in $H_2S_2O_7$

$= 2 \times (+1) + 2x + 7 \times (-2)$
$= 2 + 2x - 14 = 2x - 12.$

(c) Putting $2x - 12 = 0$ as above we have $x = +6$ or the oxidation number of $H_2S_2O_7$ is +6

(d) Sum of the oxidation numbers of various atoms in $Na_2S_2O_3$ is

$= 2 \times (+1) + 2x + 3 \times (-2)$
$= 2 + 2x - 6$
$= 2x - 6$
$= 2x - 4$

Putting $2x - 4 = 0$, we have $x = +2$ or the oxidation number of $Na_2S_2O_3$ is +2.

Try to find the oxidation number of Mn in, $KMnO_4$, MnO and K_2MnO_4.

Eample 2 : What is the oxidation number of Cr in $Cr_2O_7^{2-}$

or, $2x + 7 \times (-2) = -2$ (rule 7)

or, $2x - 14 = -2$

or, $2x = 12$ or, $x = +6$ or the oxidation number of $Cr_2O_7^{2-}$ is +6

Try to find the oxidation number of P in PO_4^{3-} and HPO_3^{2-}.

Valency Oxidation State and Oxidation Number

Valency of an element is defined as a number indicating its combining capacity. For example, it represents the number of hydrogen atoms, which can combine with a given atom. It also represents the number of single bonds, which an atom can form. It is also defined as a number of electrons its atom is able to lend, borrow or share. *In any case valency is a pure number and has no plus or minus sign associated with it.*

Oxidation number is defined as the charge which an atom appears to have when electrons are counted. It is positive or negative. For example, in ammonia the valency of nitrogen is +3 but its oxidation number is –3.

In ionic compounds the oxidation state of an element is the same as the charge on the ion formed from an atom or an element. For example, in potassium bromide potassium is said to be in the +1 oxidation state and bromine in –1 oxidation state. It ionises as

$$KBr \longleftrightarrow K^+ + Br^-$$

Oxidation state of aluminium is Al_2O_3 is +3 and the total oxidation number of two aluminium atoms is +6. Thus oxidation state of an element is its oxidation number per atom.

There may actually be a difference between the magnitude of valency and the oxidation number. For example, consider the following compounds of carbon:

CH_4 Methane

CH_3Cl Methyl chloride

CH_2Cl_2 Methylene chloride

$CHCl_3$ Chloroform

CCl_4 Carbon tetrachloride

In each case one atom of carbon shares a total of four pairs of electrons, with other atoms. Carbon atom, therefore, tetravalent in each case.

Oxidation number for carbon id CH_4, CH_3Cl, Ch_2Cl_2, $CHCl_3$ and CCl_4, is –4 –2, 0, +2 and +4 respectively.

Thus while valency of carbon remains constant (=4) in each one of the five compounds, its oxidation number varies from –4 to +4

Oxidation and Reduction in terms of Oxidation Numbers : The term oxidation refers to any chemical change involving increasing in oxidation number whereas the term reduction applies to any chemical change involving decrease in oxidation number.

Consider the following chemical changes:

(*i*) $$2H_2 + O_2 \longrightarrow 2H_2O$$

The oxidation number of hydrogen changes from 0 (in H_2) to +1 (in H_2O). It is therefore, a case of oxidation of hydrogen.

(*ii*) Sugar ($C_{12}H_{22}O_{11}$) burns to give CO_2 and water. In this oxidation number of carbon increases from 0 (in $C_{12}H_{22}O_{11}$) to +4 in CO_2. The sugar is therefore, said to have undergone oxidation.

(*iii*) When oxygen reacts with hydrogen to give water [example (*i*)] the oxidation number of oxygen decreases from 0 (in O_2) to –2 (in H_2O). It is therefore, a case of reduction of oxygen.

In the same reaction, oxidation number of hydrogen increases, and that of oxygen decreases, i.e., hydrogen undergoes oxidation while oxygen undergoes reduction. Thus oxidation and reduction occur together. This is because in oxidation and reduction change of oxidation number occurs which pulls the electrons from one atom A to other atom B that pulls the electrons towards it. As a result of this electron shift oxidation number of A increases and that of B decreases. Oxidation and Reduction therefore must occur together.

An oxidising agent is a substance, which brings about oxidation. It contains an atom, which undergoes a decrease in oxidation number. It can also be defined as a substance, which picks up electrons and thus brings about de-electronation.

Potassium chlorate ($KClO_3$) is an oxidising agent and during the reaction it is converted to KCl. In this oxidation number of chlorine decreases from +5 (in $KClO_3$) to –1 (in KCl). This amounts to its picking up of six electrons from other atoms.

A reducing agent is a substance, which brings about reduction. It contains an atom, which shows an increase in oxidation number. It may also be defined as a substance, which supplies electrons.

Sulphurous acid (H_2SO_3) is a reducing agent and during the reaction in which it acts as a reducing agent, itself gets oxidised to sulphuric acid (H_2SO_4). In this reaction the oxidation number of S increases from +4 (in H_2SO_3) to +6(in H_2SO_4).

Various terms used in oxidation-reduction are summarised in the following table:

Term	Oxidation Number	Change	Electron Change
Oxidation	Increases		Loss of Electrons
Reduction	Decreases		Gain of Electrons
Oxidising Agent	Decreases		Picks up Electrons
Reducing Agent	Increases		Supplies Electrons

Activity Series

While discussing the preparation and properties of hydrogen and oxygen we noticed that:

(*i*) Some metals displace hydrogen from water at room temprature.

(*ii*) A number of metals displace hydrogen from acids.

(*iii*) Some metals burn brilliantly in oxygen while others do not.

On the basic of these differences in chemical activity, the metals have been arranged in order of their decreasing tendency to undergo oxidation, i.e., decreasing tendency of their atoms to give electrons. Table (activity series of metals) gives the arrangement of some common metals in such activity series or electrochemical series. (table is shown below)

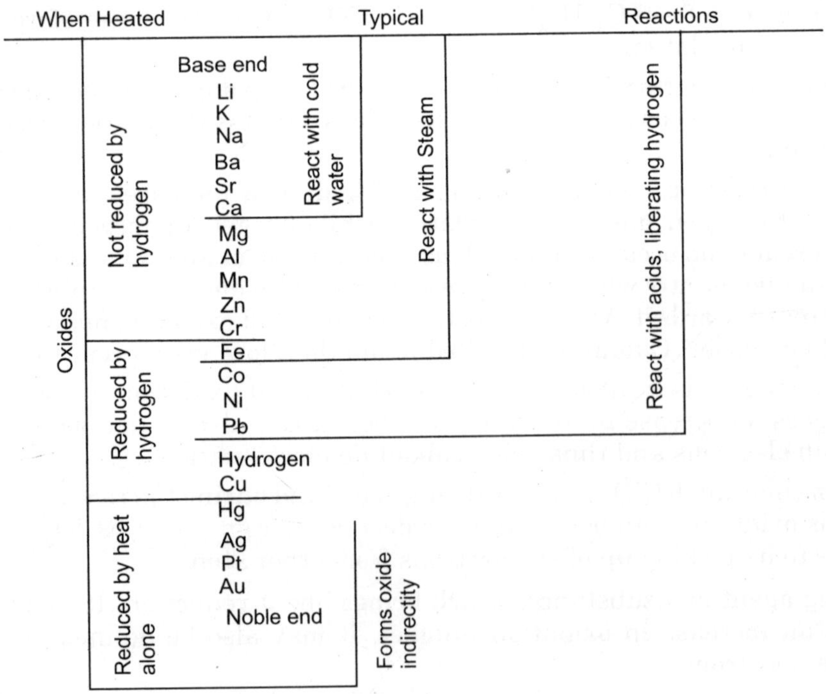

In this series element at the top has the greatest tendency to give up electron (undergo oxidation) and the one at the bottom has the least. This helps in deciding whether a certain reaction will occur or not. For example, let us consider the possible reactions of dilute hydrochloric acid with zinc and copper.

$$Zn + 3HCl \longrightarrow ZnCl_2 + H_2 \qquad \textit{(Will this reaction occur?)}$$

$$Zn \longrightarrow Zn^{2+} + 2e^- \qquad \textit{(Oxidation)}$$

$$2H^+ + 2e^- \longrightarrow H_2 \qquad \textit{(Reduction)}$$

Zinc being above hydrogen in activity series is expected to lose electrons readily to hydrogen ion. Hence, the reaction will occur.

In the reaction of copper and hydrochloric acid the copper being below the hydrogen in activity series, therefore the reaction will not occur.

$$Cu + 2HCl \longrightarrow CuCl_2 + H_2$$

$$Cu \longrightarrow Cu^{2+} + 2e^-$$

$$2H + 2e^- \longrightarrow H_2$$

Try to find whether the reaction between silver and hydrochloric acid will occur or not.

In general

(*i*) Metals placed above the hydrogen in activity series liberates from the acids.

(*ii*) Metals higher up in the series will displace those below, which are less active from their solutions. For example, zinc displaces copper from solutions.

(*iii*) Only very active metals react with water at room temperature. The reduction of H^+ ions.

Now let us consider the possibility of preparing oxygen by heating the two metallic oxides, CaO and HgO.

$$2HgO \longrightarrow 2Hg + O_2 \qquad \text{(will this occur?)}$$

$$2Hg^{2+} + 4e^- \longrightarrow 2Hg \qquad \text{(Reduction)}$$

$$2O^{2-} \longrightarrow O_2 + 4e^- \qquad \text{(Oxidation)}$$

Mercury being towards the bottom of the activity series loses electrons with difficulty and would, therefore, gain electrons readily. We would, therefore, expect this reaction to occur. Mercuric oxide decomposes on heating and gives oxygen.

In general

(*i*) Metals lying above hydrogen in activity rust easily, those below it does not

(*ii*) Oxides of manganese and other metals below it cannot be reduced to the metallic state by heating in the stream of hydrogen.

(*iii*) Oxides of iron and other metals lying below it decompose when heated.

Ionic Equations

Molecular and Ionic Equations : On adding silver nitrate solution to a sodium chloride solution we get a white precipitate of silver chloride. Sodium chloride (a salt) is produced on mixing sodium hydroxide and hydrochloric acid in equivalent proportion. These reactions are usually represented by the following chemical equations:

$$NaCl + AgNO_3 \quad AgCl + NaNO_3$$
$$NaOH + HCl \quad NaCl + H_2O$$

These equations are termed as molecular equations. If all strong bases and acids and strong bases are completely ionised, it follows that molecular equations given above are not strictly appropriate when no molecules are actually present. Chemists differ somewhat in their views on writing such equations. Some argue that since it is simpler to write molecular equations, there is no harm in writing such equations in the molecular form as long as the chemist is fully aware of the ionic character of the reacting substance. Others insist on expressing highly ionised compounds only in the form of ionic symbols, which is obviously more exact. It is of course essential to know the ionic or non-ionic nature of the compounds in advance.

A few simple rules needed for writing ionic equations are given below :

(*i*) All *strong electrolytes* are expressed in ionic symbols if they are soluble in water and all *weak electrolytes* and *covalent substances* are expressed in molecular form.

(*ii*) Any electrolyte, which is highly insoluble in water, is generally written in the molecular form to indicate its insolubility. Actually, this is incorrect since many such solids are ionic in crystal from as well. This is however, traditional.

(*iii*) In addition to the atoms which must balance on both sides of the equations for the given charges must also balance.

The following example will illustrate the method of writing ionic equations for the given molecular equations.

1. *Potassium chloride and silver nitrate react to produce potassium nitrate (soluble) and silver and silver chloride (insoluble).*

   Molecular : $KCl + AgNO_3 \longrightarrow AgCl\downarrow + KNO_3$

   Ionic : $K^+ + Cl^- + Ag^+ + NO_3^- \longrightarrow AgCl\downarrow + K^+ + NO_3^-$

   Since K^+ and NO_3^- ions appear equally on both sides of the equation these may be cancelled and the ionic equation may simply be expressed as

   $$Ag^+ + Cl^- \longrightarrow AgCl\downarrow$$

   This ionic equation express all that actually happens.

2. Potassium sulphate and barium chloride solutions react to produce barium sulphate (insoluble) and potassium chloride (soluble).

   Molecular : $K_2SO_4 + BaCl_2 \longrightarrow BaSO_4 + 2KCl$

   Ionic : $2K^+ + SO_4^{2-} + Ba^{3+} + 2Cl^- \longrightarrow BaSO_4 + 2K^+ + 2Cl^-$

   Cancelling the ions which are common to both sides $SO_4^{2-} + Ba^{3+} \longrightarrow BaSO_4$

3. Sodium Hydroxide and hydrochloric acid solutions react to produce sodium chloride (soluble) and water.

   Molecular : $NaOH + HCl \longrightarrow NaCl + H_2O$

   Ionic : $Na^+ + OH^- + H^+ + Cl^- \longrightarrow Na^+ + Cl^- + H_2O$

The Ion-Electron Method for Balancing Equations for Oxidation-Reduction Reactions By use of Half-reactions

In this method the reaction is split up into two half-reactions. In one half-reaction the oxidising agent picks up electrons and gets reduced and in the other reducing agent is oxidised by supplying electrons. The two half-reactions are balanced separately and added in such a way that the electrons on the left of the one and on the right of the other cancel out.

This procedure seems to imply that electrons are produced in Redox reaction and travel through the solution from the reducing agent to the other species which is reduced. This is not true. We should, therefore, view the two half-reactions mearly as a device for obtaining the final answer rather than as a mechanism of the reaction.

Different steps involved in writing a chemical equation by the ion electron method, they are:

- Separate the oxidising and reducing agents.
- Write down one half-reaction for the oxidising agent changing into its oxidised form.
- Write down the other half-reaction showing the reducing agent changing into its oxidised form.
- Balance the atoms other than H and O for each half-reaction by adjusting co-efficients, if necessary.
- Balance the oxygen atoms on the two sides by adding H_2O to the side deficient in oxygen.
- Balance the hydrogen atoms on the two sides by adding H^+ to the side deficient in oxygen.
- Equalise the charge on both sides by adding electrons (e^-) to the side deficient in negative charge.
- If the reaction proceeds in basic solution, add enough OH^- on both sides of the half reaction to get rid of H^+ appearing there. Combine H^+ and OH^- to give H_2O and remove H_2O duplication.
- Add these two balanced half-reactions in such a way that the electrons on the right of one half-reaction and on the left of the other half-reaction cancel each other. For this each half-reaction will be multiplied by appropriate number before addition.

Following examples will illustrate the method.

Example 1 : Write balanced equation for the oxidation of ferrous to ferric ion by dichromate ion in acid solution. The dichromate under these condition yield Cr^{3+}.

$Cr_2O_7^{2+} \longrightarrow Cr^{3+}$ (reactants and products are written on the two sides)

$Cr_2O_7^{2-} \longrightarrow 2Cr^{3+}$ (balancing the atoms other than oxygen)

$Cr_2O_7^{2-} \longrightarrow 2Cr^{3+} + 7H_2O$ (balancing oxygen atoms by adding $7H_2O$ on the right)

$Cr_2O_7^{2-} + 14H^+ \longrightarrow 2Cr^{3+} + 7H_2O$ (balancing the hydrogen atom by adding $14H^+$ on the left)

$Cr_2O_7^{2-} + 14H^+ + 6e^- \longrightarrow 2Cr^{3+} + 7H_2O$(equalising the charge by adding 6 electrons on the left)

(Balancing the negative charge of the in the left hand side by $6e^-$)

The second half of the reaction :

Writing down the reactant and product of the half-reaction for the reducing agent changing into its oxidised form :

$Fe^{2+} \longrightarrow Fe^{3+}$ (I) Writing the reactant and product of the half-reaction for the reduction agent changing into its oxidised form.

$Fe^{2+} \longrightarrow Fe^{3+} + e^-$ (Equalising the charge by adding one electron on the right to equalise the charge of the Fe^{2+})

Adding two half reactions :

$$\begin{array}{rcl} Cr_2O_7^{2-} + 14H^+ + 6e^- & \longrightarrow & 2Cr^{3+} + 7H_2O \\ 6Fe^{2+} & \longrightarrow & 6Fe^{3+} + 6e^- \\ \hline Cr_2O_7^{2+} + 6Fe^{2+} + 14H^+ & \longrightarrow & 2Cr^{3+} + 6Fe^{3+} + 7H_2O \end{array}$$

Balance the following equations :

1. $H^+ + MnO_2 + Fe^{2+} \longrightarrow Fe^{3+} + Mn^{2+}$
2. $H_2C_2O_4 + H_2O_2 \longrightarrow CO_2 + H_2O$

CHAPTER

24

Radioactive Chemistry

Introduction

In this chapter, we shall discuss the radioactivity, the radioactive properties of different particles, the nuclear energy, the chain reaction, the different instruments for detecting radioactivity. The instruments for different experiments on radioactivity. The radioactive elements called Lanthanides and Actinides. The nuclear chain reaction, has also been explained, with the creation of atom bomb.

Radioactivity

The word radio comes from the word radiation, Thus the word radioactivity means the activity or the physio-chemical reaction of those substances which are capable of producing the energy in the ionic form.

In 1896 *Henry Becquerel* found that a uranium compound affected a photographic plate wrapped in lightproof paper. He called the phenomenon as radioactivity.

Those elements that are regarded as radioactive elements they have a very large atomic numbers, these orbits are fully packed with electrons; they also have many orbits. This decreases the attraction force between the nucleus and the electrons and the outer most orbit that is the valence orbit, has the least attractive force, the electrons can easily radiate from their orbit making the substance radioactive. We shall now discuss few properties of radioactivity.

Characteristics of Radioactivity

Investigations have shown a number of characteristics of radioactive elements, which are stated below:

(*i*) All radioactive elements are of high atomic number, generally they have atomic number more than 83.

(*ii*) Radioactivity is an atomic phenomenon, i.e.; it is a characteristic property of the atom of any radioactive element, no matter whether it is in the free state or compound with any other element. Thus, the rays out of the uranium or its salt are exactly similar to those produced by radium, thorium, polonium etc and their salts. Emission of rays occurs at a rate development only upon the nature of the element concerned.

It is seen that the intensity of radioactivity of 1 mg of radium or any of its compounds like radium nitrate or radium bromide is the same.

(*iii*) Radioactivity is a spontaneous process, which is never affected by any chemical reagents or physical conditions. The process goes on exactly at the same time rate in the presence and in the absence of light also. No sign of dimension or acceleration of its rate is detected by extreme variations in the external conditions of temperature and pressure.

(*iv*) The emission of radiation of a radioactive element is accompanied by the liberation of an enormous quantity of energy in the form of heat.

(*v*) With the emission of radiation, an atom of radioactive substance element breaks down and forms the atom of a different element having a lower atomic weight. The emission continues till atom through a series of changes became non-radioactive.

In fact, radioactivity should be regarded as a nuclear phenomenon. Generally, the atom of a radio active substance having neutron-proton ratio N:P greater than a certain fixed value possesses an unstable nucleus capable of spontaneous disintigration showing radioactivity. It involves the change of the nucleus of a radioactive atom to produce the nucleus of the atom of the other element.

The phenomenon mentioned above are all properties of a natural radioactive substance and the radioactivity is therefore called Natural Activity.

Radioactivity is the phenomenon in which the atoms of certain heavy elements (like uranium, thorium radium polonium etc) undergo spontaneous disintegration or breaking down with the emission of insoluble, penetrating radiations at a rate absolutely independent of the external or environmental conditions and transform into the atoms of new elements.

Nature and Properties of Radiations Emitting from a Radioactive Source

The radiations from a radioactive source attack a photographic plate even in dark. They are capable of passing through very thin layers of metallic sheets and induce phosphorescence on a fluorescence substance like zinc sulphide, barium platinocyanide. They also ionise air or any other gas in the neighbourhood by separating electrons from the gas molecules. It also affects the living bodies by altering the internal biochemical reactions and causing carcinogenic effects.

Alpha, Beta and Gamma Rays

Investigations made by Rutherford show that radiations are composite in nature and consist of three different kinds of rays, denoted by Greek letter *Alpha (α) Beta (β) and Gamma (γ)*. These rays are characterised by its deflection in a magnetic field.

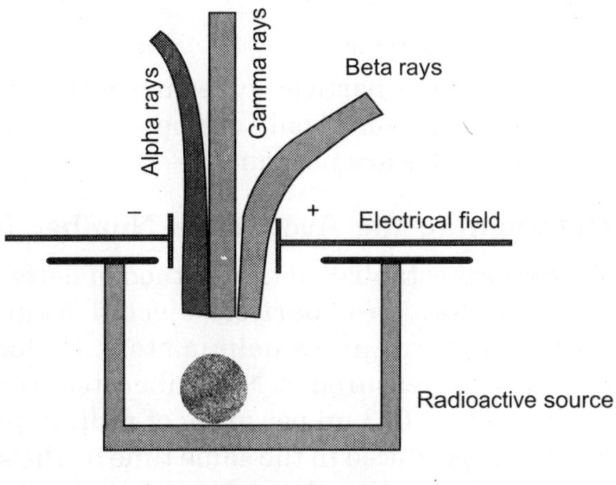

Fig. 24.1. The different radioactive rays in the electrical field.

When subjected to the influence of a strong electrical field (or magnetic field) the alpha rays are slightly deflected towards the North Pole. The β rays are maximum deflected in the opposite direction of the α rays and the γ rays do not respond to magnetic fields see Fig. 24.1.

Alpha Rays : It has been proved that the alpha rays consists of highly positive charged small material particles. Each alpha particle has a mass of 4 (H =1) and a charge of +2 units (twice as large as that of a proton). That is to say, an alpha particle consists of two protons (each carrying a unit positive charge and a mass 1). We know that helium atom consists of two protons, two neutrons and two electrons. Thus an alpha particle is identical with a doubly charged helium ion (He^{++}) i.e. a helium atom without its two orbital electrons.

For a better expression, an α-particle often symbolised as ${}^4{}_2H^{++}$. It is observed that when the emission of an alpha particle, (mass 4 on the atomic weight scale) in a radioactive charge, will make the resultant element 4 units less than that of parent element, for example :

Radium	$\xrightarrow[226]{-\alpha}$ Radon (a radioactive element)	+	α particle
Atomic Weight	222		4

The alpha particles at the moment of their emission from a radioactive source has a very high velocity of about 1.609×10^9 cm/sec.

Due to high speed and comparatively large size, they possess great momentum and considerable kinetic energy (= mv^2). The alpha particles produce phosphorescence on zinc sulphide screen and each particle on striking the screen produces a bright flash or scintillation of light. The power of the particles gives high kinetic energy.

The alpha particles also have a great ionization effect on the gasses through which they pass. The high-speed particles cause ionization by knocking electrons out of the gas molecules they collide with.

Being relatively large in size, their penetrating power is low. They are stopped or completely absorbed by an ordinary piece of paper or by an aluminium foil of less than 0.1 mm thick. While passing through a thin layer of air, they loose their charge and the air is ultimately converted into helium atoms by picking up electrons from the surrounding air.

$$\underset{\alpha\text{-particle}}{He^{++}} + \underset{\text{electron}}{2e} \longrightarrow \underset{\text{helium atom}}{He}$$

When the α-particle comes in contact with skin, causes severe burns on the skin, but the burns are caused only on the surface of the skin because the penetrating power of the alpha particles are not much.

Determining the Avogadro's Number from Radioactivity

Rutherford established the method of determination of Avogadro's number on the basic of the fact that an α-particle ejected from a radioactive source ultimately turns into a helium atom. He found that the volume of helium gas measured at NTP, liberated from the disintegration of radium was 0.043 ml per gram of radium per year. The number of α-particle produced in the same time by the same amount of radium was calculated directly and was found to be 11.6×10^{17}.

This value of N agrees with the same determined by other methods.

Now each alpha particle is transformed into one helium atom or molecule of helium atom. One mole of gas occupies 22.4 litres or 22400 ml. Hence the Avogadro's number (N), which is the number of molecules in one mole, can be obtained as

$$N = \frac{11.6 \times 10^{17} \times 22400}{0.043} = 6.05 \times 10^{23}$$

Beta Rays : Deflection experiment in an electrical or magnetic field shows that the β-rays are made up of negatively charged very minute particles, called β-particles. Determination of charge, mass and the study of the properties of the particles indicate that the β-rays is the stream of fast moving electrons. A beta particle is expresses by the symbol ${}^{0}_{-1}e$, which indicates its mass nearly zero and charge –1. Thus, when a beta particle or an electron with negligible mass is emitted in a radioactive change, the resulting element (called the daughter element) has an atomic weight identical with that of its parent. These particles are emitted from the radioactive source atoms of tremendous velocity ranging between 40 to 90% that of light. Of course, the velocity of β-particles very much dependant to the source from which they are emitted.

Although the velocity of the beta particles is very high, it's mass is so negligibly small, that its kinetic energy is much less than that of alpha particle.

Beta particles are capable of passing through the aluminium sheet of 5mm thick. On the other hand the ionising power of the β particles is much less than that of the α particle, β-rays have a very less effect on the zinc sulphide screen.

Gamma Rays : Gamma rays does not consist of material particles. In fact these rays are electromagnetic radiations, similar to that of light. They resemble X rays but have much shorter wavelength. The γ-rays have much higher velocity and greater penetrating power than those of alpha and beta rays. They move with a velocity nearly equal to that of light that is 1,86,000 miles per second. They are capable of penetrating through an iron plate of 25 cm thick, an aluminium sheet of 60 cm thick and a lead sheet of 8 cm thick. Gamma rays are known as hard rays because their penetrating power through matter is much greater than that of X-rays.

They are weak ionizers of gases; their effect on the living matter is slow but highly injurious.

The penetrating power of the alpha, beta and the gamma rays are shown in the Fig. 24.2.

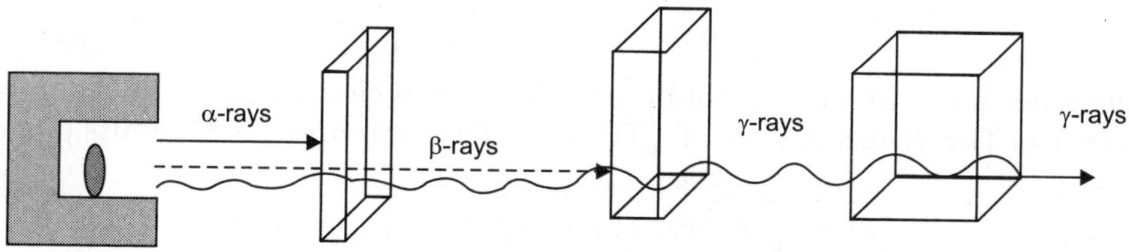

Fig. 24.2. The Penetration of alpha, beta and Gamma rays.

The radioactive rays explained above can be summarised in the table below:

Property	Alpha Rays	Beta Rays	Gamma Rays
Mass	4.03314 amu	5.487915×10^{-4} amu	No mass
Charge	9.60432×10^{-9} esu	4.80216×10^{-10} esu	No mass
Velocity	1.609×10^{7} cm/sec	1.5×10^{11} cm/sec	No mass

The Theory of Radioactivity

It is true that during a radioactive charge, a radioactive source atom breaks down spontaneously to produce the atom of a new element with different or same atomic weight depending upon the ejection of the α or β particle. Now, the new element may in turn be radioactive and may emit a particle resulting in the formation of yet another element. This is the summary of the theory of the radioactive disintegration, which was established by *Rutherford* and *Shoddy.* According to this theory, the radioactive charges occur in series. At each stage, a charge is accompanied by the emission of an α or β particle are never emitted together in one single charge. After a number of successive charges each involving either α or β activity, a non-radioactive end product (stable atom of lead). It is to be also to be noted that the emission of the number of alpha and beta particle are also fixed which.

For example, an atom of uranium ($_{92}U^{238}$) or an atom of radium ($_{88}Ra^{226}$) will individually eject α particle. No below emission is observed in the disintegration of Uranium-238 or radium-226. It should be bear in mind that, $_{92}U^{238}$ on emission of α-particle is transformed into thallium $_{90}Tl^{239}$. The disintegration of $_{92}U^{238}$ into $_{90}Tl^{239}$ can be represented by the following equation.

$$_{92}U^{235} \longrightarrow \, _{90}Tl^{239} + \alpha$$

or

$$_{92}U^{235} \longrightarrow \, _{90}Tl^{239} + \, _{2}He^{4} \text{ [since } \alpha = \text{He nucleus]}$$

On the other hand, an atom of thallium ($_{90}Tl^{239}$) will eject a β-particle. It can never eject α-particle. The disintegration of $_{90}U^{239}$ into Protactinium $_{91}Pa^{239}$ can be represented by the following equation.

$$_{92}U^{235} \longrightarrow {}_{90}Tl^{239} + \alpha$$

or, $$_{92}U^{235} \longrightarrow {}_{90}Tl^{239} + {}_{2}He^{4} \text{ [since } \alpha = \text{He nuclei]}$$

On the other hand, an atom of thallium ($_{90}Tl^{239}$) will eject a β-particle. It can be never eject α-particle. The disintegration of $_{90}Tl^{239}$ into Protactinium $_{91}Pa^{239}$ emits β particle.

$$_{90}Tl^{239} \longrightarrow {}_{91}Pa^{239} + \beta$$

or, $$_{90}Tl^{239} \longrightarrow {}_{91}Pa^{239} + {}_{-1}e^{0}$$

Thus, a radio active of uranium ($_{92}U^{238}$) changes ultimately to a non-radioactive atom of lead $_{82}Pb^{206}$ with the emission of α-particle in 8 stage a β-particles in 6 stages, the overall process is shown below :

$$_{92}U^{235} \longrightarrow {}_{82}Pb^{206} + \underbrace{8\,{}_{2}^{4}He^{++}}_{8\text{-}\alpha \text{ particle}} + \underbrace{6\,e_{-1}^{0}}_{6\text{-}\beta \text{ particle}}$$

$$_{92}U^{238} \xrightarrow{\alpha\text{ (emission)}} {}_{90}U^{238} \xrightarrow{\beta\text{ (emission)}} {}_{91}Pa^{238} \xrightarrow{\beta} {}_{92}U^{239} \xrightarrow{\alpha} {}_{90}Tl^{230} \xrightarrow{\alpha} {}_{88}Ra^{226} \xrightarrow{\alpha} {}_{86}Ra^{222}$$

$$\xrightarrow{\alpha} {}_{84}Po^{218} \xrightarrow{\alpha} {}_{82}Pb^{214} \xrightarrow{\beta} {}_{83}Bi^{214} \xrightarrow{\beta} {}_{84}Po^{214} \xrightarrow{\beta} {}_{82}Pb^{210} \xrightarrow{\beta} {}_{83}Bi^{210} \xrightarrow{\beta} {}_{82}Pb^{206}$$

The numerical figure placed just above right hand side of the symbol of the element in the above table indicates the mass number (number of protons and neutrons in the nucleus). The numerical figure on the left expresses the atomic number i.e. the positive charge on the nucleus of the atom of the concerned element.

In a balanced nuclear equation the sum of the mass number of the different species in the left side and the sum of the mass numbers of the different species in the right side must be equal to each other. Similarly the sum of the atomic numbers of the different species on the left side must be equal to the sum of the atomic numbers of the different species on the right side :

- It is to be remembered that the liberation of heat energy in a radioactivè charge results from the interaction between·the neutrons (i.e. the constituents of the nucleus - the protons and the neutrons). Alpha or beta particle does not mean the pre-existence of the particles or such in the atomic nucleus of a radioactive element. It has been logically explained that in a heavy nucleus and α or β-particles formed just at the moment of its emission.
- α, β and γ rays are never found to come out from the same radioactive element. But a sample of radioactive substance may contain more than one radioactive element or it may successively produce various radioactive disintegration. That is why it is seen that a single radioactive source in emitting three different rays.

Causes of Radioactivity

As the radioactivity is a nuclear phenomenon, it must be connected with the unstability of the nucleus. The factor upon which the stability of the nucleus depends is the N: P in the case of N: P=1 the nucleus is stable. The elements having N: P ratio more than 1.54 are said to be unstable and are regarded as radioactive. For example, the state of $_{11}N^{23}$ the number of neutrons is (23-11) = 12 or N : P = 12:11 or 1.09 so the element is non-radioactive.

But $_{92}U^{238}$ number of neutrons is (238 – 92) = 146 or the N:P = 146:92 or 1.58, so the element is radioactive.

In the case of Radium $_{88}Ra^{226}$ the number of neutrons is (226 - 88) = 138

$\therefore$ N : P = 138:88 or 1.56 so it is radioactive.

The elements having atomic number greater than 83 and the numbers of neutrons greater than 126 are unstable. They do not fall within the belt of stability (see Fig. 24.3) and are all radioactive. The nucleus of such elements tend to attain stability by lowering their N : P ratio to a favourable value through spontaneous radioactive disintegration.

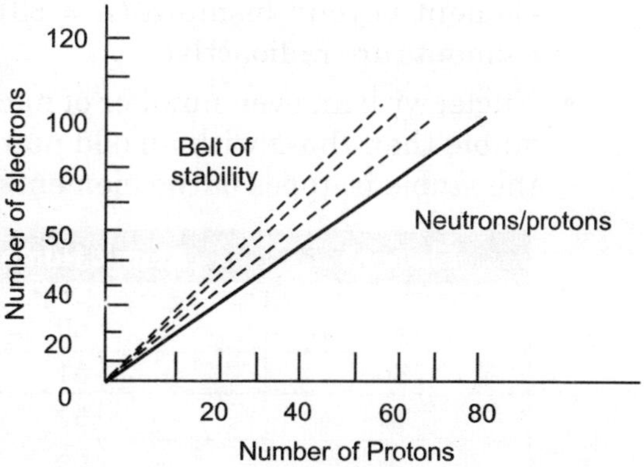

Fig. 24.3. The graph of Nuclear Stability.

It is to be remembered that there are three types of forces, which operate in the nucleus. They are proton-proton repulsion, is mainly responsible for the unstability of the nucleus. So to stand against this proton-proton repulsion the neutron-proton attraction and the neutron-neutron repulsion is to be increased. In fact this is the only possible for those elements which are having atomic number below 83 or up to 83. After bismuth with atomic number 83, as the atomic number increases, the proton-proton repulsion becomes faster. The increase in value of N/P ratio cannot be minimised due to the fast growing repulsion.

As a result the nuclei of such elements become unstable and start to disintegrate through a radioactive process so that they may reach a stable configuration. According to the Heisenberg's Uncertainty Principle, nothing can remain unstable, so to obtain the stable state the radioactive disintegration takes place.

Nuclear Stability

According to classical electrostatics, we should accept that in a nucleus to repel one another (due like charges), causing the nucleus to fly apart. It turns out however, that at the very short distance of separation characteristics of atomic nuclei, about 10^{-15}m, there are strong attractive forces between nuclear particles. The stability of the nucleus depends upon the balance between these forces and these of electrostatic repulsion. Our knowledge of nuclear stability today lives a great deal to be desired. All we have a series of empirical rules for predicting which nuclei will be most stable. The more important of these rules are listed below :

- The neutron to proton ratio required for the stability varies with atomic number (see Fig 24.3) below for light elements, this ratio is close to one for example.

  Stable isotopes (dots in the diagram) have neutron-proton ratios that fall within a narrow range known as belt of stability. For light isotopes of small atomic number, the stable ratio is 1.0, with heavier isotopes it increases to about 1.5. There are no stable isotopes for the elements of atomic number greater that 83.

  The isotopes $^{12}{}_{6}C$ $^{14}{}_{7}N$ and $^{16}{}_{8}O$ are stable, as atomic number increases, the ratio increases, the belt of stability sifts to higher numbers of neutrons. With very heavy isotopes such as $^{206}{}_{82}Pb$, the stable neutron-proton ratio is 1:5 (206 -82)/82 = 124/ 82 = 1.51, or 1.5 : 1.51.

- Nuclei combining more than 83 protons are unstable, putting it another way, no element beyond bismuth (Z = 83) has a stable isotope. All the isotopes of such elements are radioactive.
- Nuclei with an even number of nucleus, either protons or neutrons tend to be more stable than those with an odd number of nuclear particles. Consider, for example, the stable isotopes of the elements of atomic number 50 - 53.

Element	Atomic Number	Mass Number of the Stable Isotope
Sn	50	112, 114, 116, 117, 118, 119, 120, 121, 122, 124
Sb	51	121, 123
Te	52	120, 122, 123, 124, 125, 126, 128, 130,
I	53	127

  Of the 21 stable isotopes of these elements, 18 have an even number of protons, 16 have an even number of neutrons, 13 of 21 have both even numbers of proton and neutrons.

- Certain numbers of protons and neutrons appear to be particularly stable 2, 20, 50, 82 and 126. These magic numbers were first suggested by Maria Mayor, who developed a nuclear shell model, Mayor got Nobel Prize in 1936 for this.

Difference between Radioactive Change and Ordinary Chemical Change

Radioactive Change	Chemical Change
1. A radioactive change is independent of all the central factors like pressure temperature, etc. It is spontaneous process, the rate of which can neither be accelerated nor retarded.	1. A chemical process is dependent on lots of factors such as, temperature pressure etc and it can also be accelerated or retarded as well.
2. In a radioactive process an element is changed to another fundamental element.	2. In chemical change an element is never converted into a new element.
3. A radioactive change is always irreversible.	3. A chemical change is not always irreversible.

Radioactive Change	Chemical Change
4. A radioactive change causes a release of much heat, as atomic energy.	4. The chemical change does not contain so much energy.
5. In a radioactive change α, β and γ rays are produced.	5. In a chemical change no such rays are emitted.
6. Other chemical substance or reaction does not influence the radioactive changes.	6. Chemical reactions do change always under the influence of the other substance or compound.

Detection of Radioactivity

There are many apparatus for the detection of radioactivity, the description of some of them are following:

(A) By using phosphorescence of Zinc Sulphate : This method is used for detecting radioactive substances but the process can detect only the presence of the alpha particles. The apparatus is called spintheriscope, see Fig. 24.4.

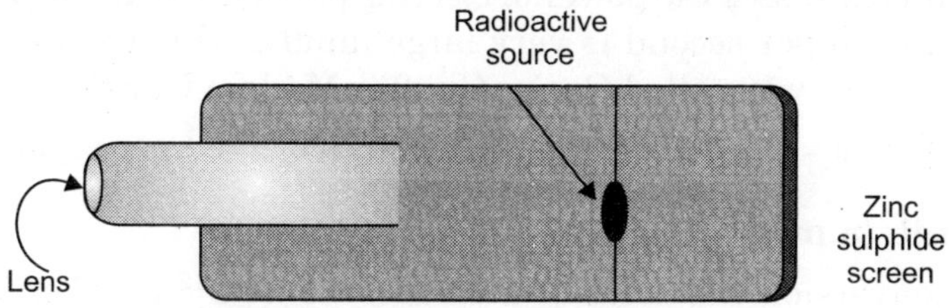

Fig. 24.4. The Spinthariscope.

The working process of the apparatus is very simple, the alpha rays from the radioactive source is emitted and it hits the zinc sulphide screen, making the screen phosphorescent thus, the tiny blinks of light will be formed at each hit, thereby enabling us to detect the alpha rays.

(B) By Ionisation of Gas : Causing discharge of an electroscope. This method for detection and measurement of radioactivity depends upon the ionization of gas through which the α or the β particle pass. These fast moving particles knock out the orbiting electrons from their valence orbits of the gas molecules to which they collide. As a result the molecules are converted into ions and become the conductor of electricity. Therefore, it becomes very easy to detect the radioactivity by using the gold leaf electroscope. An electroscope is initially charged by an external source which is opposite in polarity of the charge of the conducting gas. Now if the ionised gas is brought near the electroscope it will be discharged due to unlike charges. Thus we can prove the presence of the radioactivity.

(C) By the use of Geiger Counter, or Geiger Tube : The basic operating principle of the Geiger counter is that it consists of an ionization chamber shaped in the form of a tube (there the name has derived from.)

This ionization tube is subjected to a high voltage, through a load resistor R as shown in the Fig. 24.5. The output of the ionization chamber is fed into the linear amplifier via a capacitor of very low capacitance; the output of the linear amplifier is read in the oscilloscope.

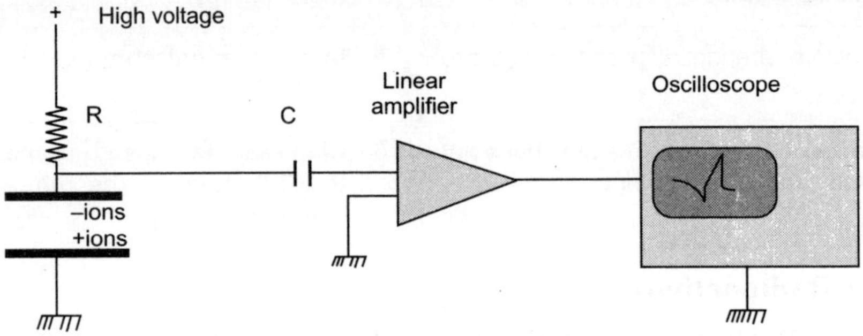

Fig. 24.5. The basic circuit of Geiger Muller Tube.

The Units of Radioactivity

The primary unit used for the detection of radioactivity is called the disintegration per second dps. This unit expresses the power of ejecting particles (α and β particles per secor.d). Since the disintegration per second is very large number this has been converted into a practical unit of radioactivity called Curie (C) after Madam Curie.

$$1 \text{ Curie} = 3.7 \times 10^{10} \text{ (}\alpha \text{ or } \beta \text{ particles ejected per second)}$$

Atomic Mass Unit (a.m.u)

According to the conversion the mass of a carbon atom ${}_6C^{12}$ is taken to be 12 amu, this 1 amu is equal to 0 of the carbon ${}_6C^{12}$ atom.

$$\begin{aligned} 1 \text{ a.m.u.} &= 0\text{th of the mass of } {}_6C^{12} \text{ atom} \\ &= 0\ 12/N_0 \text{ GM where } N_0 \text{ is Avogadro's number} \\ &= 0 \times 12/6.022 \times 10^{23} \text{ or } 1/6.022 \times 10^{-24} \text{ g} \end{aligned}$$

or $$= 1.6605779 \times 10^{-24} \text{ g}$$

or, $$= 1.66 \times 10^{-27} \text{ kg}$$

Electron Volt : According to Einstein's mass energy relationship,

$$E = mc^2$$

Where m is mass in grams, C is the velocity of light in centimetre per second and E is energy in ergs. Therefore, energy equivalent or 1 amu is given by.

$$E = 1.66 \times 10^{-27} \times (3 \times 10^{10} \text{ cm/sec})$$

$$= \frac{1.494 \times 10^{-3} \text{ joule}}{1.602 \times 10^{-19}} \text{ [Since 1 electron volt} = 1.602 \times 10^{-19}\text{]}$$

or $= \frac{1.494\times10^{-3}}{1.602\times10^{-13}}$ million electron volt

or $= 932.59$ MeV

[Electron volt is defined as the energy acquired by an electron, (change = $1.602 \times 10^{-19} \times 1) = 1.602 \times 10^{-19}$ joule].

Isotopes Isotones and Isobars

According to Dalton's atomic theory, atoms of the same element are exactly alike and equal in weight. But latter investigations have proved that equality of weight is not the characteristics of all atoms of the same element. From a study of radioactivity, Shoddy recognised that an element might have atoms of identical chemical properties but of different atomic weight. But such atoms of any given element are always found to contain the same number of protons in their nuclei. The phenomenon leading to the existence of varieties of atoms (with different masses) of the same element is known as isotopy and the varieties of atoms are called isotopes.

> *Isotopes are defines as the atoms of two elements having different atomic mass but same atomic number*

How Isotopes are Formed

The change involved in the elimination of an α or β particle may be depicted as follows:

- X $\xrightarrow{-\alpha}$ $_{Z-Z}X_1^{4-4}$ (where $x = a$ radioactive element, A= mass number Z = atomic number X_1 = new element).
- E $\xrightarrow{-\beta}$ $_{Z+1}E_1^{A}$ (E = and element E_1 = new element formed).

It is obvious that by successive expulsions of one α and two β particles, we arrive at a product, which is having a same atomic number as the original atom. In other words, the original atom and the product are isotopes.

Thus by emission of an α particles, Uranium –I produces Uranium X_1, which on successive expulsion of the two β-particles, one after another, giving Uranium X_2 and Uranium–I since both Uranium–1 and Uranium–II have the same atomic number, 92.

Uranium I	$\xrightarrow{-\alpha}$ Uranium X_1	$\xrightarrow{-\beta}$ Uranium A_2	$\xrightarrow{-\beta}$ Uranium II
At Number 92	At Number 90	at Number 91	At number 91
At Mass 238	At mass 234	At mass 234	At mass 234
(ISOTOPE)			(ISOTOPE)

Thus, isotopes may also be defined as the different atomic phenomenon, of the same element with different atomic weights (mass) but with the same atomic number. Invariably they occupy the same space in the periodic table.

Although isotopes were first detected in the radioactive sources but now a day it has been seen that the non-radioactive sources also stay with the mixtures of their isotopes.

Examples of Isotopy

A well-known example of isotopy is found in chlorine and magnesium

"Cl $_{17}{}^{37}$Cl, *Mg $^{25}{}_{12}$Mg $^{26}{}_{12}$Mg

The isotopes of hydrogen has been discussed in the chapter 9 to tome 1.

Isotones and Isobars

The atoms of elements, which have the same mass numbers but different atomic numbers, are called isobars.

The isotopes of argon (at no 18), potassium (at number 19) and calcium (at number 20) are the e.g. of isobars.

$_{18}Ar^{40}$, $_{19}K^{40}$ and $_{20}C^{40}$ are examples of isobars.

Isobars have different atomic numbers of protons in their nuclei but the sum of the protons and the neutrons in the nuclei is the same, except that they have the same atomic mass, the isobars differ in their physical as well as the chemical properties.

The emission of the β particles from a radioactive element does not change the atomic mass but changes the atomic number. Hence, the element is the always an isobar of the previous element.

Let us consider the following radioactive changes :

	Ra – A ⟶	Ra–B ⟶	Ra–C ⟶	Ra–C
Atomic Number	82	80	81	82
Mass Number	218	214	214	214

It is evident from the above that Ra–B, Ra–C and Ra–C are isobaric with one another but Ra–A is isotopic with Ra–C.

Some elements having different mass numbers may have same number of neutrons in the nuclei, they are called isotones.

Isotones display different chemical properties

Examples : $_{14}Si^{30}$ (protons 14 neutrons 16) $_{15}P^{31}$(proton 15 neutrons 16) $_{16}S^{32}$ (protons 16 neutrons 16).

Fajan's Russel and Soddy's Group Displacement Law

The emission of an alpha or a beta particle from a radioactive element produces a new element, which will have the atomic number (i.e. positive charge on the nucleus) different from that of the original element. With this change in atomic number, the daughter element corresponds to an element in a different group of the periodic table than the parent element. This is obvious as the position of an element in the periodic table is fixed by its atomic number.

The loss of an alpha particle (mass 4 and charge +2) from a radioactive element gives an element of nucleus of which has the positive charge less by 2 units and mass by 4 units than that of parent element. Therefore, the atomic number of the element produced by emission of an alpha particle becomes 2 units smaller and the daughter element shifts two places (groups) to the left of the parent in the periodic table.

For example, in the table, radium (atomic number 88) is in the group II while radium (at no 86) derived from its parent due to the loss of an α-particle in the zero group.

$$_{88}R^{226} \xrightarrow{-\alpha} {}_{86}Rn^{222} + {}_{2}H^{4}\ (\alpha)$$

Again, thorium (at no 90) belongs to group IV in the periodic table but the product radium (at number 88) generated from it by emission of an alpha particle is a group II element.

Emission of β-particle (mass 0 charge –1) Or electron from a nucleus does not cause any appreciable charge in mass (atomic weight) but the positive charge on the resulting nucleus is increased by one unit. Hence, the atomic number of the product is one unit greater than that of parent element formed, shifts to next group in periodic table one place right to the parent.

For example:

Radium (in group II) (at number 88 mass 228) $\xrightarrow{-\beta}$ (Actinium at number 89 (in group III)

$$_{82}Pb^{210} \xrightarrow{-\beta} {}_{83}Bi^{210} + {}_{-1}e^{0}\ [\beta\text{-particle}]$$

Thus, if the atomic number of the parent radioactive element is known, atomic number of all its disintegration products can be calculated by subtracting 2 for every alpha particle emitted and adding 2 for every beta particle emitted.

The sift of an element in the periodic table in radioactive changes is summarised in the group displacement law.

Emission of an alpha particle in a radioactive change generates an element, which occupies a position in the periodic table two groups to the left of the parent element. While the emission of a β-particle produces an element that belongs to the next group to the right of the parent.

Some Atomic Particles Besides Proton Electron and Neutron

Hideki Yukawa (1935) a Japanese scientist suggested that besides protons and neutrons another particle oscillates between them, which he called meson, having a velocity closed to light.

Mesons have been detected in some radioactive experiments they are 273 times heavier than electrons and may electrically positive, negative or neutral. They interact with proton and neutron therefore, the proton is converted into neutron and the neutron is converted into proton temporarily by exchanging their charges from meson.

Proton + Meson (–) $\longleftrightarrow$ Neutron

Neutron + Meson (+) $\longleftrightarrow$ Proton

So the meson starts oscillating between the protons and neutrons.

Anderson (1932) discovered positron. He discovered the existence of the positive electrons. Positrons collide with electrons and both disappear being transformed into an electron.

Binding Energies Regarding the Stability of Nuclei

Binding energy per nucleon of the number of isotopes are plotted in the graph in Fig. 24.6.

It is seen that the maximum occurs at about mass number 56(iron). Thus nucleus of iron is thermodynamically much stable. It is also seen that the points for helium (mass number = 4), carbon (mass number = 12) and oxygen (mass number = 16) lie quite high in the graph. This shows that the nuclei of these elements are exceptionally stable. The maximum binding energy is about 8.7meV. This is the energy required to remove a proton or a neutron from most stable nucleus.

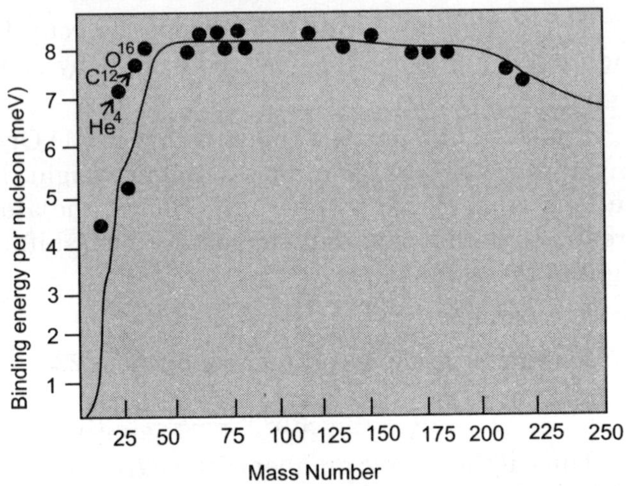

Fig. 24.6. The graph of binding energy per nucleon.

The graph near the maximum is almost flat from A = 50 to about A = 100. This shows that the addition of a single nucleon to any nucleus in this region increases the binding energy by roughly the same amount.

Half Life

We often hear about the half-life of a radioactive substance, the time required for half life is important in several ways. It determines how long a sample of an unstable or dangerous isotope must be stored before it decays to a level that does not represent a hazard. For example, cobalt 60, an isotope that emits gamma used to treat cancer has a half-life of 5.2 years. The intensity of the radiation produced by a cobalt 60 source decreases by one half-life every 5.2 years. Thus cobalt 60 source must be replaced regularly. The disposal of unwanted plutonium -239 presents a problem because it has a half-life of 24, 000 years and at the end of the 24,000 years the mass of the plutonium will be reduced to half and so on.

The half-life of radioactive elements vary widely for example, the half-life of a recently reported element (information received from IUPAC) element 111, is 1.5×10^{-3} seconds. The half-lives of some common elements are listed below :

Isotopes	Half-lives	Types of emission
$^{233}U_{92}$	1.6×10^5 years	α
$^{234}U_{92}$	2.45×10^5 years	α
$^{238}U_{92}$	4.51×10^9 years	α
$^{239}U_{92}$	23.54 months	β
$^{239}Np_{93}$	2.3 days	β

Artificial Radioactive Isotopes Used in the Medical Diagnosis

$^{3}_{1}H$	Tritium (hydrogen-3)	Measure water content of body
$^{11}_{6}C$	Carbon-11	Brain scan with position emission transverse tomography (PET) to trace glucose pathway
$^{14}_{6}C$	Carbon-14	Radioimmunoassay
$^{32}_{15}P$	Phosphorus-32	Detection of eye tumours
$^{51}_{24}Cr$	Chromium-51	Albumism diagnosis, size and shape of spleen, gastrointestinal disordes
$^{59}_{26}Fe$	Iron-59	Bone marrow function, diagnosis of anemias
$^{60}_{27}Co$	Cobalt-60	Treatment of cancer
$^{67}_{31}Ga$	Gallium-67	Whole-body scan for tumours
$^{75}_{34}Se$	Selenium-75	Pancreas scan
$^{81m}_{36}Kr$	Krypton-81m	Lung ventilation scan
$^{85}_{88}Sr$	Strontium-85	Bone scan for bone diseases, including cancer
$^{89m}_{93}Tc$	Technetium-99m	Brain, liver, kidney, bone marrow scans; diagnosis of damaged heart muscles
$^{131}_{53}I$	Iodine-131	Diagnosis of thyroid malfunction; treatment of hyperthyroidism and thyroid cancer
$^{197}_{80}Hg$	Mercury-197	Kidney scan

RADIOACTIVE DECAY (detection of half life)

Radioactive decay is the first order process. The rate constant k can be expressed in terms of the half-life, t_1 by the equation

> Ln is the natural log only found in the scientific calculator, this would be available in the *fx*-82lb model.

$$\ln \frac{[A_0]}{[A]} = kt$$

Where A_0 = initial concentration of the radioactive element A is the concentration of the element in the give time period T.

or $$t_1 = \ln \frac{[A_0]}{[A]} \times \frac{1}{k}$$

If we set the time equal to the half-life as t_1 of the initial concentration hence at time

$t_1[A] = 1[A_0]$, therefore, $t_1 = \ln = \frac{[A_o]}{[A]} \times \frac{}{k}$

$t_1 = \ln \frac{[A_0]}{1[A_0]} \times \frac{}{k}$ $\times$ [since A = 1 $[A_0]$ thus, t_1 $_{= \text{natural}}$ log of $2/k$ or 0.693/K

The initial amount of a radioactive isotope and the amount remaining after a given period of time are expressed by the logarithmic relationship that applies to all first order reactions.

$$\ln = \frac{C_0}{Ct} = kt$$

Where C_0 is the mass number of moles of the isotope and C_t is the mass number of moles remaining at time t.

Example A : What is the rate constant for the radioactive disintegration of cobalt 60 isotope? $_{27}Co^{60}$ decays with half life of 5.2 years to produce $_{28}Ni^{60}$.

$K = 0.693/\ t_1 = 0.692\ /5.2$ or $.13\ \text{years}^{-1}$

Example B : Calculate the fraction of a sample of the $_{27}Co^{60}$ isotope that remain after 15 years.

Solution: $\ln\ = \frac{C_0}{C_t}$

$= (0.13\ y^{-1})\ (15\ \text{years}) = 1.95$

$\therefore$ C_0/C_t = antilog of ln = 7.03 (use e^x button of your calculator)

the fraction that will remain is equal to the amount at time t divided by the initial amount, or C_t/C_0 which is equal to 1/7.03 = 0.14.

Therefore, the fraction will remain after 15 years is 0.14, or the 14% of the original cobalt 60 sample.

Example C : How long does it take for a sample of cobalt 60 to disintegrate to the extent that only 20% of the original amount remain?

Solution : $C_t = 0.020 \times C_0$

Therefore, $C_0/C_t = 1/0.020 = 50$

Or $\ln C_0/C_t = kt$

$\text{Ln } 50 = (0.13\ y^{-1})t$

$T = \ln 50/\ 0.13^{-1} = 3.9112\ /0.13 = 30$ years.

Try Radon 222 has a half-life of 3.823 days. How long will a sample of radon with mass 0.750 g to decay to other elements living only 0.0100 of radon 222? [**Ans.** 11.1 days]

Radioactive Disposal

It is already known to us that the radioactive substance like plutonium-239 has a half life of 24,000 years, so we must be careful about the disposing them. Otherwise the radiations will endanger our normal life, the picture below shows the careless disposal of the radioactive plutonium-239 in the North Sea, the worn barrel, is almost decayed and the radioactive plutonium is getting contaminated with the water. It is an information from IUPAC that there are 28,390 more barrels has been drowned in the North Sea as disposal.

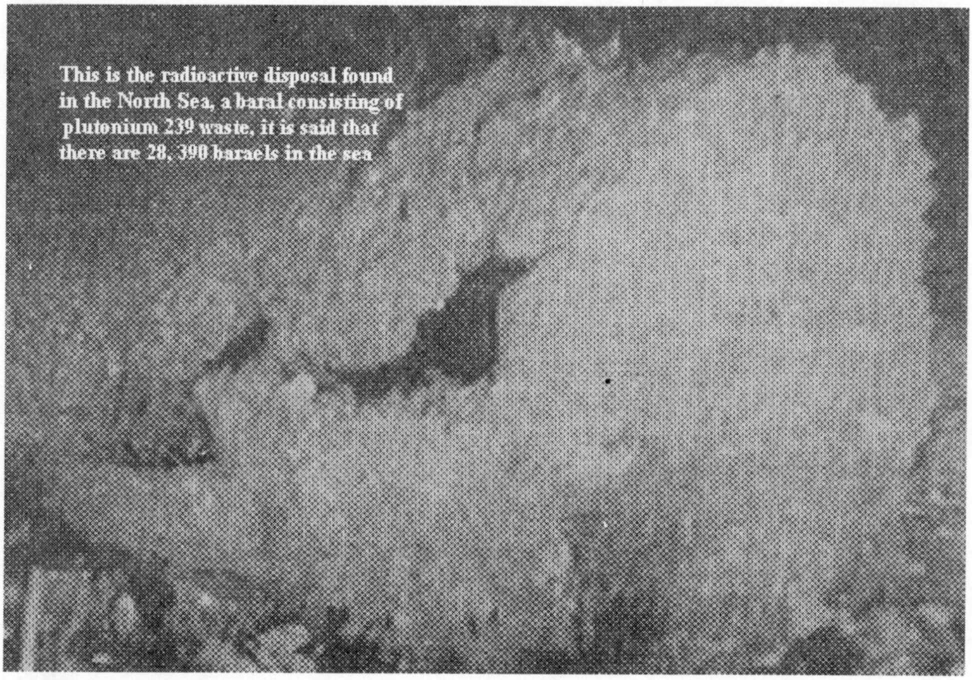

Nuclear Reactions

Changes in nuclei that result in changes in their atomic number, mass number and energy states are called nuclear reactions.

To describe a nuclear reaction, we use an equation that identifies the nuclides in the reaction, their mass numbers, atomic numbers and the other particles involved in the reaction. These particles include :

1. Alpha particle (nHe, or α), helium nuclei consisting of two neutrons and two protons
2. Beta particles ($-^{0}{}_{1}e$ or β)
3. Positrons ($+^{0}{}_{1}e$ or β^{+}), particles with the same mass as an electron but with 1 unit of positive charge.
4. Protons (eH or P), nuclei of hydrogen atom
5. Neutrons, (bn or n), particles with a mass approximately equal to that of a proton but no charge.

Nuclear reactions also involve gamma rays (γ). Gamma rays are electromagnetic radiations like X-rays but with grater energies and shorter wavelengths. The equations of several nuclear reactions that have played an important role in the history of nuclear chemistry follow :

1. The first naturally occurring radioactive element has been isolated, polonium, which was discovered by Madam Curie in 1898. It decays as follows:

$$_{84}Po^{212} \longrightarrow {}_{82}Pb^{208} + \frac{4}{2}He\ (\alpha)$$

2. Technetium was first prepared in 1937. This was the first artificial element on earth. It decays by beta emission.

$$_{43}Tc^{98} \longrightarrow {}_{44}Ru^{98} + -{}^{0}{}_{1}e \text{ or } \beta$$

3. The first controlled nucleus chain reaction was carried out in the University of Chicago.

$$_{90}U^{235} + 3\frac{1}{0}n \longrightarrow {}_{35}Br^{87} + {}_{57}La^{157} + 3\frac{1}{0}n$$

4. One of the reaction is involved in the thermonuclear weapon and one prepared for the peaceful production of energy by nuclear fission is

$_1H^3$ + ♐H ⟶ ♐He + $\frac{1}{0}n$

5. The first element prepared in 1919 by bombarding nitrogen atoms with α particles.

$$\frac{14}{7} + \frac{4}{2}He \longrightarrow \frac{17}{8}O + \frac{1}{1}H$$

Each number reactions is balanced in two ways:

1. The sum of the mass number of the reactants equals the sum of the mass number of products.
2. The sum of the atomic numbers of the reactants equals to the sum of the atomic numbers of the reactions equals the sum of the atomic numbers of the products.

If the atomic number and the mass number of all but, one of the particles in a nuclear reaction is known. We can identify the particle by balancing the reaction. For example, we could determine that $^{17}_{0}O$ is a product of the reaction of $\frac{4}{7}$ and $\frac{14}{7}N$. He if we know that a proton $\frac{1}{1}H$, was one of the two products.

Example : The reaction of a α particle with $_{12}Mg^{25} + \frac{4}{7}He \longrightarrow \frac{A}{Z}X + H\frac{1}{1}$

The A is the mass number and Z is atomic number

$\therefore$ $25 + 4 = A + 1$

or, $A = 29 - 1$ or $A = 28$

$\therefore$ $12 + 2 = 2 + 1$ or $A = 4 - 1 = 13$.

The element having mass number = 28 and atomic number 13 is aluminium.

Try The nuclide $_{53}I^{125}$ captures an electron $(-{}^{0}{}_{1}e)$ to produce a new nucleus and no other positive particle. What is the equation of the reaction?

Artificial Radioactivity

In the year 1934 Curie and Joliot were studying the emission of positrons by aluminium which had been bombarded by alpha particles. They were amazed to see that, the piece of aluminium was still emitting positrons after the bombardment has stopped and the rate followed the same law, which governed the release of radiations by naturally radioactive substances. It is clear that the bombardment has transformed the aluminium atoms into an isotope which was unstable in character and which under went subsequent disintegration just like radium. The name thus given to this phenomenon as *Artificial Radioactivity.*

Subsequent analysis disclosed what happened, an atom of aluminium, which absorbed an alpha particle emitted a neutron and was transformed into an atom of phosphorous. But this new atom of phosphorous was unstable and ejected a positron, thereby changing to an atom of silicon. Many radioactive isotopes known today were discovered after lot of experiments.

Nuclear Fission

All most as soon as *Curie* discovered radium in 1898 physicists were impressed by the fact that the radioactive elements gave off a steady stream of energy in the form of various radiations. Some scientists explained that if imagined a mechanism by which these atoms will extract energy from their surroundings. *Becquerel Rutherford* and *Shoddy* among others were convinced that the radioactive atoms were themselves storehouse of energy. Rutherford's Experiment which was pioneer in atom smashing the experiments made clear that every atom was such a reservoir of energy. By 1932 there was general agreement that the release of atomic energy was in accordance with the equation for mass energy equivalence, namely $E = mc^2$ which Einstein had postulated in 1905. Though atomic energy would become available for practical use. The day was far off in efficiency of atomic bombardments and of the fact that there was no way of controlling the result are not of many kinds. It will be possible to create mew elements not known to exist in nature, by bombarding uranium. In 1934 *Fermi* and his colleagues tried this experiment, he reported that after capturing one neutron U_{238} has become U_{239}. However the subsequent emission of an electron transformed one of the neutrons in its nucleus into a proton. Thus the atom changed from element number 92 to element number 93. The subsequent emission of another electrons transformed this element number 93 to element number 94. These elements were known as *Transuranium Elements.*

Hann and *Strassmann* found that one of the radioactive products of the reaction could not be separated chemically from barium and so reluctantly came to the conclusion that it was barium. Until then they have spoken of these products as *radium isotopes.*

The reaction of the uranium in the fission reaction is as follows :

$$^{235}_{92}U + \frac{1}{0}N \longrightarrow {}_{56}Ba^{190} + {}_{36}K^{93} + 3\frac{1}{0}n$$

Thus *fission is defined as the bombardment of the heavier atomic nucleus with a neutron and splitting it into lighter elements spontaneously.*

Here it is necessary to say that on the left half on the entry of the neutron in the equation above actually sets the process of splitting a barium nucleus and a krypton nucleus is produced, with three more neutrons. It this three neutron hit another uranium nuclei, then in turn they will generate three more neutrons, so that the initial neutron in the first step has produced nine electrons. Such a spontaneous process is called chain reaction.

It is just like a game of nine pins, suppose if the ball knocks down three pins and these three pin will knock down other three more and in turn all nine pins will lie on the floor.

The uranium fission is spontaneous process much like this. The isotope of uranium having mass 235 makes up a little less that 1% of natural uranium ore. When a U-235 nucleus is hit by a neutron, it decompose and gives out neutrons three at a time; but the chances are very less that they will hit another U-235 nuclei. They collide frequently with more abundant U-238 nuclei and terminate the chain reaction.

However, it is possible to concentrate the rarer isotope and a make a ball of uranium metal that is practically pure uranium. If these nuclei are brought sufficiently close together, the density of the neutron emission becomes so high that the chain reaction sets in. The chain reaction was first shown experimentally in the Chicago University.

Chain Reaction

When the uranium 235 underwent fission it released one to three neutrons, U_{235} will undergo fission with fast neutron as well as slow neutrons. An atomic power plant is based on the fission of uranium 235 with slow neutrons.

A chain reaction using fast neutrons will be an explosive reaction, in other words an atom bomb. The Principle chain reaction is shown in fig 290. U_{238} with fast neutron a chain reaction could not be set up in U_{238} with fast neutrons.

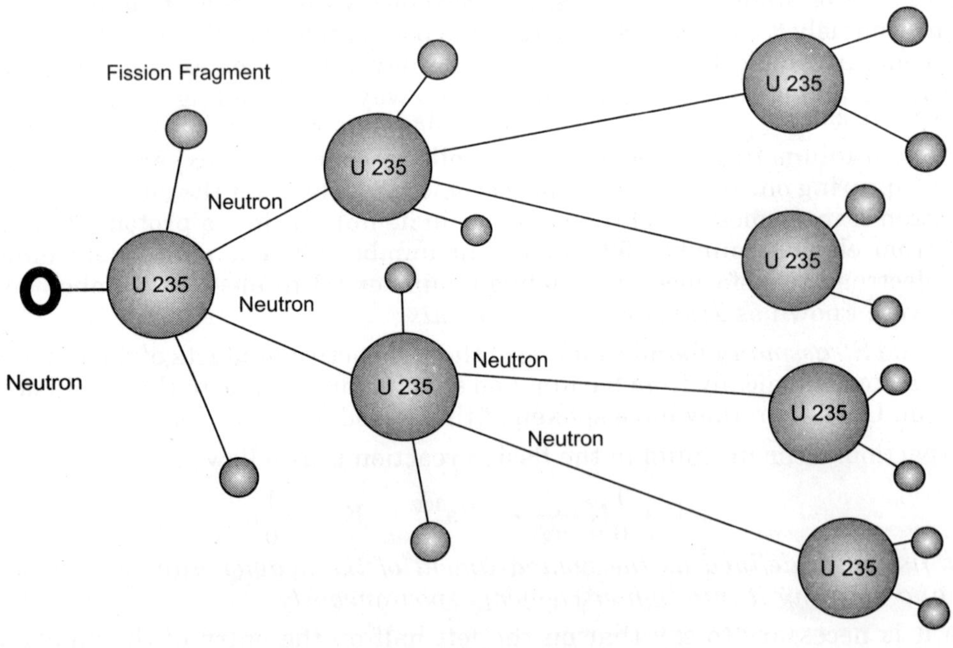

Fig. 24.7. The Chain Reaction.

A sample of Uranium 235 was finally reached the threshold of the era of the atomic energy.

Nuclear Fusion

In 1932 two physicists *Cockcroft* and *Walton* found that a hydrogen nucleus and a lithium nucleus can combine with them and an enormous energy was released.

$$\frac{1}{1}H + {}_3L^{37} \longrightarrow \frac{4}{2}He + \frac{4}{2}He + \text{ energy}$$

These reactions are called fusion reaction; thus *it can be defined as the transformation of lighter elements like hydrogen into helium by adding another heavier element is called fusion.*

The best natural example of the fusion is in the sun, where the hydrogen gas is getting converted to helium producing the solar energy.

Nuclear Reactors

Synthesis of Nuclides : The first nucleus prepared in the laboratory was made by Rutherford in 1919 by a transmutation reaction. The bombardment of nuclei of one type with other nuclei or with neutrons. Rutherford bombarded nitrogen atoms with α particles emitted from natural radioactive isotope of natural radium and radioactive isotope of radium and observed protons resulting from the reaction.

$$\frac{14}{7}N + \frac{4}{2}He \longrightarrow \frac{17}{8}O + \frac{1}{1}H$$

The ⊠O and ♏H nuclear are stable, so no further reaction takes place.

Before 1940 the heaviest element was uranium, where atomic number is 92. In 1942 *McMillan* and *Abelson* were able to make element 93 neptunium (Np), by bombarding uranium 238 with reactions. The reaction gives unstable uranium-239 with a half-life of 2.3 days, and decays to plutonium (Pu) element 94. The reactions are as follows :

$$^{238}_{92}U + \frac{1}{0}n \longrightarrow {}^{235}_{92}U$$

$$^{235}_{92}U \longrightarrow {}_{93}Np^{239} + {}_{-1}e^0$$

$${}_{93}Np^{239} \longrightarrow {}_{92}Pu^{239} + {}_{-1}e^0$$

Since 1940 hundred of tons of plutonium have been prepared for atomic purpose.

The Production of Plutonium

Long before that historic moment December 2 1942, it was realised that if the nuclear reactor attained a self-sustaining chain reaction, it would be a device for the conversion of Uranium 238 into plutonium (see Fig. 24.8).

For this reason studies of plutonium were instituted in the Metallurgical Laboratory at the time of its organisation even earlier in the summer of 1940 *Lawrence* and his colleagues at the University of California had carried one experiments with minute amounts of plutonium produced with the aid of a cyclotron. The first experimental confirmation was that Uranium 238 could be converted into neptunium was obtained by *McMillan* and Abelson. However the amount was not sufficiently large to confirm the subsequent conversion of the neptunium into plutonium.

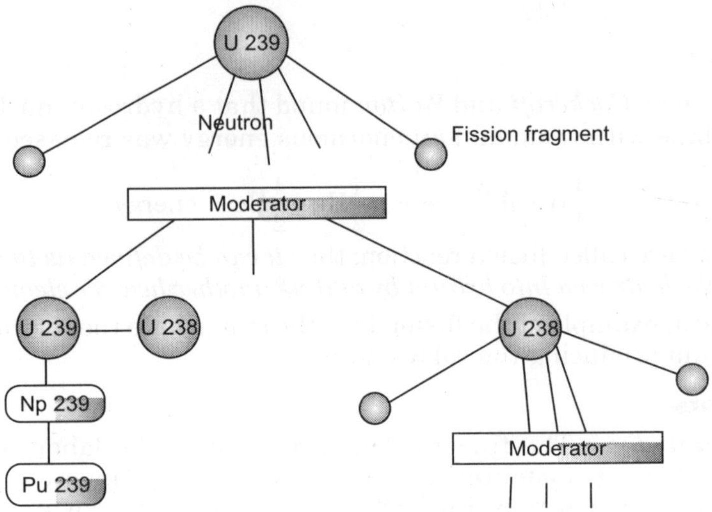

Fig. 24.8. The Production of plutonium for chain reaction.

As a reactor starts to operate, heavier isotopes of plutonium (Pu-240, Pu-241 and Pu-242) are, produced by subsequent capture of neutrons by higher plutonium nuclei. The longer the fuel stays in the reactor higher the concentration of bearer plutonium isotopes. It is estimated that about 1650 tones of plutonium will exit in the year 2000. Most of this will be a component of used nuclear reactor fuels.

Nuclear Power Reactors

Fission of U-235 nucleus produces on average 2.5 neutrons as well as fission fragments. These neutrons may cause the fission of other uranium 235 atoms, which in turn provides mere neutrons, setting up a chain reaction. Nuclear fission becomes self sustaining when the number of neutrons produced by the fission equals or exceeds the number of neutrons absorbed, by splitting nuclei plus the number that escape into the surroundings. *The amount of a fissionable material that will support a self-sustainable chain reaction is called critical Mass (CARI).*

The critical mass of a fissionable material depends upon the shape of the sample and as well as the type of the material. We shall discuss the construction of atom bomb later on.

Nuclear Reactors

Any nuclear reactor must contain five stages; they are as follows, (1) Fuel, (2) Moderator (3) cooling system and (4) A shield and a containment system (see Fig. 24.9).

Nuclear Fuel : Fissionable isotopes (commonly found in a mixture of U235 and U238 must be present in sufficient quantity to provide a self-sustaining chain reaction. Most reactors in the USA uses pellets of uranium oxide, U_3O_8 (also called urano-uranic oxide),

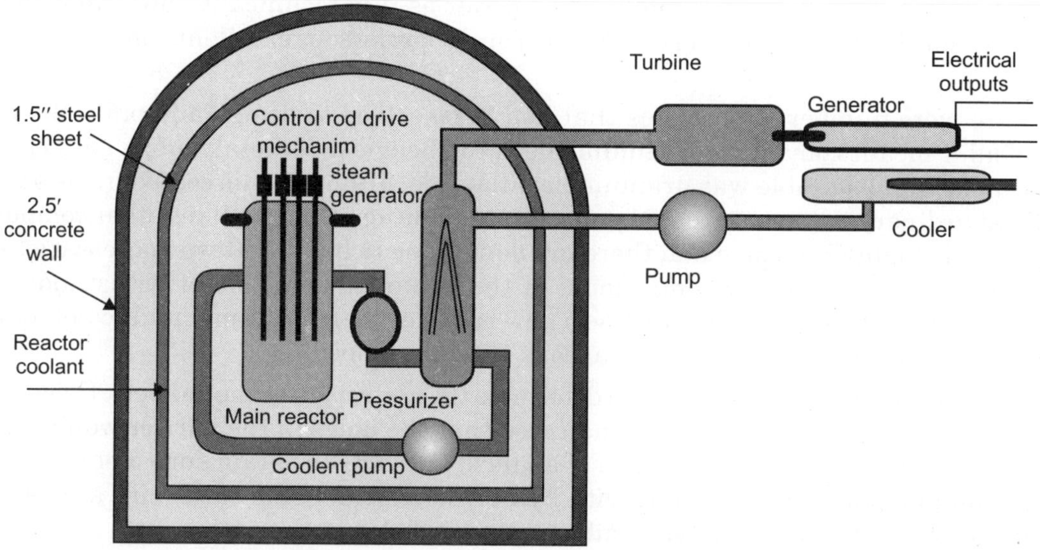

Fig. 24.9. The Diagram of a Nuclear Reactor.

in which the concentration of uranium 235 has been increased from its initial 1% to 3%. The U_3O_8 are contained in a tube (a fuel rod) of protective material, usually a zirconium alloy. The core of a typical nuclear power reactor in the U.S. (National Radioactive Laboratory) has about 40,000 kg of enriched U_3O_8 contained in several thousand fuel rods.

Naturally occurring uranium is a mixture of several isotopes; uranium 238 is the most abundant. About every 140 uranium atom there is only one U-235 isotope. To obtain heavier concentration of U-235 it is necessary to separate it from the U-238. The most successful method uses fractional diffusion of gaseous UF_6 (uranium hexafluoride). At a low pressure through porous barriers to separate $^{235}UF_6$ from $^{238}UF_6$. This method is based on the fact that the slightly porous barrier $^{235}UF_6$ molecules diffuse through porous barriers slightly faster than the heavier $^{238}UF_6$ molecules (Graham's Law). When a number of gases are kept in a container with porous walls, effusion of gases occurs through the small openings of the walls. The lighter gases effuse much rapidly through these small openings than the heavier ones. In 1832 Thomas Graham stated that the rate of effusion of gases are inversely proportional to the square root of their densities (or their molar masses)

$$\frac{\text{Rate of effusion of gas A}}{\text{Rate of effusion of gas B}} = \frac{\sqrt{\text{Density B}}}{\sqrt{\text{Density A}}} = \frac{\sqrt{\text{Molar Mass B}}}{\sqrt{\text{Molar Mass A}}}$$

Effusion through a porous barrier from a region of higher pressure to one of lower pressure was used for the large scale separation of gaseous $^{235}UF_6$ from $^{238}UF_6$ is done.

Gas Diffusion Method : As long as 1832 Thomas Graham has shown that a mixture of two gases can be separated by permitting it to diffuse through minute holes of porous sheet of clay or some similar substance. The lighter gas because of higher speed of its molecules tends to pass more rapidly. The barrier as it is technically called in effect is a kind of atomic sieve. By using a succession of such barriers an excellent separation can be made.

There were number of problems that had to be solved to design and build the plant, first of all it meant converting Uranium one of the heaviest of metals into a gas. The only compounds found suitable was uranium hexafluoride, although this compound is actually a solid at ordinary temperature and pressure. In addition, this gas likes other compounds of fluorine is highly corrosive and therefore dangerous to handle. It was necessary to find materials for the parts and components of the apparatus capable of withstanding the corrosive action of gas. For this purpose a new series of chemical compounds combinations of carbon and fluorine known fluorocarbons were developed.

The next important problem was to determine the materials for the barriers. Calculations based on the kinetic theory of gases indicated that the holes in the barrier would have to be in the neighbourhood of a millionth of an inch. One way to obtain such a barrier was to etch a thin plate of silver zinc alloy with hydrochloric acid. Some of the zinc atoms would be carried off by the acid, behaving submicroscopic holes in the plate.

The method of operation of the plant is that of multistage recycling. When about half of the gas in any stage has passed through the barrier, what remains is drawn off and reintroduced into previous stage, the flow of gas is indicated in Fig 24.10. Thus as uranium 235 moves towards one end of the plant Uranium 238 moves towards the other. It takes the given sample of uranium hexafluoride several months to pass through all the stages. Operation of the plant required the development of several pumps, which were leak proof and able to with stand the corrosive nature of the gas.

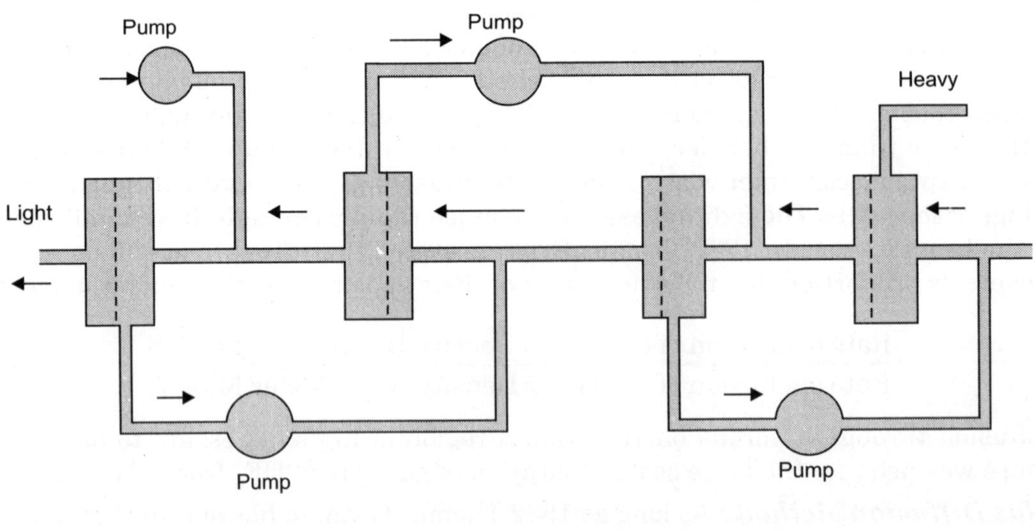

Fig. 24.10. The Separation of Isotopes by Multistage Recycling.

Electromagnetic Separation : Each unit consists of a gigantic electromagnet, which created the necessary magnetic field. The uranium, in the form of a volatile salt was vapourised and ionised then introduced into this field. The two isotopes described divergent paths in the magnetic field because of their difference in mass (see Fig. 24.11). The uranium 235 in the bomb that exploded over Hiroshima came from the electromagnetic separation plant.

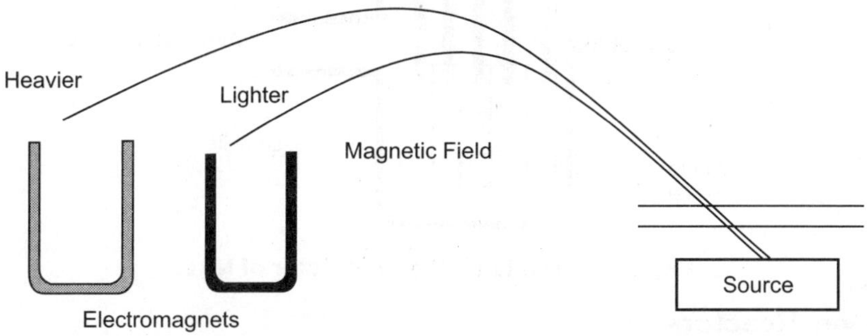

Fig. 24.11. The separation of uranium 235 by Electromagnetic Process.

Moderator : Neutrons produced by nuclear reactions moves too fast to cause fission. They must be slowed down in order to stop the atomic explosion and also to protect it from producing additional reactions. In a reactor neutrons are slowed by creating collision with nuclei of a moderator such as heavy water D_2O, graphite, carbon dioxide or light (ordinary) water, H_2O. These materials are used because they do not react with or absorb neutrons. Most reactors in USA use light water as the moderator. Heavy water and graphite is used in other countries.

A Coolant : The coolant carries the heat produced by the fusion reaction to an external boiler or turbine, where it is turned into electrical energy. The coolant is a gas or a liquid that is pumped through the reactor core. Some coolants also serve as moderators liquid sodium is used as a coolant and moderator.

A Control System : A nuclear reactor is controlled by adjusting the number of slow neutrons present to keep the rate of the chain reaction at a safe level. Control is maintained by special rods that absorb neutrons. In control rods containing cadmium or boron-10 are often used. Boron-10 for example absorbs neutrons by a reaction that produces lithium-7 and α-particles. The chain reaction can be stopped by inserting all the control rods into the core between the fuel rods.

A Shield and a Prevention System : A nuclear reactor produces a good amount of radiations, even it is shut down, the waste disposals are also highly radioactive. So a reactor operators or people working near to a reactor must be protected from its radiations.

A reactor often consists of three major protective parts.

1. A reactor vessel, that is made of a steel sheet 3-20 cm thick which absorbs most of the radiations.
2. A main shield, of 1-3m concrete wall which surrounds the whole reactor system.
3. A personal shield of lighter material which protects the operators from X-rays and γ-rays.

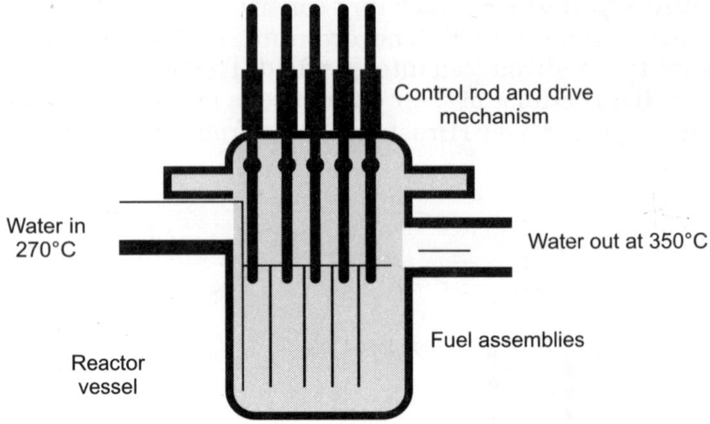

Fig. 24.12. The Light Water Reactor of NRL.

Nuclear Fusion Reactors

The process of combining very light nuclei into heavier nuclei is called fusion reaction. The principal source of energy in the sun is the main fusion reaction, in which four hydrogen nuclei react to form one helium nucleus and 2 proton.

$$4\frac{1}{1}H \longrightarrow \frac{4}{2}He + 2_{+1}e^0 \text{ positron}$$

A helium nucleus has a mass of 0.7% less than that of four hydrogen nuclei; this lost mass is converted into energy during the fusion.

It has been determined that deuterium $4\frac{1}{1}H$ and tritium ^3_1H, the heavy isotopes of hydrogen, undergo fusion at extremely high temperatures (thermonuclear fusion). They form helium nuclei and neutrons.

$$\frac{1}{1}H + \frac{3}{1}H \longrightarrow \frac{4}{2}He + \frac{1}{0}n$$

The change proceeds with a mass of 0.0188 amu, corresponding to the release of 1.69×10^9 kJ per mole of $\frac{4}{2}He$ formed. The very high temperature is necessary to give the nuclei enough kinetic energy to overcome the very strong repulsive forces resulting from the positive charges on their nuclei so they can collide.

Uncontrolled fusion is the power source of hydrogen bomb. However, if we could control fusion reactions of heavy isotopes of hydrogen at reasonably low temperatures, hydrogen from the water of the oceans could supply immense amount of energy for future generations.

A fusion reactor is a nuclear reactor where the fusion reactions of the nuclei are controlled.

Useful fusion reactions appear to require very high temperatures for their initiation in about 10^8K. At this temperature all molecules dissociate into atoms and the atoms ionise, forming the forth state of matter, plasma, because no solid material can exist in that temperature in stable condition. Plasma consists and circulates in a circular chamber called plasma arc controlled by magnetic field of 1.5 Tesla and the plasma current of 250kA. A practical application of the plasma arc is TOKAMAK, the little sun developed in Saha Institute of Nuclear Physics.

Creation of Atom Bomb

British Canadian Scientists as well as American physicists were the main mastermind behind these Atomic Explosives.

Atomic Energy for military purpose, the possibility of a chain reaction that might lead either to the controlled release of atomic energy or to the creation of an atomic bomb because the subject of immediate speculation.

Nuclear fission sometimes occurred in three elements Uranium, Thorium and Protactinium, releasing large amount of energy. Only neutrons caused the fission of Thorium and Protactinium but only fast neutrons could cause the fission of Uranium but fast or slow neutrons could cause the fission of Uranium.

A chain reaction could be obtained in ordinary uranium if one neutron released by every fission of Uranium 235 would cause another fission.

If the reaction fell below 1, the reaction will die out and if the reaction exceeds 1 the reaction will be accelerated. This ratio is known as *multiplication factor* and was represented in the calculations by *K.* the nuclear physicists sometimes refer to it as the great God K.

It is impossible to prevent stray neutrons from starting a chain reaction when the size of the charge exceeds the critical size. The problem of designing a bomb, therefore, was to create a mechanism, which would bring together quickly and firmly the number of pieces of uranium 235 or plutonium of sub-critical size. If the assembled pieces form a mass of critical size a chain reaction will begin and the bomb will explode. The obvious method of very rapidly assembling an atom bomb was to shoot one part as a projectile in a gun against a second part as shown in Fig. 24.13.

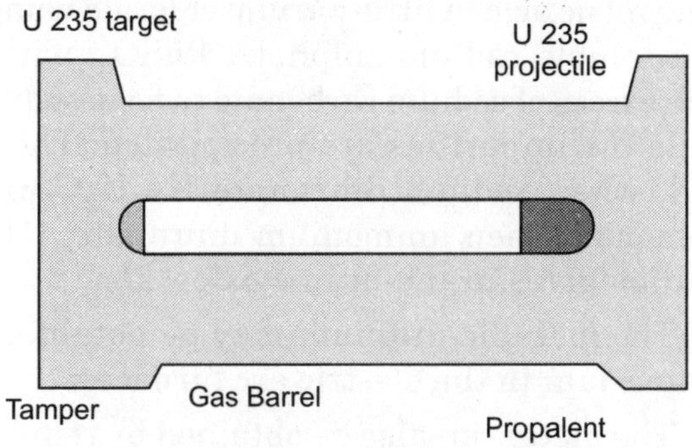

Fig. 24.13. The assembling of Atom Bomb.

Radioactive Elements

We shall now study about the radioactive elements, their properties, compounds etc. These are divide into two groups, the *Actinides* and the *Lanthanides.*

URANIUM

Melting Point : 1132°C
Boiling Point : 3818°C

Atomic Number : 92
Atomic Weight : 238.029
Isotopes : U-234, U-235, U-238

History and Occurrence

Uranium was discovered in pitchblende in 1789. Becquerel made the discovery of the phenomenon of radioactivity in 1896 from an examination of the uranium minerals and Madam Curie discovered radium in pitchblende in 1898.

All uranium minerals contain radium in minute amounts in equilibrium, as it is a product of uranium disintegration. The important uranium containing mineral is *pitchblende,* UO_2, or *uranite*, *Torbernite* the copper uranite, $Cu(UO_2)_2P_2O_8 12H_2O$. Pitchblende is found in Joachimthal Hungary, Cornwall of England, Caucasus, Australia and in Canada where rich ore occurs. Variety in pitchblende rich in thorium and rare earth's found in Norway and U.S.A. Many thorium minerals also contain uranium.

Extraction

Uranium minerals were formerly treated mainly for the extraction of radium and uranium was mainly obtained as bi-product. Since the development of atom bomb, it has acquired importance and being produced in large amounts.

The powdered uranium mineral is roasted and fused with sodium carbonate and nitre, the mass is boiled with water and the residue is digested with dilute sulphuric acid and some nitric acid. A little barium chloride being added to the extract. Radium remains behind as insoluble radium sulphate, $RaSO_4$, with the resulting $BaSO_4$. The solution is boiled with excess of sodium carbonate to form sodium uranyl carbonate, which remains dissolved, while the impurities are precipitated. The solution is neutralised with dilute sulphuric acid, when sodium diuranate $Na_2U_2O_7 6H_2O$ is precipitated as a yellow solid. With ammonia, when ammonium diuranate, $(NH_4)_2U_2O_7$ is precipitated. The precipitate on ignition gives urano-uranic oxide, U_3O_8.

The metallic uranium may be obtained by reducing the oxide U_3O_8, with carbon or magnesium in the electric arc furnace.

The metal can also be obtained by reducing uranium tetrachloride, UCl_4, with sodium on a large scale, pure uranium is prepared by heating tetrafluoride UF_4 with metallic calcium in an electric arc furnace, or by electrolysis of potassium uranyl pentafluoride, mixed with sodium chloride and calcium chloride using molybdenum cathode.

Compounds of Uranium

Oxide and Hydroxide of Uranium : Uranium forms three oxides; uranous oxide, UO_2, which is basic and forms uranous salts, urano uranic oxide, U_3O_8 and uranous trioxide, which behaves mainly as acidic oxide.

Physical Properties

Melting Point	1132°C
Boiling Point	3818°C
Emission	α-particle
Density	19.0 g/cc
Heat of Fusion	4.7 Kcal/mol
Heat of capacity	6.6 cal/mol
Electrical Resistivity	25 ohm cm
Isotopes	U-233, U-234, U-235, U-238
Half Life	
U-233	1.62×10^5 years
U-234	2.48×10^5 years
U-235	7.13×10^8 years
U-238	
Particle Energy (U-238)	4.2 MeV
Standard Entropy	12.03 deg^{-1} mol^{-1}
Isotope Mass	
U-235	235.12
U-238	238.12
Mass Defect	
U-235	+0.12
U-238	+0.12
Packing Fraction	
U-235	+5.1
U-238	+5.0
Binding Energy per Nucleon	
U-235	7.6MeV
U-238	7.5 MeV
Radioactive Abundance	
U-235	0.71%
U-238	99.28%
N/P Ratio	
U-235	1.554
U-238	1.587

Uranous Oxide (UO_2): It is obtained as a black solid by heating urano uranous oxide in hydrogen

Urano uranous Oxide + Hydrogen ⟶ Uranous Oxide + Water

$$U_3O_8 + H_2 \longrightarrow 3UO_2 + H_2O$$

It dissolves in nitric acid with the formation of uranyl nitrate.

Uranous Oxide + Nitric Acid ⟶ Uranyl Nitrate + Water

$$UO_2 + 2HNO_3 \longrightarrow UO_2(NO_3)_2 + H_2O$$

It behaves like a reducing agent and reduces ammonia. Uranous hydroxide is precipitated from uranous salt solution by the addition of alkalis. This dissolves in acids from the uranous salts.

Urano Uranic Oxide (U_2O_8): It is the naturally occurring uranium compound. It is formed when uranous oxide, uranium trioxide or ammonium diuranate is gently heated in air. It is the most stable salt of uranium and is found in dark green powder. It gets slowly dissolved in sulphuric acid yielding uranous sulphate.

Urano uranous Oxide + Sulphuric Acid ⟶ Uranous Sulphate + Uranyl Sulphate + Water

$$U_2O_8 + 4H_2SO_4 \longrightarrow U(SO_4)_2 + 2UO_2SO_4 + 4H_2O$$

Uranium Trioxide (UO_3): This compound of uranium is obtained as an orange coloured amorphous powder by heating uranyl nitrate $UO_2(NO_3)_2$ at 250°C temperature. On further heating, it is changed to U_3O_8. It forms hydrate with water. It dissolves in hydrochloric acid forming uranyl chloride,UO_2Cl_2.

Uranium Trioxide + Hydrochloric Acid ⟶ Uranyl Chloride + Water

$$UO_3 + HCl \longrightarrow UO_2Cl_2 + H_2O$$

The uranates are insoluble in water including, sodium potassium and ammonium uranates. The alkali di uranates are precipitated as yellow amorphous substance from solutions of uranyl salts by the addition of alkali hydroxides.

Uranyl Nitrate + Potassium Hydroxide ⟶ Potassium Di Uranate + Potassium Nitrate + Water

$$2UO_2(NO_3)_2 + 6KOH \longrightarrow K_2U_2O_7 + 4KNO_3 + 3H_2O$$

Halides : Uranium forms uranium tetrafluoride, UF_4 as green insoluble amorphous powder and uranium hexafluoride, UF_6 as a colourless crystalline solid, by the action of fluorine on uranium pentachloride, UCl_5

Uranium Pentachloride + Fluorine ⟶ Uranium Tetrafluoride + Uranium Hexafluoride + Chlorine

$$2UCl_5 + 5F_2 \longrightarrow UF_4 + UF_6 + Cl_2$$

Uranium hexafluoride sublimes at 56°C and reacts with water to give yellow uranyl fluoride, UO_2F_2. The tetrafluoride is precipitated from a solution of uranous salts by the addition of solution of fluorides.

Uranium forms three chlorides namely the tetrachloride, which is found in the dark green crystal, delequescent and is readily soluble in water and alcohol. The aqueous solution of the tetrachloride reacts strongly in acid, due to hydrolysis.

The tetrachloride is a powerful reducing agent, as it is readily oxidised to uranyl compounds. When the tetrachloride is heated in a current of hydrogen it is reduced to uranium trichloride. UCl_3 which dissolves in water to form a red solution with strong reducing properties. The trichloride in aqueous solution decomposes with evolution of hydrogen and forms ruby-red crystals by the action of chlorine on UCl_4. Uranyl chloride, $UOCl_2H_2O$ is obtained in yellowish-green crystals from a solution of uranium trioxide in hydrochloric acid. The crystals are very soluble in water. The anhydrous uranyl chloride is formed by the action of chlorine upon uranium dioxide UO_2.

Uranyl Nitrate ($UO_2(NO_3)_26H_2O$): The commonest and the most stable compound of uranium is uranyl nitrate. It is obtained as a hexahydrate $UO_2(NO_3)6H_2O$ in the form of yellow delequescent crystals with a green fluorescence. The crystals are readily soluble in water and ether. The aqueous solution contains the uranyl cation UO_2^{++}, but it is considerably hydrolysed.

Sulphates: Uranous oxide dissolves in dilute sulphuric acid and from the dark green uranous sulphate.

$U(SO_4)4H_2O$, can be crystallised out. The solution of the sulphate is readily hydrolysed and acts as reducing agent to form Uranyl sulphate, $UO_2SO_43H_2O$ by evaporation. It is prepared by dissolving uranic acid in concentrated sulphuric acid, or by heating uranyl nitrate with sulphuric acid.

Uranyl Acetate ($UO_2(CH_3COO)_22H_2O$): The compound is a yellow coloured crystal with strong green fluorescence. It is prepared by dissolving uranium trioxide, UO_3 or the hydroxide $UO_2(OH)_2$ in acetic acid.

Uranyl Hydrogen Phosphate ($UO_2HPO_44H_2O$): This can be obtained as yellow sparingly soluble substance from uranyl nitrate and disodium phosphate.

Uranyl Sulphide (UO_2S): This is obtained as a brown prepcipitate from a solution. It is soluble in dilute acids.

Detection and Estimation

Uranium compounds, when heated in a borax bead, colour of the bead in the oxidising flame, is yellow and green in the reducing flame. The green colour is due to the formation of uranous salts. From the solution of uranyl salt, potassium ferrocyanide gives a brown precipitate of uranyl ferrocyanide. This is very sensitive reaction of uranium.

Uranium is estimated by the precipitation as ammonium dichromate and the then by igniting the precipitate to U_3O_8.

RADIUM

Melting Point : 700°C
Boiling Point : 1140°C

Atomic Number : 88
Atomic Weight : 226.025
Isotopes : 14, radium-226
Half-life : 1620 years

History and Occurrence

Madam Curie discovered radium in the year 1898 in the uranium pitchblende. It is an extremely rare element, present in minute quantities in uranium ores. It is produced from the disintegration of uranium. Radium is present in milligrams per ton of uranium ore.

Physical Property

Melting Point	700°C
Boiling Point	1140°C
Density	5 g/cc
Standard Entropy	17 deg^{-1} mol^{-1}
Particle Energy	4.78 MeV(α) 4.79(β)
Emission	α-particle, β-particle and γ-particle
Neutron Proton Ratio	1.567
Mass Defect (Radium C)	+12
Binding Energy	7.8 MeV
Packing Fraction(Radium C)	5309.734513274

Extraction

Radium is obtained from pitchblende; a separation is of the uranium from the radium and the other alkline earths by means of an acid-sulphate treatment. This yields a relatively small precipitate of the alkline earth sulphate, which is washed free of uranium. This greyish precipitate is known, as the *white cake* ands is the raw material of the radium industry.

White cake contains radium together with a large amount of barium (often added before sulphate treatment to insure maximum recovery of the radium), some calcium little or no strontium and a significant amount of radiolead (inactive lead and radium D, Pb^{210}). In addition, the white cake contains a large amount of silica. Other elements such as iron, silver and rare earths are present in variable amounts. The overall plan of the separation is to convert the radium and barium to bromide so that, by fractional crystallisation, the radium can be finally obtained as *radium bromide* of at least 90% purity.

The flow diagram is shown in the Fig. 24.14.

Radium Extraction

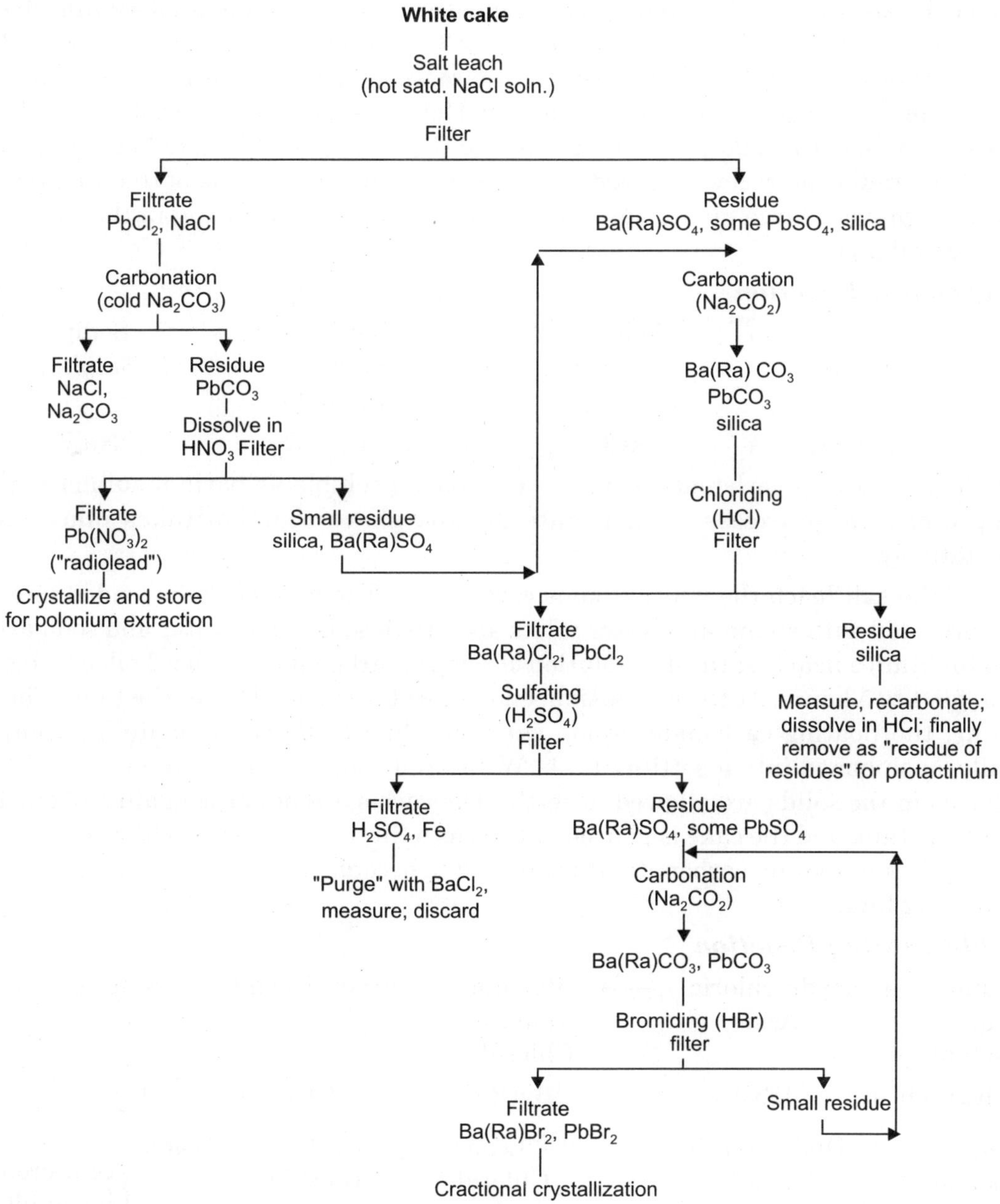

Fig. 24.14

(1) Salt Leach Reaction

$$\text{Lead Sulphate} + \text{Sodium Chloride} \longrightarrow \text{Lead Chloride} + \text{Sodium Sulphate}$$

$$PbSO_4 + 2NaCl \longrightarrow PbCl_2 + Na_2SO_4$$

The white cake is placed in a rubber-lined tank equipped with steam inlet. Water and sodium chloride are added so that on heating with steam a concentrated sodium chloride solution is in contact with the white cake. After 3 or 4 hours the contents of the tank are rapidly dropped on a cloth filter covering a square wooden support or filtration. Filtration must be rapid and is generally done in vacuum. Hot water is used for two washes. The salt leaches the bulk of the lead out. The filtrate and the washes are collected in large wooden tank for the radiolead recovery speed is necessary for the operations of the filtration and washing to prevent large quantities of lead chloride from crystallising in the filter cloth, filter bed and pipes.

(2) Carbonate Reaction

Barium (radium) Sulphate	+	Sodium Carbonate	$\longrightarrow$	Barium (radium) carbonate	+	Sodium Sulphate
$Ba(Ra)SO_4$	+	$NaCO_3$	$\longrightarrow$	$Ba(Ra)CO_3$	+	$NaSO_4$

Although barium carbonate is about ten times a soluble as barium sulphate, in the presence of a large excess of carbonate ion the above reaction takes place almost quantitatively.

After the salt leach the white cake is carbonated. The white cake is loaded into steel tanks provided with steam and water inlets and stirrers. Enough water and soda ash are added to obtain a nearly saturated solution of sodium carbonate, which is boiled by blowing steam, for 4 to 5 hours. At the conclusion, enough water is added to fill the tank. The next morning, the floating carbonate liquor, carrying almost all the sulphate ion contained initially, is siphoned into a settling tank. Water and sodium carbonate are added to the residue as in the solids are allowed to settle. The carbonate liquor is again siphoned into the settling tank and the cake is poured onto a cloth filter and filtered. The cake is washed relatively free of sodium carbonate. It is allowed to become filter-dry and brought into the chlorinating tank.

(3) Chlorinating Reaction

Barium (radium) carbonate	+	Hydrochloric Acid	$\longrightarrow$	Barium (radium) Chloride	+	Carbon Dioxide	+	Water	
$2Ba(Ra)CO_3$	+	$4HCl$	$\longrightarrow$	$2Ba(Ra)Cl_2$	+	$2CO_2$	+	$2H_2O$	
Lead Carbonate	+	Hydrochloric Acid		Lead Chloride	+	Carbon Dioxide	+	Water	Silica remains insoluble.
$PbCO_3$	+	$2HCl$	$\longrightarrow$	$PbCl_2$	+	CO_2	+	H_2O	

The carbonate cake is treated with hydrochloric acid until complete dissolution of the carbonate has occurred. A rubber lined steel tank equipped with steam inlet is used for this purpose and the contents are filtered with a cloth filter and washed free of chloride ions.

The filtrate is poured into a glass-lined steel tank. The residue is subjected to another recarburization and chlorinating process.

The residue from the second chlorinating is so-called "Rückrückstand" or "residue of residues," which is measured for radioactive content. It is starting material for the extraction of protactinium.

(4) Sulpahte Reaction

Barium (radium) Chloride	+	Sulphuric Acid	$\longrightarrow$	Barium(radium) Sulphate
$Ba(Ra)Cl_2$	+	H_2SO_4	$\longrightarrow$	$Ba(Ra)SO_4$

The chloride liquor is treated with excess sulphuric acid and the system is agitated for the working day and allowed to settle. The sulphate liquor is pumped into a settling tank and the sulphate cake is washed. After removal of the wash, the sulphate is put into crocks and allowed to settle. The wash is pumped into a special carbonation tank for the final carbonation.

(5) Final Carbonation : The same procedure as in step (2) used. The finished carbonate is washed free of carbonate ion. It is then ready for dissolution with hydrobromic acid.

(6) Brominating Reaction

Barium (radium) carbonate	+	Hydrogen Bromide	$\longrightarrow$	Barium (radium) bromide	+	Carbon Dioxide	+	Water
$Ba(Ra)CO_3$	+	2HBr		$Ba(Ra)Br_2$	+	CO_2	+	H_2O

The finished carbonate is treated with hydrobromic acid, either fresh or regenerated in a tank similar to the chloride tank, until all the carbonate has been converted into bromide. The bromide solution is filtered through a cloth filter (supported on chemical stoneware) and the small residue is washed and returned for recarbonation with subsequent charges of white cake.

1. Fractional Crystallisation: When a solution of radium and barium bromide, acidulated with hydrochloric, is evaporated, crystals will form in the surface when a certain concentration is achieved. If the system is allowed to cool, two phases will develop, a crystal phase and a liquor phase. The ratio of radium to barium is moved and the crystal phase is dissolved in water and recrystallised, the new ultimate crystal phase will have a heavier radium-barium ratio than the previous crystal phase. By many successive fractional crystallisations it is possible to achieve radium bromide crystals whose purity exceeds 90%.

The Mt Kisco, refinary New York uses glass-lined, steamed-jacked steel evaporators, uses quartz basin and Pyrex and quartz laboratory ware. Fractionation is continued until a stage is reached where in the liquor contains so low a concentration of radium that further fractionation is useless.

The product crystals contain 90% RaBr, are dried, crushed and introduced into weighted glass tubes. The tubes are reweighted and are sealed off in a flame, the sealing time being noted for each tube. The tube is then stored in a numbered lead cylinder. The radium value of each tube is determined by direct gamma comparison with a suitable standard.

2. Regeneration of Hydrobromic Acid : The end liquor is essentially barium bromide solution plus impurities such as iron and lead. This solution is first made alkline with barium hydroxide, precipitating the base metals, which are filtered off. This residue is the starting material for actinium extraction. The alkline barium bromide solution is with excess sulphuric acid, which serves to precipitate al the barium and leave a solution of hydrobromic acid and sulphuric acid. The excess of sulphuric acid is removed by careful addition of slightly alkline barium bromide liquor. This leaves hydrobromic acid, containing a slight amount of barium, which may be used as regenerated acid in the brominating step. Barium sulphate is used as a carrier in other operations.

Extraction of Radium D

The needles are cut and heated with excess sulphuric acid until fumes of sulphur trioxide appear. This serves to dissolves the sulphate and extract most of it form the needles. After

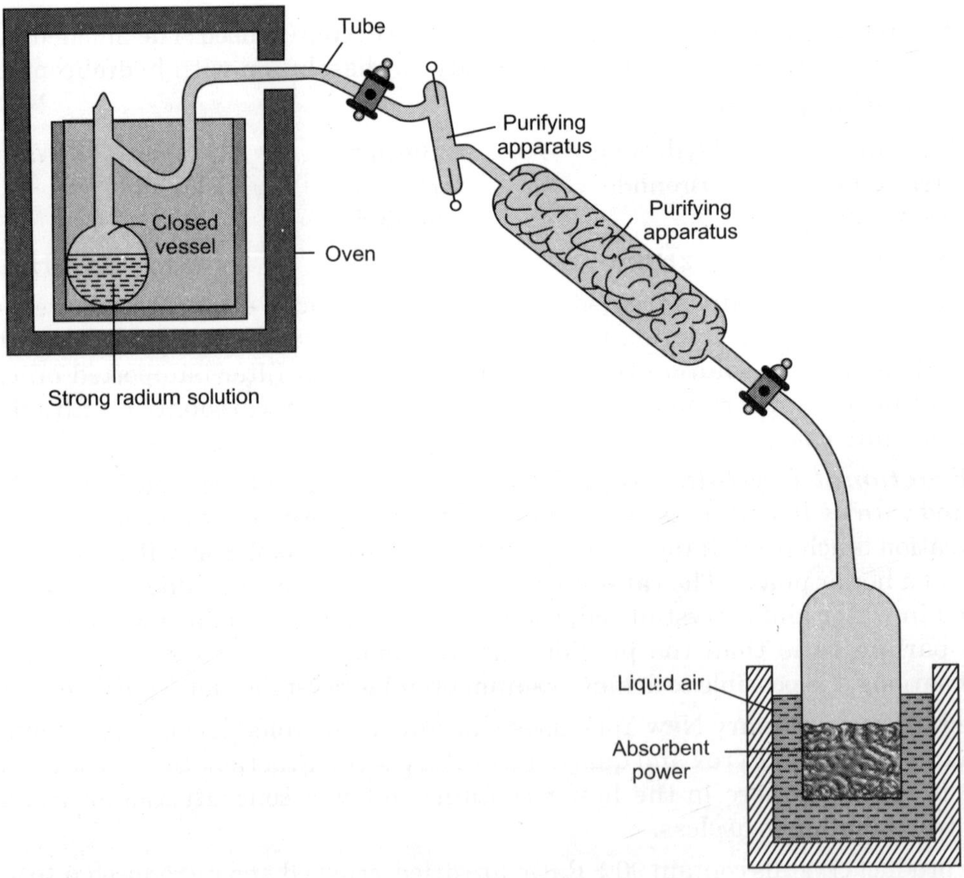

Fig. 24.15. The Extraction of Radium-D.

cooling, the sulphuric acid, containing radium sulphate, both dissolved and reprecipitated, is decanted away from the containers which are retreated with sulphuric acid, after addition of barium sulphate to act as carrier. The sulphuric acid liquors are combined and evaporated to expel most of the sulphuric acid, after which the residue of sulphates is carbonated with soda ash, to be converted into bromides, from which the radium (Ra^{226}) and barium bromide are separated by fractional crystallisation. Before the first crystallisation the solution is purified by adding barium hydroxide and hydrogen sulphide to precipitate the sulphide of radium D. A part of the original radium D is recovered from this precipitate.

The carbonate liquor from the carbonation of the sulphates contains most of the radium D (13 milligrams per 100 milicuries) dissolves in the large excess sodium carbonate. The carbonate liquor contains little radium (Ra^{226}) To recover the radium D, the following steps are to be followed as shown in Fig. 24.15.

Compounds of Radium

We shall now study about the compounds of radium, they are as follows :

Radium Bromide ($RaBr_2$) : It is available as the refined compound from white cake as explained in the extraction of radium. It has a melting point of 728°C and sublimes at 900°C and is soluble in alcohol and water. It is used in medicine and physical research.

Radium Sulphate ($RaSO_4$): It is prepared by the action of radium bromide and sulphuric acid

$$\text{Radium Bromide} + \text{Sulphuric Acid} \longrightarrow \text{Radium Sulphate} + \text{Hydrgen Bromide}$$

$$RaBr_2 + H_2SO_4 \longrightarrow RaSO_4 + 2HBr$$

It is found in white crystal as a pure salt, but sometimes due to impurities it becomes yellow, or orange or pink.

THORIUM

Melting Point : 1700°C
Boiling Point : 4500°C

Atomic Number : 90
Atomic Weight : 232.038

History and Occurrence

Berzelius in 1828 isolated thorium from the Norwegian deposit *thorite, $ThSiO_4$*. The important and the chief source of thorium is monazite, it contains mixture of thorium phosphate crystals about 4 to 18% of thorium phosphate. This is easily found in the Kerala (India) (India is the largest thorium producing country in the world) and beach and river sands of Brazil.

Importance of Thorium to India

In recent years, the monazite sands of Trivantapuram have acquired a great importance. The factory for the production of rare earth compounds owned by the Indian rare earth Private Ltd went into production of thorium in 1952.

Extraction of Thorium from Monazite

The first step in the extraction of thorium is that of concentration of monazite sand. The monazite sand being much heavier than quartz crystal gets caught into the riffles, while the quartz sand and the other light materials are washed off. The heavier materials after being separated are send to magnetic separators. The monazite being non-magnetic easily gets separated form the other magnetic impurities such as ferrous deposits.

The next step is the opening up of the minerals consists in heating it with an excess of concentrated sulphuric acid in cast iron pots fitted with stirrers.

The reaction, being exothermic, has to be carefully controlled to prevent its violence. The product consisting of the anhydrous sulphates of all the materials present is run into cold water to bring the sulphates into solution.

The solution contains the sulphates, free phosphoric acid and the unused excess of sulphuric acid. The method of precipitating thorium from this solution consists in general addition of alkali; so as to make a precipitate of thorium phosphate[1]. This is filtered off, dissolved in minimum quantity of acid and the process of fractional precipitation repeated. Most of the thorium is thus concentrated in the precipitate.

The precipitate of thorium phosphate (still containing same rare-earth phosphates) is dissolved in concentrated hydrochloric acid. Oxalic acid is then added to precipitate thorium as the acid insoluble oxalate.

The oxalate precipitate is washed and extracted with sodium carbonate solution. The thorium passes into solution as the complex sodium pentacarbonatothorate, $Na_6[Th(CO_3)_5]$, while the rare earth component is left undissolved.

The solution containing sodium carbonatothorate is treated with sulphuric acid and the resulting thorium sulphate purified by crystallisation. The process is repeated for a number of times by dissolving the sulphate crystals in water, precipitating thorium as the hydroxide, dissolving the same in sulphuric acid and recrystallizing the sulphate, thorium is thus obtained practically pure as *thorium sulphate octahydrate,* $Th(SO_4)_2.8H_2O$.

The nitrate and other salts of thorium are obtained from the sulphate by precipitating the hydroxide; $Th(OH)_4$ and dissolving the well-washed precipitate in the appropriate acid.

Purification : Thorium of the highest degree of purity can now be obtained by thermal decomposition of thorium iodide by the filament growth process.

Other Properties : A freshly polished surface of thorium is platinum-white in appearance, but gets tarnished in air to a dark grey coating. The metal is soft, ductile, malleable and low tensile strength. Water and alkalis do not have any action on thorium. Thorium acts with chlorine, bromine and iodine at a high temperature of 500°C to form tetrahalides.

It burns in oxygen with evolution of much heat, forming the dioxide, ThO_2. It also combines with nitrogen to form, thorium nitride, Th_3N_4. When thorium, in the form of powder is heated in the hydrogen at 400°C it reacts with incandescence to give a black powder of the composition TH_3, the thorium hydride.

1. Thorium phosphate is very sparingly soluble in dilute acids, so the process of fractional distillation is used.

Physical Properties

Melting Point	1700°C
Boiling Point	4500°C
Density	11.7 g/cc
Isotope Mass	232.11
Mass Defect	+0.11
Packing Fraction	+4.7
Binding Energy	7.5 MeV
Relative Abundance	100%
Neutron Proton Ratio	1.577
Half Life	1.39×10^{10} years
Energy	4.20 MeV
Emission	a-particle
Isotope	Thorium A, Thorium B, Thorium C, Thorium C", Thorium D (stable isotope)

Compounds of Thorium

Thorium Hydroxide [Th(OH)$_4$]: When ammonia or an alkali hydroxide solution is added to a solution of a thorium salt, a white precipitate is formed, which is usually assumed to be thorium hydroxide, $Th(OH)_4$. It is however, amorphous and does not give X-Ray interference rings. After prolonged heating under the solution from which it has been separated, it exhibits X-Ray interference rings, but these then correspond to thorium oxide, ThO_2. It may, therefore, be concluded that the precipitate most probably consists of hydrous thorium dioxide.

Thorium hydroxide is insoluble in alkalis, but dissolves readily in acids to form salts of thorium. It is fairly strong base and quickly absorbs carbon dioxide from air. Though insoluble in alkali carbonate solutions due to the formation of pentacarbonatothorate complex, $[Th(CO_3)_5]^{6-}$.

Thorium Oxide, Thorium Dioxide, Thoria (ThO$_2$): This compound is prepared by the action of thorium sulphate and alkali compound, followed by ignition of well-washed precipitate of the hydrous oxide. It is also obtained by roasting a power of thorium nitrate. Thorium dioxide is a white powder, on fusion with borax a crystalline product can be obtained.

After being strongly ignited thorium dioxide is practically insoluble in acids, but may be "opened up" by fusion with sodium hydrogen sulphate or by heating with concentrated sulphuric acid, the product in each case, is of course thorium sulphate.

Thorium oxide does not have of any acidic character and also remains unaffected when fused with alkalis. It differs in these respects from the dioxides of titanium, zirconium and hafnium.

Uses of thorium Oxide, (*i*) Incandescent Gas Mantels, (*ii*) Electric light filament (*iii*) in metallurgy, and as catalyst in organic chemistry.

Thorium Chloride [ThCl$_4$]: It is obtained by heating a mixture of chlorine and carbon monoxide (or the vapour of carbon tetrachloride) over thorium oxide heated to redness.

Thorium Dioxide	+	Carbon Monoxide	+	Chlorine	$\longrightarrow$	Thorium Chloride	+	Carbon Dioxide
ThO_2	+	2CO	+	$2Cl_2$	$\longrightarrow$	$ThCl_4$	+	$2CO_2$

Thorium Dioxide	+	Carbon Tetrachloride	$\longrightarrow$	Thorium Chloride	+	Carbon Dioxide
ThO_2	+	CCl_4	$\longrightarrow$	$ThCl_4$	+	CO_2

It forms colourless needles, melts at 765°C and boils at 922°C. It dissolves in water with evolution of heat and undergoes considerable hydrolysis during dissolution. The aqueous solution is therefore strongly acidic.

Thorium Fluoride [ThF$_4$]: It is prepared as anhydrous powder by passing hydrogen fluoride vapour over thorium chloride at about 400°C. The tetrahydrate is obtained from thorium hydroxide and aqueous hydrofluoric acid. The salt may also be prepared by precipitation from a thorium salt solution by adding hydrofluoric acid or ammonium fluoride. The precipitate is gelatenous, but becomes granular on standing in contact with the solution from which it has been produced.

Thorium Sulphate [Th(SO$_4$)$_2$]: The anhydrous may be obtained by careful dehydration of the commercial octahydrate or by heating thorium oxide with concentrated sulphuric acid and evaporating to dryness until no more fumes of the acid are evolved. It is a white powder, readily soluble in water.

Thorium sulphate is obtained as the first commercial product from monazite and continues as the starting material in the production of other thorium compounds.

Thorium Nitrate [Th (NO$_3$)$_4$] : It may be obtained by crystallisation from the hydroxide in nitric acid. When crystallisation is carried out in the cold, it is the decahydrate that separates out. At temperatures higher than 20°, the hexahydrate and the tetrahydrate separate out.

The commercial salt used in the gas mantel industry, is partly dehydrated and may also contain 2 to 3 percent of thorium sulphate intentionally added to it.

Detection and Estimation of Thorium

There is no characteristic dry test for thorium. The thorium compounds reacts with alkali hydroxide to form thorium hydroxide. Ammonium carbonate reacts with thorium hydroxide to precipitate thorium carbonate, but it is dissolved in the excess of the ammonium carbonate, due to the formation of carbonatothorate, e.g. $(NH_4)_6[Th(CO_3)_5]$. On heating the solution to 50°C a basic carbonate is precipitated, which redissolves on cooling the liquid.

The carbonatothorate complex, however, is sufficiently stable not to permit precipitation by the addition of ammonia. The formation of the stable carbonatothorate is used for the separation of thorium from rare earths, with which it is associated in monazite.

PLUTONIUM

Melting Point : 639°C
Boiling Point : 3235°C

Atomic Number : 94
Atomic Weight : 244

History and Occurrence

Glenn T Seaborg, Joseph W Kennedy, and *Aurther C Wahl,* who produced plutonium, in deuteron bombardment of U-238 in the 60 inch cyclotron at Berkeley California. The sequence of the reaction is shown below :

$$^{238}_{92}U + \frac{1}{0}n \longrightarrow {}_{92}U^{239} \xrightarrow{-\beta} {}_{93}Np^{239} \xrightarrow{-\beta} {}_{94}Pu^{239}$$

The production is carried out on the controlled base in a nuclear chain reactor. Plutonium-239 is recovered from the spent nuclear fuel by solvent extraction with tributyl phosphate dissolved in kerosene.

Plutonium-239 is used as nuclear fuel. It is also employed as an explosive ingredient for nuclear weapons.

Extraction of Plutonium

Redox Process : This process is based on the ability of hexone (methyl isobutyl ketone) to extract uranyl nitrates and plutonium nitrates from aqueous solution. Plutonium and fission product nitrates on the other hand, remain practically unextracted by this solvent. Two separate extraction columns as shown in Fig. 24.16.

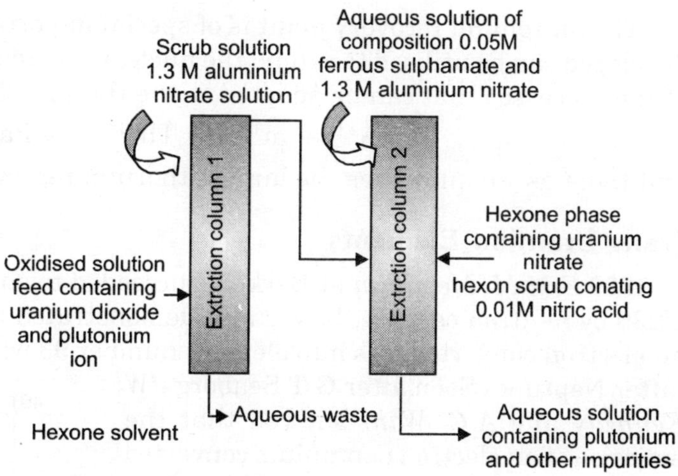

Fig. 24.16. The Extraction of Plutonium by redox process.

Physical Properties

Melting Point	639°C
Boiling Point	3235°C
Specific Gravity	19.84
Energy	0.02MeV
Half Life	2.5×10^4 years
Neutron to Proton Ratio	1.548
Binding Energy	7.6MeV
Isotope Mass	244.12
Mass Defect	+0.12
Emission (Pu-241)	β-particle
Critical Mass(Pu-239)	400g

Production of Plutonium India : India has set up a big plutonium plant at Trombay, Maharashtra where the nuclear wastes are recycled to produce uranium. The process used here consists of dissolution of the nuclear waste, followed by solvent extraction of the plutonium and the fission product streams are separated from one another. The uranium and the plutonium streams are further purified the former by a third cycle of solvent extraction and the latter by ion exchange. The uranium is fabricated into fuel rods for reuse in an atomic reactor, while plutonium is kept for future nuclear power programme of the community.

The Trombay plutonium plant, called 'project phoenix', is the first one to be set up in Asia. The construction was completed in 1964.

The plutonium recovery plant is of special importance to India as the technology here developed can be utilised to extend the nuclear fuel reserves of the country by development of the processes that can be adopted to the thorium-233/uranium-233 breeding cycle.

$$Th^{232} + {}_0n^1 \longrightarrow Th^{233} \longrightarrow Pa^{233} \longrightarrow U^{238}$$

and India as we know, has the largest thorium reserves in the world.

Trans Uranium Elements

In 1939 *E.M.M. Mcmillan* at Berkely succeeded to show that U238 was transformed into U239 by neutron capture. It was also demonstrated by *P.H. Abelson* that the emission of an electron converted this into element number 93 which Mcmillan named as Neptunium (after Neptune) Soon after G.T Seaborg *JW Kennedy* and *A.C. Wahl* showed that the emission of an electron neptunium converted it into another element named Plutonium (after the name of Pluto).

In 1944 *Seaborg, Albert Ghiorso* and *L.O. Morgan* succeeded in producing element Number 95 by bombarding plutonium with alpha particle they named it as Americium, they also found element number 96 which they named as Curium (after Pierre and Madam Curie). In 1950, Seaborg and his colleagues obtained element number 92 Berkelium (after the name of Berkely) by bombarding Americium with alpha particles. In similar fashion they created the element number 98 by bombarding curium with alpha particles. They named it as Californium (after California).

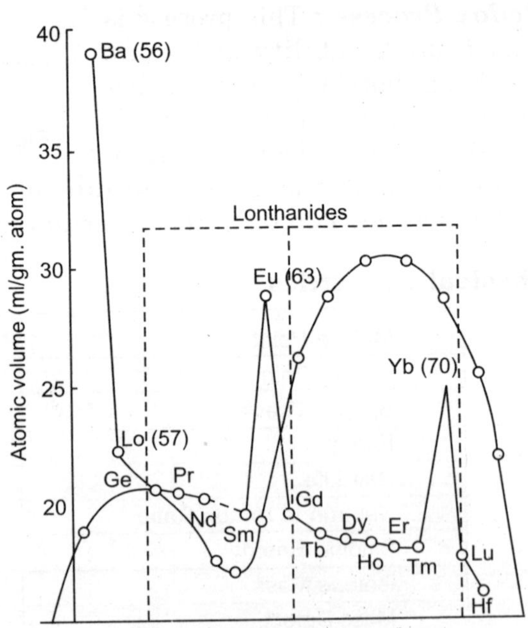

Fig. 24.17. The different properties of rare earths compared in the graph

Elements Number 99 and 100 were discovered during the analysis of the radioactive dust collected after the first thermonuclear explosion at Eniwetok UN atoll in Pacific in 1952. This element was named as Einsteinium (after Albert Einstein) and number hundred as Fermium (after Enrico Fermi).

Element number 101 was discovered by Seaborg and his team in 1955 by bombarding Einsteinium with alpha particles, Einsteinium which they used consisted of a microscopic film deposited on a target of gold foil. The new element was named as Mendelevium (after Mendeleef).

Element Number 102 was created at the University of California in 1958, with the aid of new heavy ion accelerator. This was named as Nobelium (after Alfred Nobel).

Albert Ghiorso, Torbjorn Sikkand Alnon E Larsh and Robert M Latimer created element number 103 in 1961 at the University of California. This element was produced by bombardment of Californium with Boron ions. Number 103 were named Lawrencium (after E.O Lawrence).

CERIUM

Melting Point : 795°C — Atomic Number : 58

Boiling Point : 3257°C — Atomic Weight : 140.12

History and Occurrence

Cerium was discovered in 1803, by *Klaporth, Berzelius* and *Hisinger*. The element was named after the asteroid Ceres discovered on 1801.

Cerium, the first member of Lanthanide series, occurs in many minerals, the important source being monazite, found in Kerala State.

Extraction and Purification

Separation of cerium from rare-earth mixtures, and its purification is done by adding concentrated nitric acid to the hot mixture of the ore consisting cerium. To this mixture ammonium nitrate is also added, when orange crystals of ammonium hexanitratoocerate $(NH_4)_2[Ce(NO_3)_6]$ separate out. After removing the crystals by filtration, the mother liquor is further concentrated to get an additional crop of crystals.

A more recent method of separation consists in *solvent extraction* of cerium solutions using tributyl phosphate.

Metallic Cerium : The metal is commercially prepared by the electrolysis of fused chlorides of cerium, calcium and sodium (or potassium). The mixture is melted in an iron pot, which acts as a cathode during electrolysis. The anode is an iron or carbon rod. Cerium, in the molten condition, collects at the bottom of the pot, where it is tapped and cast into plates, weighing 250kg.

Physical Properties

Melting Point	795°C
Boiling Point	3257°C
Density	6.78

Cerium is chemically active, a freshly cut surface of cerium tarnishes at room temperature in a few hours. Cerium will ignite it is filed or put under a heavy friction. It burns in air to form a dioxide, CeO_2. It decomposes with water, setting free hydrogen and forming cerium (iii) hydroxide. Dilute, non-oxidising acids react with the metal to form cerium (iii) salts. Concentrated sulphuric acid, however, has no action.

Mischmetall and Ferrocerium : Mischmetall, or mishmetal, is an alloy of cerium prepared from a mixture of rare earth chlorides, which contain the metals in the ratio in which they occur in the ore. Its average composition is Ce, 50%; La, 40%; other rare-earth metals 3% and Fe, 7%. Mischmetall is pyrophoric; small particles filed off the metal catch fire in air and burn, therefore it is used in the cigarette lighters. Sometimes it is mixed with excess of iron to increase its hardness; it is then called *Ferrocerium*.

(1) The principal use of the cerium alloy is in the cigarette lighters.

(2) Addition of a small proportion of cerium or mischmetall to stainless steel improves. the quality and also removes the hot-rolling defects.

(3) Addition of the mischmetall increases the efficiency of the steel by 25%.

(4) Mischmetall increases the castibility and strength of magnesium alloys considerably. Certain alloys of this class, containing a small portion of zinc in addition to rare earth metals finds its use in casting and wrought iron alloys.

(5) Magnesium alloys containing mischmetall (3%) and zirconium (1%) have sufficiently high strength at elevated temperature and have been found particularly useful in making jet engine parts.

Compounds of Cerium

Cerium Compounds are divided into two groups, the *cerous* and *ceric* types. Among these, cerous compounds are mostly stable, by the ceric salts are not stable like the cerous, they are extremely hydrolysed in aqueous solution and on diluting the solution, basic salts often get precipitated, Ceric salts often get reduced to cerous salts. Ceric salts also form a number of complexes that are much more stable than the simple ceric ion.

Cerous Chloride [$CeCl_3$. $7H_2O$]: It is prepared by dissolving cerous carbonate or hydrous oxide in hydrochloric acid, the mixture is evaporated and transferred to a syrupy mass and then allowed to cool.

$$\text{Cerous Carbonate} + \text{Hydrochloric Acid} \longrightarrow \text{Cerous Chloride} + \text{Carbonic Acid}$$

$$Ce_2(CO_3)_3 + 6HCl \longrightarrow 2CeCl_3 + 3H_2CO_3$$

Cerous Nitrate [$Ce(NO_3)_3$]: This compound is obtained by the addition of nitric acid to cerous carbonate, and the mass is allowed to crystallise.

$$\text{Cerous Carbonate} + \text{Nitric Acid} \longrightarrow \text{Cerous Nitrate} + \text{Carbonic Acid}$$

$$Ce_2(CO_3)_3 + 6HNO_3 \longrightarrow 2Ce(NO_3)_3 + 3H_2CO_3$$

Cerous Sulphate [$Ce_2(SO_4)_3$]: It is prepared by reducing ceric sulphate with sulphuric acid in the presence of hydrogen peroxide. On evaporating the solution under various conditions, as many as five hydrates, containing, 4, 5, 8, 9 and $12H_2O$ respectively, have been prepared. The anhydrous salt is obtained by heating the hydrated salt at about 350°C.

Cerous Oxalate [$Ce_2(C_2O_4)_3$]: It is the only oxalate[1] of cerium that is known. It separates as a white crystalline precipitate on adding oxalic acid to a cerous salt solution. It is practically insoluble in water and dissolves only to a slight extent in dilute acids, behaviour that is typical to all the rare-earth oxalates.

Ceric Chloride [$CeCl_4$]: It is known only as a deep red solution of the complex, *hexachloroceric acid*, $H_2[CeCl_6]$, which, on standing looses chlorine and changes to the colourless cerous chloride solution.

$$\text{Hexachloroceric Acid} \longrightarrow \text{Cerous Chloride} + \text{Hydrogen Chloride} + \text{Chlorine}$$
$$2H_2[CeCl_6] \longrightarrow 2CeCl_3 + 4HCl + Cl_2$$

Ceric Sulphate [$Ce(SO_4)_2$]: The anhydrous salt is prepared by heating cerium dioxide with concentrated sulphuric acid. It dissolves in water, giving a brown-yellow solution, from which the salt separates below 3°C as the decahydrate, between 3°C and 42.5°C as the octahydrate and above the latter temperature as the tetrahydrate.

On diluting an aqueous solution of ceric sulphate, hydrolysis occurs and ultimately, sparingly soluble basic salts get precipitated. Hence, for use in volumetric analysis, the salt is dissolved in dilute sulphuric acid (0.5N or stronger).

Cerium sulphate is widely used in volumetric analysis as an oxidising agent. The salt is used in volumetric analysis as an oxidising agent. The equivalent weight of the salt, in acid solution, is the same as its molecular weight, as may be seen from the equation.

$$Ce^{++} + e^- \longrightarrow Ce^{3+}$$

The salt employed for the titration of Fe^{2+}, $C_2O_4^{\,2-}$, $NO_2^{\,-}$, Cu^+ Mo^{5+}, Te^{4+} and U^{4+} ions. It is also used as an oxidiser in preparative organic chemistry.

Ceric Ammonium Sulphate [$Ce(SO_4)_2.2(NH_4)_2SO_4.2H_2O$]: More correctly called *Ammonium terasulphatocerate*, $(NH_4)_4[Ce(SO_4)_4].2H_2O$, being more soluble than ceric sulphate itself, it is used in the volumetric reagent. It is prepared by adding ammonium sulphate to a solution of the simple salt having excess of dilute sulphuric acid, to prevent hydrolysis and then crystallisation is done.

Ceric Nitrate [$Ce(OH)(NO_3)_3 3H_2O$]: It is obtained as red crystals by dissolving ceric oxide in concentrated nitric acid and evaporating the solution. The alt dissolves in water to give a yellow solution having an acid reaction due to hydrolysis. On adding nitric acid to a freshly prepared solution of the salt, a bright red coloured develops owing to the formation of nitrato complexes, such as $[Ce(NO_3)_6]^{2-}$.

Addition of an alkali (or ammonium) nitrate to a solution of ceric nitrate, followed by crystallisation, yields bright red anhydrous nitrato salts. Ammonium hexanitratoocerate $(NH_4)_2.[Ce(NO_3)_6]$, is an important example. It is freely soluble in water but sparingly in nitric acid. Its formation affords a simple method for the separation of cerium from other rare-earth metals.

---

1. This is because the ceric oxalate is unstable due to the presence of strongly oxidising Ce^{++} and the reducing $C_2O_4^{\,2-}$ ions destryong each other.

Oxides and Hydrous Oxides of Cerium : Cerium forms the following oxides and hydrous oxides.

Ceric Oxide (CeO_2): This compound is also known as cerium dioxide, prepared by igniting cerium salts of volatile acids in air. It is a white powder with a slight yellowish tint and turns brown-yellow on heating. It is completely insoluble in hydrochloric or nitric acid. Concentrated sulphuric acid converts it into ceric sulphate.

$$\text{Ceric Oxide} + \text{Sulphuric Acid} \longrightarrow Ce(SO_4)_2 + 2H_2O$$

Which is soluble in water but entirely insoluble in concentrated sulphuric acid.

Ceric oxide employed in highly efficient glass-polishing compositions.

Cerium Hydrous Oxide (CeO_2xH_2O): It is obtained as a yellowish, precipitate on adding ammonium hydroxide, or an alkali metal hydroxide, to a ceric salt solution. It dissolves in concentrated nitric acid to form a red coloured solution containing hexachloroceric acid.

Cerous Oxide (Ce_2O_3): It is prepared by complete reduction of cerium dioxide by heating in a current of hydrogen. It is pyrophoric and burns spontaneously in air to form ceric dioxide.

Cerous Hydrous Oxide : This compound is obtained as a white precipitate by adding an alkali solution to a cerous salt solution. It has a powerful tendency to undergo oxidation on exposure to air, it gets oxidised to deep violet-coloured intermediate hydrous oxide and ultimately changes to yellow cerous hydrous oxide.

Cerous Peroxide Hydrate : When a cerous salt solution is treated with ammonium hydroxide and hydrogen peroxides a red-brown precipitate of cerous peroxide hydrate is formed. Very small quantities of cerium can be detected by making use of this reaction. The indication in such case is the formation of yellow colouration.

Detection and Estimation of Cerium

When a solution of a cerous salt is added with ammonium and sodium hydroxide and then a little hydrogen peroxide is added, a red brown precipitate of cerous peroxide occurs.

POLONIUM

Polonium was discovered by Madam Curie and the name was give after her motherland Poland.

Polonium occurs three times in the uranium series and twice each in the thorium and the actinium series, but the only isotope with more than a very short half-life is Po^{210} in the uranium series, with a half-life of 140 days. One milicuries of polonium weighs 0.22 microgram.

Special techniques are required for the investigation of the properties of polonium because it is only obtainable in very small quantities.

There are two allotropic forms of polonium, with transition temperature between 65 and 85°C, depending upon the thickness of the sample. Keeping the sample in low temperature from –190 to 25°C produces shrinking cracks, which is actually the negative temperature co-efficient of expansion. High temperature form melts at 254°C.

Although polonium is in the same family as sulphur selenium and tellurium, it is more closely related chemically to its horizontal neighbour in the periodic table, bismuth. It is amphoteric and behaves as a cation with pH below 5. With proper carriers, it is insoluble in excess ammonia. Polonium lies below silver and above gold in the electromotive series.

Extraction from Radioactive Lead Chloride

The process begins properly with the extracted radiolead from the white cake. This is for the non-radioactive lead, but with a greater portion of Pb^{206}, Pb^{207} and Pb^{208} than normal lead and containing a fraction of about 3×10^{-8} (or 1 milicuries per pound) of radium D (Pb^{210}). The radioactive lead is in the form of lead chloride (solid) in equilibrium with a saturated solution of lead chloride in concentrated sodium chloride. The lead is converted to lead carbonate by sodium carbonate with agitation at room temperature, the reaction being:

$$\text{Lead Chloride} + \text{Sodium Carbonate} \longrightarrow \text{Lead Carbonate} + \text{Sodium Chloride}$$
$$PbCl_2 + Na_2CO_3 \longrightarrow PbCO_3 + 2NaCl$$

The lead carbonate is filtered, washed free of chloride and dissolved in nitric acid in glass vessel:

$$\text{Lead Carbonate} + \text{Nitric Acid} \longrightarrow \text{Lead Nitrate} + \text{Water} + \text{Carbon Dioxide}$$
$$PbCO_3 + 2HNO_3 \longrightarrow Pb(NO_3)_2 + H_2O + CO_2$$

The lead nitrate solution filtered and evaporated to crystals of lead nitrate. These are filtered and evaporated to crystals of lead nitrate. These are filtered, washed briefly with cold water, dried and stored in boxes, from this lead nitrate which the polonium is periodically extracted.

The quantity of polonium produced by a given quantity of radium D is a function of time. The expression for this function is given by Rutherford, Chadwick and Ellis.

$$N_{Po} = N_{OraD} (a \exp \{-\lambda_1 t\} + b\{\exp - \lambda_2 t\} + c \exp\{ -\lambda_3 t\}$$

Where, N_{OraD} = Number of radium D initially,

N_{Po} = Number of atoms of polonium at time t.

$$\lambda_1 = \text{Disintegration constant of radium D} = \frac{0.693}{22}\,\text{year}^{-1}$$

$$\lambda_2 = \text{Disintegration constant of radium E} = \frac{0.693}{5.0}\,\text{day}^{-1}$$

$$\lambda_3 = \text{Disintegration constant of Polonium} = \frac{0.693}{138}\,\text{day}^{-1}$$

This is a very important equation and students must understand it properly.

and
$$a = \lambda_1\lambda_2/(\lambda_2 - \lambda_1)(\lambda_3 - \lambda_1)$$
$$b = \lambda_1\lambda_2/(\lambda_1 - \lambda_2)(\lambda_3 - \lambda_2)$$
$$c = \lambda_1\lambda_2/(\lambda_1 - \lambda_3)(\lambda_2 - \lambda_3)$$

Calculation indicates that, in 100 days, 1 milicuries of radium D yields 0.39 milicuries of polonium. The half-life of radium D is 22 years and the amount of polonium rises to a maximum of 0.87 milicuries after 2.3 years.

According to the method of *Horn, Pavay* and *B. Pregel,* polonium can be separated from radioactive lead nitrate by the following steps :

- Separation of polonium from the bulk of radiolead with an iron carrier by precipitation as hydroxide in approximately neutral solution.
- Separation of polonium from the large bulk of radioactive lead with an iron carrier by precipitation as hydroxide in approximately neutral solution.
- Separation of polonium from nickel by dissolution of the nickel and polonium in nitric acid, followed by precipitation of polonium along with a small amount of the iron, carrier, using ammonia. The nickel forms a soluble complex ion.
- Reduction of the amount of iron by deposition of the polonium on a nickel foil. The nickel foil is dissolved in nitric acid, a very small amount of iron is added and represipitation is carried out with ammonia. This yields the polonium together with an extremely small amount of iron oxide. The solids are dissolved in hydrochloride acids and the solution is calibrated.

Measurement of Plutonium

Polonium is measured by its alpha emission. In the polonium refining process. Polonium is deposited on nickel foils in practically molecular thickness. The alpha emission is measured by the spintheriscope consisting a photomultiplier and a photocell, which is connected to DC voltage reading device. The light produced by the fluorescent screen is multiplied by the multiplier and these light emissions produces a proportional voltage in the photocell, which is indicated by the meter. Blau and Dreyfuss have described this method. Microcurie values of polonium can be measured in alpha ionization chamber with the Lindemann electrometer. Very large sources of polonium have been measured calorimetricaly. These measurements are based on the known heat developed by alpha emission.

PROTACTINIUM

Protactinium 234 occurs in the uranium series an element of short half-life. In the actinium series, derived from uranium 235, it occurs as Pa^{231} with a half-life of 32,000 years. This isotope occurs in uranium ores to the extent of 0.8 gram per gram of radium. One milicurie of Pa^{231} corresponds to a mass of 20 milligrams. Protactinium is in the actinide series of elements. Its fluoride is soluble; the hydrous oxide and the phosphate are insoluble.

Extraction : In the separation of radium from uranium, some of the protactinium is carried along with the white cake, presumably by the silica. Protactinium is recoverable from the siliceous residue of residues. The method of extraction outlined below follows that of *Grosse* and *Agruss.*

- The silica content of the residue is reduced by caustic fusion.
- After each fusion, residue is dissolved in hydrochloric acid. Zirconium (with protactinium) is precipitated from the combined hydrochloric acid solutions by the addition of hydrogen peroxide, which converts the titanium to the proxy form.
- After the last fusion, protactinium is separated from zirconium by fractional crystallisation of the chloride. The phosphate is decomposed with sodium carbonate; the carbonate is converted to chloride. This fractionation is extremely efficient because of the fact that zirconyl chloride is not isomorphous with the chloride of protactinium.

Measurement : Protactinium is an alpha emitter. It is measurable in small amounts in alpha ionization chamber using a Lindemann electrometer arrangement. The lack of protactinium standards compels comparison with other alpha emitters such as polonium. Blau and Sinason have reported the method.

ACTINIUM

Actinium does not occur in the uranium series. In the thorium series Ac^{228} occurs, with a half-life of 6.13 hours. A more stable isotope is Ac^{227}, half-life 21.7 years, occurring in the actinium series as the product of protactinium. One milicurie of this isotope corresponds to a mass of 16 milligrams. In its chemical properties, actinium resembles lanthanum extremely closely. Consequently, actinium has the usual properties, namely insoluble hydrous oxide, insoluble fluoride, etc.

Extraction : In working up uranium ores, a fraction of actinium is carried with the white cake in the separation of radium from uranium as sulphate in subsequent radium refining. Actinium is recoverable from barium hydroxide precipitate from the end liquor of the radium fractionation. According to the methods defined by Pavey and B.Pregel, actinium can be refined as follows :

- Barium hydroxide residue is dissolved in hydrochloric acid. The lead chloride is filtered off.
- Lanthanum salt is added as carrier.
- Lanthanum, with the actinium and rare earths, is separated from iron by means of fluoride.
- The rare earth fluorides are decomposed with sulphuric acid. Alkaline earths and lead sulphide is filtered off.
- Cerium is eliminated by oxidation to the tetravalent state.
- The actinium is concentrated by separation of lanthanum from the remaining rare earths.
- The actinium is separated from lanthanum by ion-exchange methods.

Measurement : Actinium is a feeble beta emitter. Examination of the disintegration chart will show that gamma radiation is emitted by a number of its disintegration products. Direct gamma-ray comparison is thus possible. Following method is due to R.D Evans (National Radioactive Laboratory): Using a Geiger Counter in which the gamma radiation passes through 1 Inch lead and 3 inch aluminium, the radium (in equilibrium) equivalent of the actinium source is related to the actinium as follows :

13 milicuries of actinium = 1 milligram radium

War a Chemical Approach

The employment of a chemical agent directly for military purposes, that is to cause casualties by irritation, burning, and asphyxilation or poisoning. This also includes the killing by large scale by the application of poison from an aircraft. Some of these chemical agents are used for solving different types of problems are being discussed below :

Tear Gas, Diamminobutane [$(NH_3)_2C_4H_{10}$]: This is the gas used for dispersing the mob, by producing and artificial tear, by exiting their lachracimal glands.

Toxaphene ($C_{10}H_{10}Cl_8$): It is an amber coloured, white waxy solid, with a mild odour of chlorine, used for spraying killing, having a disintrigation area of 5km. The TLV (threshold limit value) in air is 0.5mg/m^3 of air.

Potassium Cyanide (KCN): It is a compound of potassium and hydrocyanic acid, produced to kill by mixing the compound in water, the toxicity level is in .005mg/m^3 of water.

Nitroglycerine ($CH_2NO_3CHNO_3CH_2NO_3$): It is used as an explosive by the application of pressure or an electrical spark. Used as time bomb charge energy. Besides this nithromethane, is also used.

CONCLUSION

Chemistry is used for the development of mankind and it is also used for the application of many inventions responsible for the improvement of human lives. So to kill persons by the application of these are not simply humanic, so we have to protest that no man should use chemistry in the destruction of human society or rather any lifeform. The actual war is the war between the disease and mankind. There is dictum kill animism and not the animal; the battle should be fought against the disease. That will be the real war in which a disease, which is the real enemy, is killed. We must stop these enemies. We must try to develop those bombs and chemical weapons, which will stop the disease and not the carrier.

CHAPTER

25

Boron

Introduction

Borax under the name of tincal was used by the Arabians as a flux. It was obtained from India, Armenia and Tibet. Davy isolated boron in 1807 by the electrolysis of moist boric acid, and in 1808 he prepared it by heating boron trioxide, B_2O_3 with potassium.

Steam and hot water issuing from the volcanic vents or fumaroles in Tuscany contain small quantities of boric acid. In the water formed there by the condensation of steam, about 0.5% of boric acid is present. This was the main source of supply of boric acid at one time. It is however of minor importance now. Boric acid is now commercially obtained from borax, $Na_2B_4O_7.10H_2O$, colamanite,$Ca_2B_6O_{11}.5H_2O$ ulexite, $Na_2Ca_2(Ba_4O_7)_3.18H_2O$. The main deposits are in California and Chilli. Boron also occurs in plants and coal ash.

Extraction

The structures of borate minerals are complex and diverse, but they characteristically contain trigonal BO_3 or tetrahedral BO_4 units in large boron-oxygen anions, some oxygen atoms in borate minerals are *monoprotonated,*[1] to give hydroxyl groups, while others are *diprotonated*[2] to give water of hydration. The cations in such minerals are usually alkali or alkaline earth cations. The structure of the borates anion in borax is shown in Fig. 25.1.

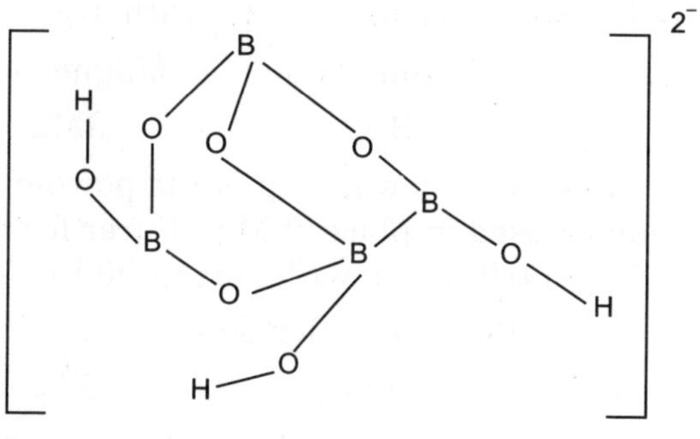

Fig. 25.1. The structure of Borate anion in Borax.

1. Nucleus consisting one Proton.
2. Nucleus consisting two Proton.

Borax also occurs in large deposits in the Mojave Desert of California, which is considered as the major source of boron. No ionic compounds involving simple B^{3+} boron cations are formed, because the ionization enthalpies cannot offset the energy required for the formation of cation. Boron does form three covalent bonds using Sp^2 unsaturated and act as a strong Lewis acid. Interaction with Lewis base (molecule or ions) gives tetrahedral adducts[3] such as BF_3. OEt_2 BF_4 and BPh_4. The formation of such Lewis acid-base adducts requires a change to Sp^3 hybridisation for boron. Another major feature of boron chemistry is the preponderance of compounds consisting of boron atoms in closed polyhedra or open basket-like appearance.

Often the structures are seen to be derivatives of fragments of the icosahedron. The frame wok of such molecules may include atoms other than boron for ex-carbon and many of those with carbon (the carboranes) form complexes with transition metals.

Among the group IIIB (13) elements it is the chemistry of boron that is unique. The chemistry of boron has a few features in common with that of aluminium they are as follows :

1. The oxide B_2O_3 and $B(OH)_3$ are acidic.
2. Borates and silicates are built on similar structural principles with sharing of oxygen atoms so that complicated chain ring or other structure result.
3. The halides of boron and silicon (except BF_3) are readily hydrolysed. The aluminium halides are solids and only hydrolysed by water. All acts as Lewis acid.
4. The halides of Boron and silicon are volatile, spontaneously flammable and readily hydrolysed. Aluminium hydride is a polymer (AH_3).

Isolation of the Elements

Boron froms a number of allotropes that are difficult to purify because of the high melting point of the solids (e.g. - 2250°C for the beta-rombhohedral form). Because of the corrosive nature of the liquid, boron is made in 95% - 98% purity as an amorphous powder by the reduction of the oxide B_2O_3 with Mg

$$\text{Boron Oxide} + \text{Magnesium} \longrightarrow \text{Boron} + \text{Magnesium Oxide}$$

$$B_2O_3 + 3Mg \longrightarrow 2B + 3MgO$$

Followed by washing of the powder with NaOH and HCl(other electropositive metals may be used in place of Mg). Purer forms of the element are available from the reduction of boron trihalides with zinc at 900°C, as the equation below :

$$\text{Boron Trichloride} + \text{Zinc} \longrightarrow \text{Boron} + \text{Zinc Chloride}$$

$$2BCl_3 + 3Zn \longrightarrow 2B + ZnCl_2$$

or from reductions with hydrogen over hot tantalum metal as a catalyst, as in the next following reaction :

$$BX_3 + 3H_2 \longrightarrow 6HX + B$$

3. A chemical addition compound.

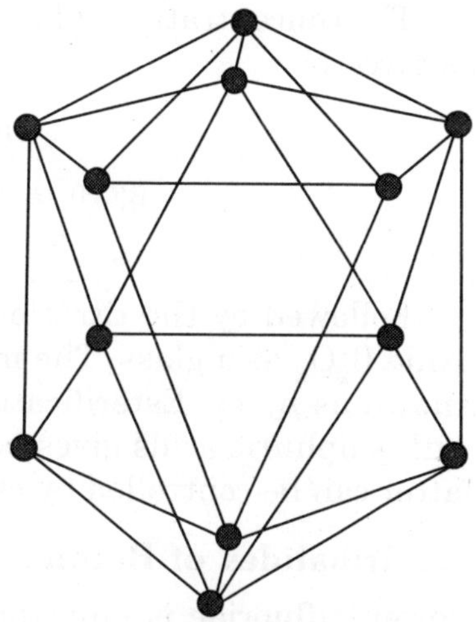

Fig. 25.2. The B12 Icosahedra

The several allotropes of crystalline boron all have structure built up of B_{12} icosahedra (as shown in Fig. 25.2.

One form is differing from another by the manner in which these icosahedra are packed into the cell (unit). Crystalline boron is very inert and is attacked only by hot and concentrated oxidising agents only. Amorphous boron is more reactive with ammonia for instance, amorphous boron at white heat gives boron nitride $(BNB)_x$ a slippery white solid with a layer structure resembling that of graphite, but with hexagonal rings of alternating B and N atoms.

Oxy-compounds of Boron

All most all the naturally occurring forms of boron are the oxygen coupling borate minerals. Endless varieties of organic derivatives containing Boron-oxygen bonds are known, the chief examples that include trigonal boron are the orthoborates $B(OR)_3$ the acyl borates $B(OCOR)_3$; the peroxo borates $B(OOR)_3$ and the boronic acids $RB(OH)_2$. It is appropriate to consider that these are the derivatives of boric acid $B(OH)_3$.

Boric Acid

From borates or by hydrolysis of boron trihalides, the acid $B(OH)_3$ can be obtained as white needles. The $B(OH)_3$ units are linked together by hydrogen bond to form infinite layers of nearly hexagonal symmetry, the layers are 3.12 Å apart and the crystals are readily cleaved along interlayer planes. Some reactions of boric acids are given in the structural figure below:

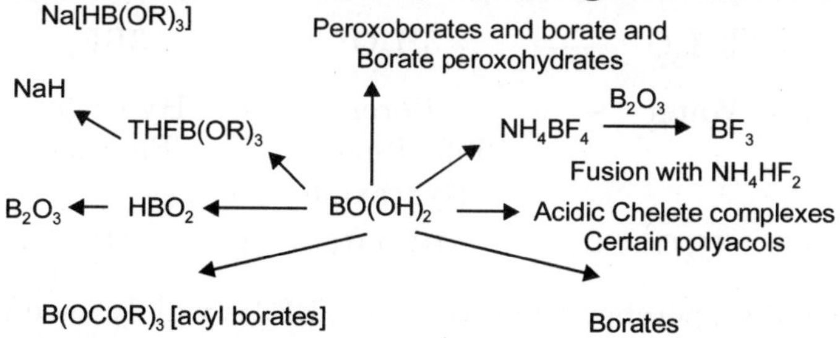

Boric Acid is moderately soluble in water where it acts as a weak Lewis acid towards OH^-

$$B(OH)_8 + H_2O = B(OH)_4^- + H^+$$

The $B(OH)_4^-$ ions occur in many borates type minerals, but most borates (especially those formed by fusion of boric acid with metal oxides)have complex structures such as the ring anion. Structure in the Fig. 25.3 and the borates form very stable complexes with 12-diols.

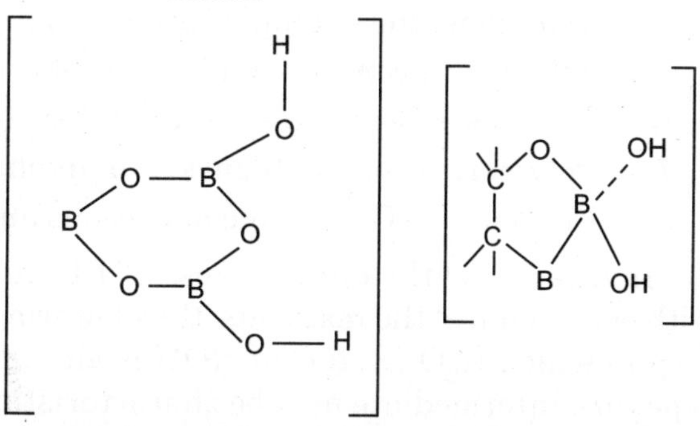

Fig. 25.3. The ring anion structure of boron.

The concentration of boric acid can be determined by the complexion with a diol such as glycerol.

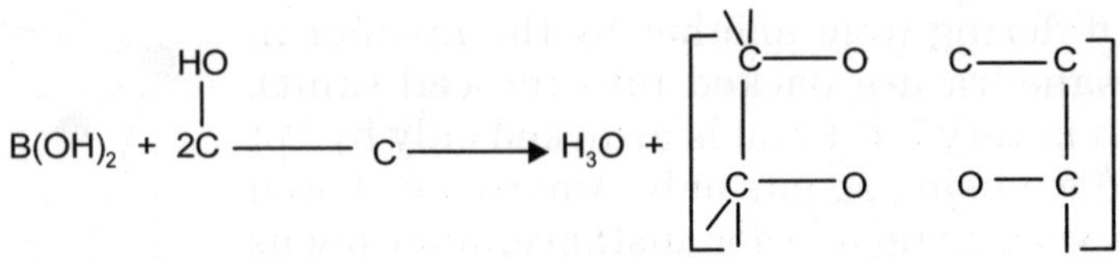

Followed by the titration of glycerol with NaOH. The fusion of the acid makes the oxide B_2O_3 as a glass. The melt readily dissolves metal oxides to give borates commonly known as *pyrex.* Esterification of the boric acid by alcohols with useful reducing agent with sulphuric acids gives boron tetraoxide $B(OR)_3$. The reactivity and selectivity of the latter can be controlled by changing the R groups.

the Trihalides of Boron

Boron trifluoride is pungent gas (boiling point 101°C) that is obained by heating B_2O_3 (boron trioxide) with NH_4BF_4 (ammonium tetrafluroborate) or CaF_2 (calcium fluoride) and concentrated H_2SO_4 (sulphuric acid) which is commercially available as oleum.

Boron trifluoride is one of the strongest Lewis acid known and reacts readily with most Lewis bases such as ethers alcohols, amines or water to give adducts and with F^- to give the tetrafluorobotate ion BF_4^-. The diethyletherate $(C_2H_5)OBF_3$ a viscous liquid is also a volatile liquid is also available commercially, unlike the other halides BF_3 only partially hydrolysed by water:

Barium Trifluoride	+	Water	$\longrightarrow$	Water	+	Boron Tetrafluoride	+	Boron Hydroxide
$4BF_3$	+	$3H_2O$	$\longrightarrow$	$3H_2O^+$	+	$3BF_4$	+	$B(OH)_3$
Boron Tetrafluoride	+	Water	$\longrightarrow$	Boron Tetrafluoride Hydroxide	+	Hydrogen Fluoride		
BF_4^-	+	H_2O	$\longrightarrow$	(BF_3OH)	+	HF		

Because of this and its potency as Lewis acids, BF_3 is widely used to promote various organic reactions, the examples are given :

(*i*) Ethers or alcohols + Acids ⟶ Esters + Water or OH

(*ii*) Alcohols + Benzene ⟶ Alakylbenzene + Water

(*iii*) Polymerization of alkenes and alkene oxides such as propylene oxide

(*iv*) Friedel ⟶ Craft's like acylations and alkylations.

In (*i*) and (*ii*) the effectiveness of BF_3 must depend on its ability to form an adducts with one or both of the reactants, thus lowering the activation energy of the rate-determining step in which H_2O (water) or ROH is eliminated by breaking C–O bonds. In reactions of type (*iii*), intermediate may be characteristics at low temperatures. Thus the interactions of benzene and C_2H_5F proceeds as shown in the equation :

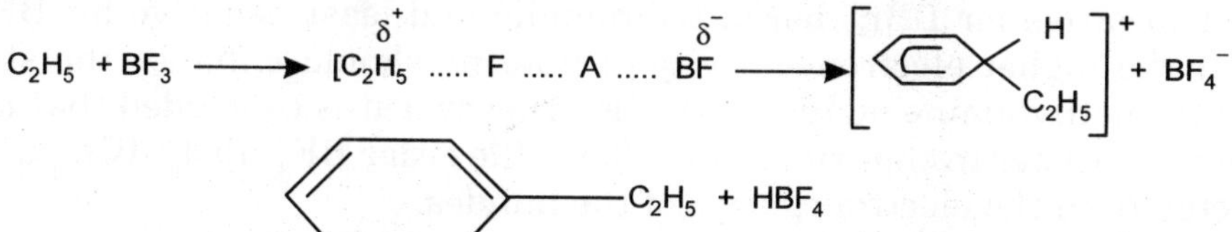

Fluroboric acid solutions are formed on dissolving $B(OH)_3$ is aqueous hydrogen fluoride.

$$B(OH)_3 + 4F_3 \longrightarrow H_3O^+ + BF_4^- + 2H_2O$$

It is clear from the above reaction equation, that the BF_3 is not actually a catalyst.

The commercial solutions contain 40% acid. Fluroboric acid as a liquid condition is a strong oxidiser and cannot of course exit as HBF_4. The ion is tetrahedral and fluroborates resemble the corresponding perchlorates in their solubilities and crystalline structures, like ClO_4^- and PF_6^- the anion has a low tendency to act as a legand towards metal ion in aqueous solutions. In non-aqueous media there is evidence for the complex formation.

Boron trichloride (boiling point 12°C) and the bromide (boiling point 90°C) are obtained by direct interaction at elevated temperature. They fume in moist air and are violently hydrolysed by water.

$$BCl_3 + 3H_2O \longrightarrow B(OH)_3 + 3HCl$$

Reactions of Trihalides of Boron

It has already been noticed that the boron trihalides are Lewis acids and they reacts readily with Lewis bases to form adducts of the trihalides when an acidic hydrogen is available.

Formation of adducts with Lewis base

Even the weakest of the base will form adducts with the trihalides of boron. Ether amines, phosphene, alcohols, carbon monoxide forms adduct with boron by donating an electron pair to boron. The rehybridization of boron that accompanies adducts formation results in a loss in BX double bond character.

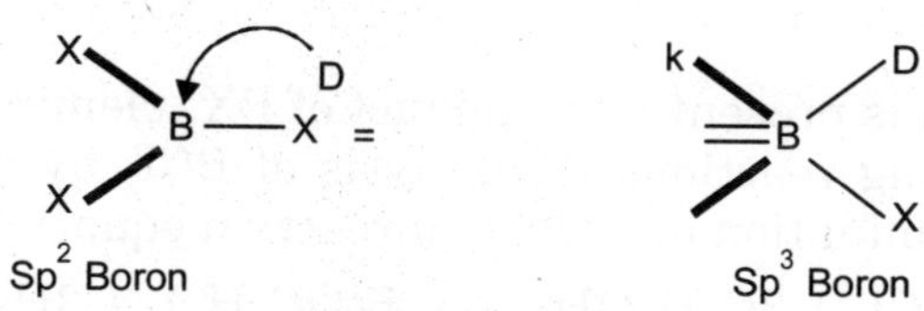

Fig. 25.4. The BX double bon character.

When the Lewis donor is trimethyl amine the enthalpy change for adduct formation is as per equation :

$$BX_3 + :H(CH_3)_3 \longrightarrow X_3B \longrightarrow N(CH_3)_3$$

Is most negative for BBr_3 (boron tribromide) and least negative for BF_3. We should except that the higher electronegativity of fluorine should enhance the stability of the compound trymethylamine adduct with BF_3. However it is concluded that the loss in BX double bond character in the trihalides follows the order $BF_3 > BCl_3 > BBr_3$ a tend opposite that expected from the electronegative of the halides.

Evidently the pi-bond system in BF_3 is especially strong because of effective overlap between the boron 2p. The fluorine 2p atomic orbitals overlap which is effective because of closely matched energy and size of the pi-donor orbitals. The 3p and the 4p atomic orbitals of Cl and Br have the proper pi-donor orbitals (3p for chlorine and up for bromine)are not well matched to those of the pi-acceptor (2p) orbital of boron.

Halide Exchange Reactions among the Boron Trihalides

Mixtures of two different trihalides of boron undergo exchange of halide as illustrated in the reaction.

$$\text{Boron Trichloride} + \text{Boron Tetrabromide} \longrightarrow \text{Boron Chlorobromide}$$

$$BCl_3 + BBr_3 \longrightarrow BBr_2Cl$$

The position of equilibrium of above reaction lies mostly to the left of the equation, but small amounts of the exchange products can be detected by spectroscopic process. The trifluoride undergoes halide exchange less readily than BBr_3 and BCl_3. No intermediates have been detected, but it is responsible to propose that the exchange involve the type of dimeric structure as shown in Fig. 25.5.

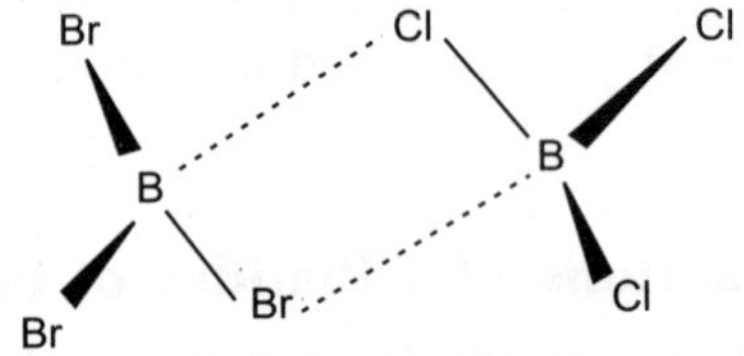

Fig. 25.5. The dimeric structure of boron tetrabromide.

Equilibria of the type illustrated by the reaction above lies mostly to the left but small amounts of the exchange product can be detected. Attempt from such systems are not successful because of the facile† nature of the equilibria. Thus no pure mixed halide of boron is known. A concentrated mechanism as illustrated in the structure (Fig 25.5 above) would be consistent with all of the facts as long as the new bridging BX bonds that form the dimmer are weak.

Elimination

When an acidic hydrogen is present in an adduct of BX_3 elemination of HX is possible as illustrated in the following reactions. Hydrolysis of BCl_3 by alcohols involves adducts formation followed by elimination of HCl as in reaction equation :

$$BCl_3 + 3C_2H_5OH \longrightarrow B(OC_2H_5)_5 + 3HCl$$

The dimethylamine adduct of BCl_3 undergoes a reaction with HCl to give an aminoborane as in the equation :

$$(CH_3)_2NH \text{ - - - - - - } BCl_3 \text{ ——— } (CH_3)_2N - BCl_2 + HCl$$

†Easily changeable.

Aminoborane and in particular, the nature of the BN bond in aminoboranes will be discussed in later subhead Boron-nitrogen compounds.

Boron forms an extensive series of molecular hydrides called *Boranes*. Typical ones are B_2H_6, B_4H_{10}, B_9H_{15}, $B_{10}H_{14}$ and $B_{20}H_{16}$. Boranes were first prepared between 1912 and 1936 by *Alfred Stock* who developed vacuum line techniques to handle the reactive materials. Stock's original synthesis now involves thermolysis of B_2H_6 under varied conditions, often in the presence of hydrogen.

The properties of some of the boranes are listed in the table below :

Important Properties of Some Boranes

Formula	Name	Melting Point °C	Boiling Point °C	Reaction with Air	Thermal Stability	Reaction with Water
B_2H_6	Diborance (6)	–164.85	–92.59	Spontaneously flammable	Fairly stable	Instant Hydrolysis
B_4H_{10}	Tetraborane (10)	–120	18	Not spontaneously flammable if pure	Decom-poses rapidly at 25°C	Hydrolysis in 24 hours
B_5H_9	Pentaborane (9)	–46.6	48	Spontaneously Flammable	Stable at 25°C Slowly decompose at 150°C	Hydrolysed only on heating
B_5H_{11}	Pentaborane (11)	–123	63	Spontaneously Flammable	Decom-poses rapidly at 25°C	Rapid Hydrolysis
B_5H_{11}	Hexaborane (12)	–82.3	87		Liquid, stable for few hours	Quantitative to give B_4H_{10}
$B_{10}H_{14}$	Decaborane (14)	99.7	213	Very stable	Stable at 150°C	Slow Hydrolysis

Diborane (6)

B_2H_6 is a gas (boiling Point –92.6°C) that is *spontaneously flammable air* and instantly hydrolysed by water to H_2 and boric acid. It is obtained quantitatively in ether, at room temperature by the reaction of sodium borohydride with BF_3 as in the reactions following:

$$\underset{\text{Borohydride}}{\text{Sodium}} + \underset{\text{Tetrafluoride}}{\text{Boron}} \longrightarrow \text{Diborane} + \underset{\text{Tetrafluoroborate}}{\text{Sodium}}$$

$$3NaBH_4 + 4BF_3 \longrightarrow 2B_2H_6 + 3NaBF_4$$

In laboratory it is best prepared by the oxidation of sodium borohydrate by iodine as following :

Sodium Borohydride	+	Iodine	$\longrightarrow$	Diborane	+	Sodium Iodide	+	Hydrogen
$2NaBH_4$	+	I_2	$\longrightarrow$	B_2H_6	+	$2NaI$	+	H_2

Industrial requirements are prepared at high temperatures by reductions with sodium hydride and BH_3

Boron Trifluoride	+	Sodium Hydride	$\longrightarrow$	Diborane	+	Sodium Fluoride
$2BF_3$	+	$6NaH$	$\longrightarrow$	B_2H_6	+	$6NaF$

Borane BH_3 has only a transient existence in the thermal decomposition of diborane.

Diborane (6)	$\longrightarrow$	Boron Trihydride	+	Diborane (9)
$2B_2H_6$	$\longrightarrow$	BH_3	+	B_3H_9

Reactions of diborane will be discussed later under this subhead. It should be stated here that diborane is an extremely versatile reagent for the synthesis. Diborane is also a powerful reducing agent for some functional groups, for example, aldehydes and organic nitrates.

Higher Boranes

The heavier boranes for examples, B_6H_{10} are mainly liquids whose flammability in air decreases with increasing molecular weight. One of the most important is decaborane, $B_{10}H_{14}$ a solid (melting point 99.7°C) that is a stable in air and only slowly important starting material of the synthesis of the $B_{10}H_{10}^{2-}$ anion and carbonate.

Structure and Bonding in Borane

The structure of the boranes is unlike those of other hydrides such as those of carbon are unique. A few of them are shown in Fig. 25.6.

There is no sufficient electron to allow the formation of conventional two electron bonds between all adjustment pairs of atoms (2c –2e) the structures of boranes that the earliest of the various concepts of a type of bonding is multicenter, which was developed.

For diborane itself (3c –2c) bonds are required to explain the B–H–B bridges. The terminal is conventional (2c – 2e) bonds to hydrogen orbitals for use in further bonding. The plane of the two remaining orbitals is perpendicular to the Bi_2 plane. When two such BH_2 groups approach each other as shown in Fig. 25.7 below:

The above figure also shows the formation of B–H–R bonds with hydrogen atoms also lying as shown in the plane orbitals where the two B–H–B (bonds) are formed. The total for four electrons required for these bonds is provided by the one electron carried by each H atom and each BH_2 group.

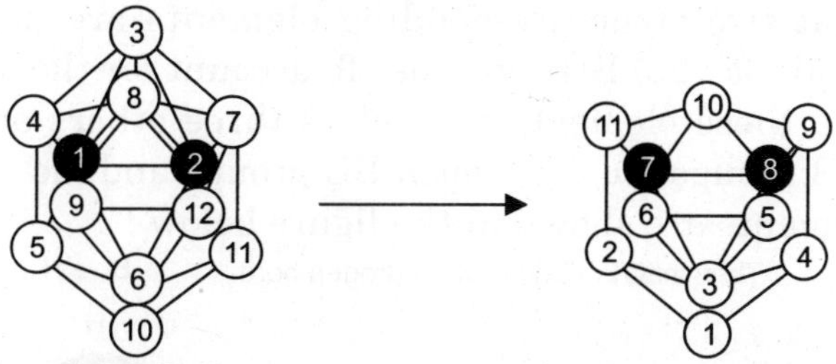

closo- 1, 2-$C_2B_{10}H_{12}$
ortho-oarborane

nido-7, 8-$C_2B_9H_{12}$

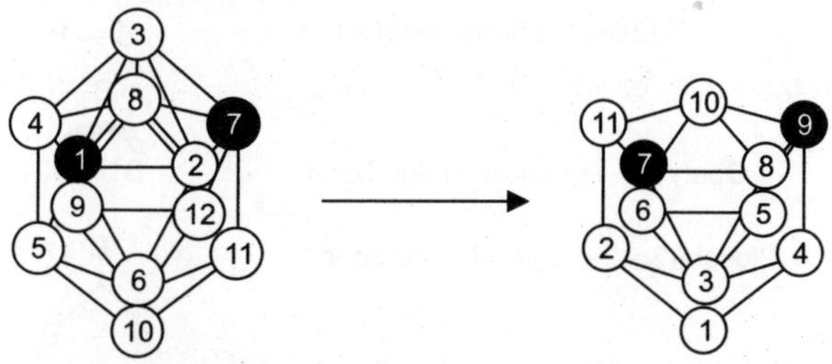

closo- 1, 7-$C_2B_{10}H_{12}$
meta-carborane

nido-7, 9-$C_2B_9H_{11}^{2}$

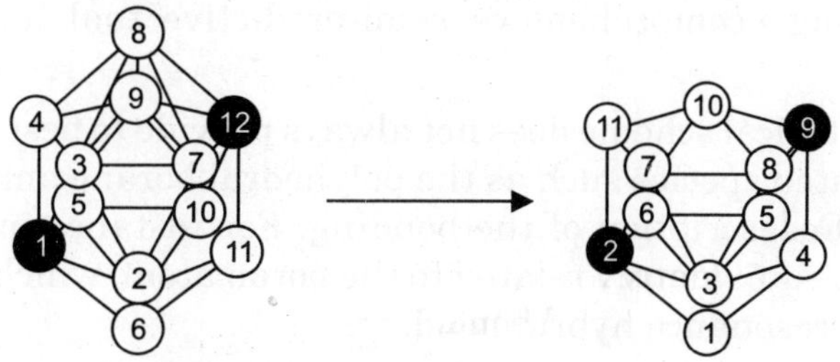

closo- 1, 12-$C_2B_{10}H_{12}$
ortho-carborane

nido-2, 9-$C_2B_9H_{11}^{2}$

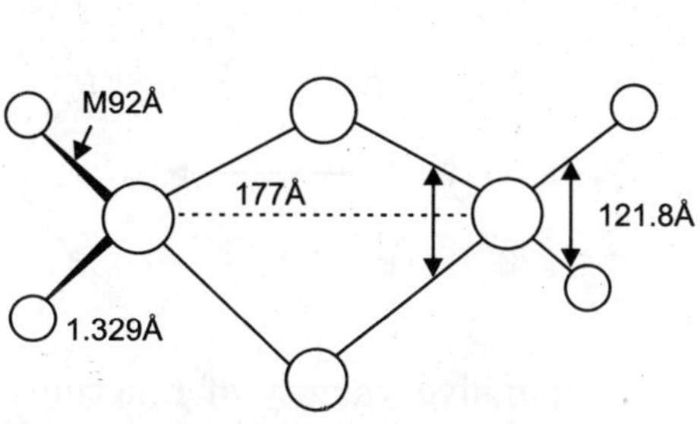

Fig. 25.6. Structures of Boranes.

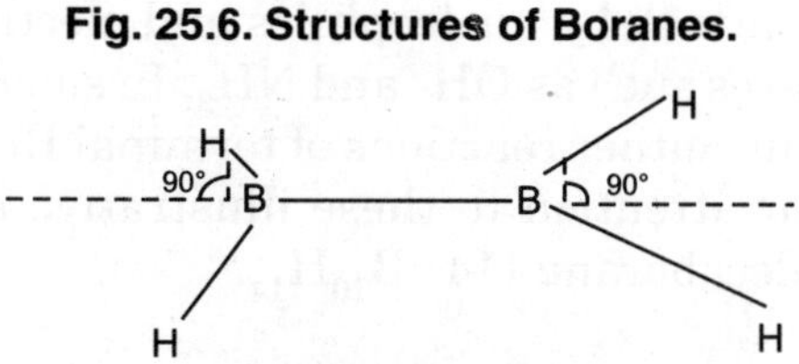

Fig. 25.7

We have seen that two structure-bonding elements are used in B_2H_6, normally (2c - 2e) BH groups and (3c - 2e) BHB groups. To account for the structures and bonding of the higher boranes, these elements as well as three others are required. The three others are (2c - 2e) BB groups (3c - 2e), open BB groups and (3c - 2e) groups. These five structure-bonding elements are shown in the figure below :

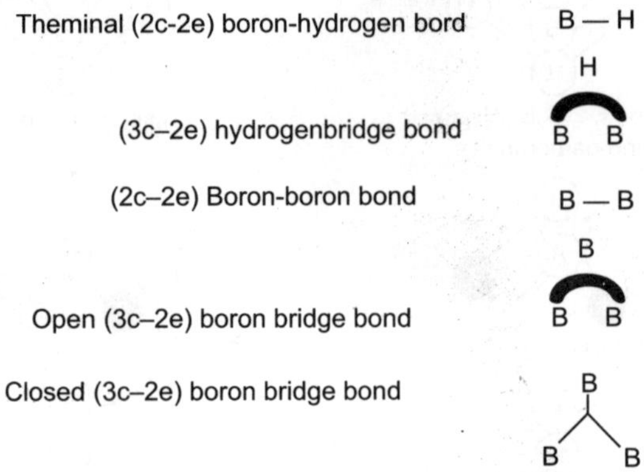

By using these five elements W, N sips comb was developed, which in term developed semi-topological descriptions of the structure and bonding in all of the boranes. The scheme is capable of elaborating a comprehensive, semi-predictive‡ tool for correlating all the structural data.

The semi-topological scheme does not always provide to best description of bonding in the borane and related species such as the polyhedral borane anions thick in terms of a highly delocalized MO description of the bonding. For instance in B_5H_9 where the four basal boron atoms are equivalently related to the boron atom, which is situated at apex. It is possible to depict a resonance hybrid bond.

and B–B element namely :

Reactions of Borane

The boranes undergo an inexpensive variety of reactions including oxidations to oxide pyrolysis to higher boranes, attack by nucleophiles and electrophils, reduction to borane anions and reactions with bases such as OH^- and NH_3. In some cases it is useful to view at least the substitutions as being either reactions of terminal BH groups or of bridging BHB groups. We shall restrict our attention to these illustrative systems diborane (6); B_2H_6 pentaborane (9) B_5H_9 and decaborane (14) $B_{10}H_{14}$.

‡To make a type of forcast.

Diborane [(6), B_2H_6]: Controlled pyrolysis of diboranes leads to most of higher boranes. Reactions with oxygen is extremely exothermic.

$$B_2H_6 + 3O_2 \longrightarrow B_2O_3 + 3H_2O \qquad \Delta H = -2160 kJ\ mol^{-1}$$

Reaction of diborane with water is intantaneous[‡]:

$$B_2H_6 + 6H_2O \longrightarrow 2B(OH)_3 + 6H_2$$

Diboranes are also hydrolyzed by alcohols as in the following reaction

$$B_2H_6 + 6ROH \longrightarrow 2B(OR)_3 + 6H_2$$

Reaction with HCl replaces a terminal H and Cl

$$B_2H_6 + HCl \longrightarrow B_2H_5Cl + H_2$$

And reactions with chlorine gives the trichloride, as in the following reactions:

$$B_2H_6 + 3Cl_2 \longrightarrow 2BCl_3 + 6HCl$$

The electron deficient three centre two-electron BHB bridges are sites of the nucleophilic attack. Small amines such as NH_3, $MeNH_2$ and $MeNH_2$ and MeNH give unsymmetrical cleavage* of diborane as in the reaction equation:

$$B_2H_6 + 2NH_3 \longrightarrow [H_2B(NH_3)_2]^+ [BH_4]^- + [H_2B_2]$$

The boronium ion products are tetrahedral and can undergo substitution by other basis as in reaction:

$$[H_2B(NH_3)_2]^+ + 2PR_3 \longrightarrow [H_2B(PR_3)]^+ + 2NH_3$$

Large amines such as MeN and pyridine give symmetrical cleavage of diborane as in the reaction:

$$B_2H_6 + 2NMe_3 \longrightarrow 2A_3B \longrightarrow NMe_3$$

The amine borane products from symmetrical cleavage of diborane are Lewis bases adducts of BH_3. Aminoboranes will be discussed under the sub head, boron-nitrogen compounds.

Reduction of diborane can be accomplished with sodium :

Diborane	+	Sodium	$\longrightarrow$	Sodium Borohydride	+	Sodium Borane
B_2H_6	+	2Na	$\longrightarrow$	$NaBH_4$	+	NaB_3H_8

Or with sodium borohydride :

Diborane	+	Sodium Borohydride	$\longrightarrow$	Sodium Borane	+	Hydrogen
B_2H_6	+	$NaBH_4$	$\longrightarrow$	NaB_3H_8	+	H_2

Reductions products of diboranes with sodium borohydride can also lead to higher borane anions as in the following equation :

Sodium Borohydride	+	Diborane	$\longrightarrow$	Sodium Borane
$2NaBH_4$	+	B_2H_6	$\longrightarrow$	$Na_2B_{12}H_{12}$

‡ Occuring suddenly.

*The quality of a crystalline substance of splitting along definite planes.

The polyhedral borane anion and carboranes will be discussed shortly.

Pentaborane (9)B_5H_9: Pentaborane has the structure shown in Fig. 25.8.

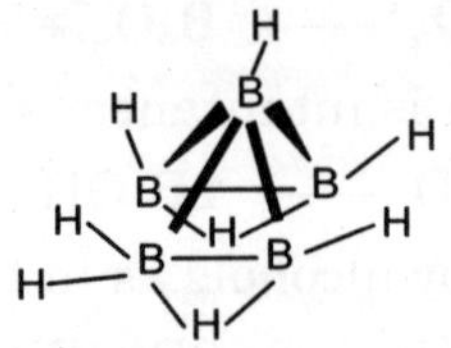

Fig. 25.8. The Structure of Pentaborane (9)

where it is clearly seen that less of one bridging hydrogen atoms. The relatively electron-rich apical group of pentaborane (9) is susceptible to attack by electrophils as in the following reaction:

$$\text{Pentaborane} + \text{Iodine} \longrightarrow \text{Boron Hydroiodide} + \text{Hydrogen Iodide}$$

$$B_5H_9 + I_2 \longrightarrow B_5H_8I + HI$$

Decaborane [(14)$B_{10}H_{14}$]: The structure of decaborane (14) is shown in Fig. 25.4.

Four electron deficient bridging BHB groups cap the top of this icosahedron fragment making this part of the molecule the preferred site for attack by nucleophiles. Like pentaborane (9) it is bridging the hydrogen atoms that are acidic

$$B_{10}H_{14} + OH^- \longrightarrow B_{10}H_{13}^- + H_2O$$

Reduction by sodium converts two of the bridging hydrogen atoms:

$$\text{Decaborane} + \text{Sodium} \longrightarrow \text{Sodium Decaborane}$$

$$B_{10}H_{14} + 2Na \qquad Na_2B_{10}H_{14}$$

Nucleophiles react to give 6, 9-distinguished products as in the following equations:

$$\text{Decaborane} + \text{Sodium Cyanide} \longrightarrow \text{Sodium Boro Cyanide} + \text{Hydrogen}$$

$$B_{10}H_{14} + 2NaCN \longrightarrow B_{10}H_{12}(CN)_2Na_2 + H_2$$

$$\text{Decaborane} + \text{Ethylene Cyanide} \longrightarrow \text{Ethylene Boro Cyanide} + \text{Hydrogen}$$

$$B_{10}H_{14} + 2CH_3CN \longrightarrow B_{10}H_{12}(NCCH_3)_2 + H_2$$

$$B_{10}H_{14} + 2PR_3 \longrightarrow B_{10}H_{12}(PR_3)_2 + H_2$$

In each reaction equation two bridging hydrogen atoms have migrated into the positions between boron atoms 5, 10 and 7, 8. Also in a forward sense the two terminal hydrogen atom at boron atoms 6 and 9 has been replaced by the incoming nucleophiles and H_2 has been eliminated.

In contrast to reactions with nucleophiles, decaborane (14) reacts with electrophils to give 2-4 or 1, 3 distributed products. An example is shown in the reaction following:

$$B_{10}H_{14} + I_2 \longrightarrow 2, 4\text{-}I_2B_{10}H_{12} + H_2$$

Changed distribution calculations using molecular orbital theory whose considerable excess negative charge should be assigned to boron atoms 1, 2, 3, and 4 with positive charge assigned to the electron deficient positions elsewhere in the molecule. It is thus gratifying that experiments show consistently that only positions 1, 2, 3 and 4 can be substituted electrophilically.

The Tetrahydroborate Ion BH_4

The tetrahydroborate ion BH_4 is the simplest of a number of borohydrate anions. It is of great importance as a reducing agent and source of H^- ion both in inorganic an organic chemistry; derivatives such as $[BH(OMe)_3]^-$ and $[BH_3CN^-]$ are also useful, the latter because it can be used in acidic solutions.

Boronhydrates of many metals have been made and some representative syntheses are:

$$4NaH + B(OMe)_3 \longrightarrow NaBH_4 + 3NaOBH_3$$

$$NaH + B(OMe)_3 \xrightarrow{THF} NaBH(OMe)_3$$

$$2LiH + B_2H_6 \longrightarrow 2LiBH_4$$

$$AlCl_3 + 3NaBH_4 \longrightarrow Al(BH_4)_3 + 3NaCl$$

$$UF_9 + 2Al(BH_4)_3 \longrightarrow U(BH_4) + 2AlF_2BH_4$$

The most important salt is $NaBH_4$. It is a white crystalline solid stable in drying air and non-volatile. It is insoluble in diethyl ether but dissolves in water, THF (tetrahydrofuran) and ethylene and glycol ethers from which it can be crystallised.

Many boronhydrides are ionic, containing the tetrahedral BH_4^- can serve as a legand, interacting more or less covalently with metal ions, by bridging hydrogen atoms. Thus in $(PH_3P)_2CuBH_4$ there are two Cu-H-B bridges, where as in $Li(BH_4)_4$ each BH_4 forms three bridges to Li as shown in Fig. 25.9.

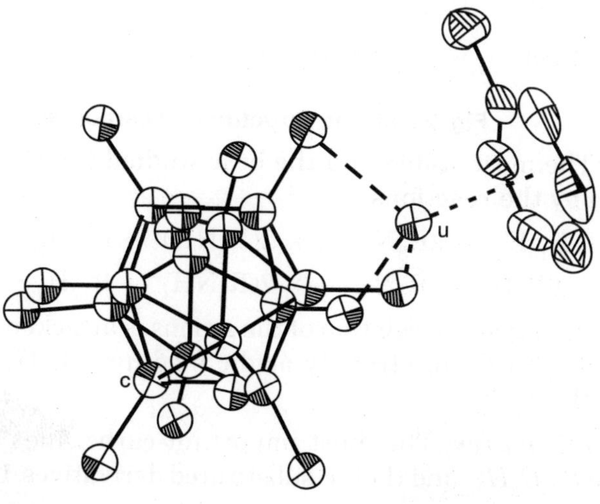

Fig. 25.9. The structure of $Li(BH_4)_4$

These M-H-B bridges are (3c-2e) bonding systems.

Polyhedral Borane Anions and Carboranes

The polyhedral borane anions have the formula BnH_2^{2-}. The carboranes may be considered to be formally derived from $B_nH_2^{-2}$ by replacement of BH^- by the isoelectric and isostructural OH. Thus two replacements lead to neutral molecule $B_{n-2}C_2H_n$. Carboranes or derivatives with n = 5 or n = 12 are known in some boron complexes. From this two or more isomers may be isolated, sulphur and phosphorous derivatives can also be obtained PH^+, for e.g.: replacing OH or BH^-.

Generally there are two broad classes of compounds:

- Those in which the atomic framework closes in on itself to form a polyhedron. These are *Closo* (in Greek closo = cage) compounds.
- Those frameworks that are open or incomplete polyhedra. These are *nido* (in Greek nido-next), which have two missing vertices. Some of the Closo carborane compounds are shown in Fig. 25.10.

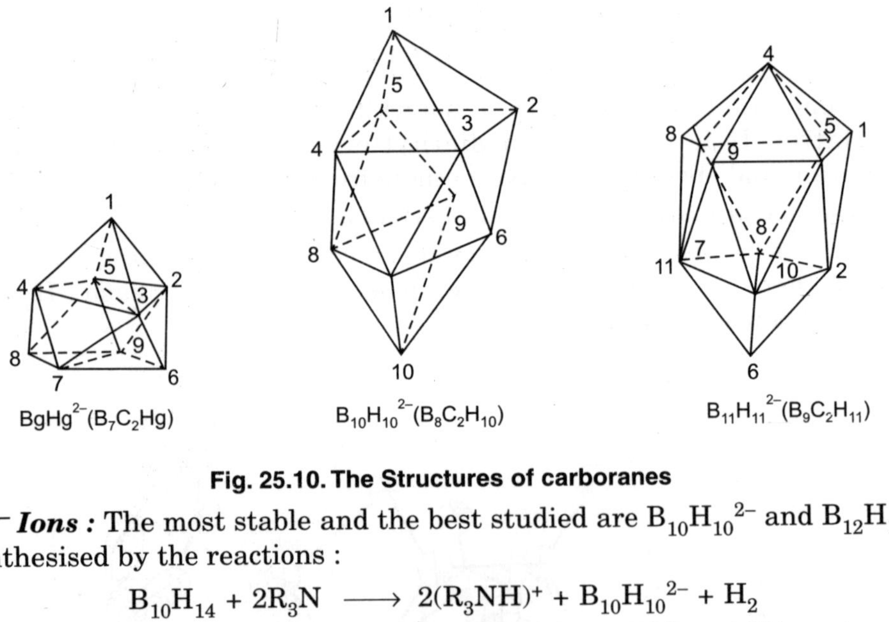

Fig. 25.10. The Structures of carboranes

$B_nH_n^{2-}$ Ions : The most stable and the best studied are $B_{10}H_{10}^{2-}$ and $B_{12}H_{12}^{2-}$ which can be synthesised by the reactions :

$$B_{10}H_{14} + 2R_3N \longrightarrow 2(R_3NH)^+ + B_{10}H_{10}^{2-} + H_2$$
$$6B_2H_6 + 2R_3N \longrightarrow 2(R_3NH)^+ + B_{12}H_{12}^{2+} + 11H_2$$

The most important general relation of the anions is attacked by electrolphillic reagents such as Br^+ $C_6H_5N_2^+$, RCO^+ in strongly acidic medium, $B_{10}H_{10}^{+2}$ is more susceptible to substitution than $B_{12}H_{12}^{2-}$.

$B_{n-2}C_2H_n$ Carboranes: The most important carboranes are *1, 2 and 1, 7 Dicabo-closo-dodecaborane $B_{10}C_2H_{12}$* and their c-substituted derivatives.1, 2 isomer may be detained by the reaction :

$$B_{10}H_{14} + 2R_2S = B_{10}H_{12}(R_2S)_2 + H_2$$

$$B_{10}H_{12}(R_2S)_2 + RC \equiv CR' = 1,2\text{-}B_{10}H_{10}C_2RR + 2R_2S + H_2$$

The $B_9C_2H_{11}^{2+}$ ions are very strong bases and readily acquire H^+ ions.

Metal Complexes of Carborane Anions

The open pentagonal forces of the $B_9C_2H_{11}^{2+}$ ions were recognised by *M.F. Huthrone* in 1964 to bear a strong resemblance structurally and electronically to the cyclopentadylenyl ion, $C_5H_5^-$, the latter strong bonds to transmission metals.

Interaction of $Na_2B_9C_2H_{11}$ with metal compounds such as those of Fe^{2+} or Co^{3+} thus leads to species isoelectronic with ferrocene, $(C_5H_5)_2$ Fe or the cobalticinium ion, $(C_5H_5)_2Co^+$, namely $(B_9C_2H_{11})Fe+$ and $(B_9C_2H_{11})_2Co^-$, respectively. The iron complex undergoes reversible oxidation like ferrocene :

$$[(C_5H_5)_2Fe^{III}]^+ + e^- = [(C_5H_5)_2Fe^{II}]^0$$

$$[(B_9C_2H_{11})_2Fe] + e^- = [(B_9C_2H_{11})_2Fe]^{2-}$$

The formed nomenclature for the $B_9C_2H_{11}^{2-}$ and its complexes are univer. The trival name "dicarboallide" ion was proposed (from alli = pot in Spanish) referring to the pot like shape of the structure as shown in Fig. 25.11 of B_9C_2.

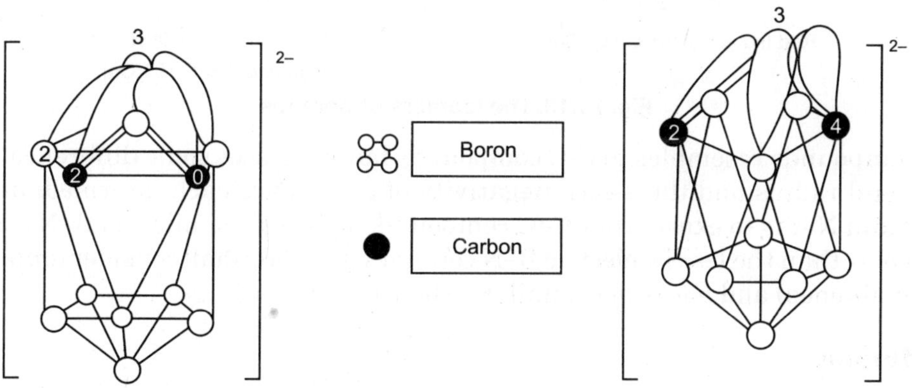

Fig. 25.11. The Structure of the isometric $B_9C_2H_{11}^{2-}$ ions.

The structures of two types of dicarboallide metal complexes are shown in Fig. 25.10, while some complexes have the metal deposited asymmetrically.

Finally, comparable with η^5 - $C_5H_5Mn(CO_3)$ there are mixed complexes with only one dicarboallide unit and other ligands such as Co, PH ,C_4, C_5H_5 and so on which is illustrated Fig. 25.12.

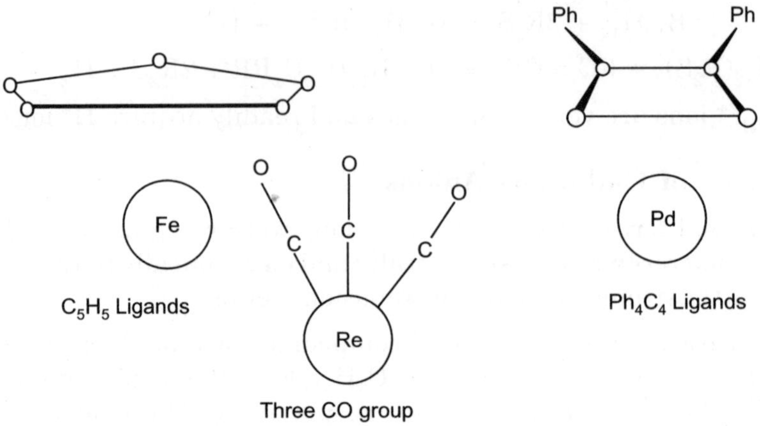

Fig. 25.12. The Dicarboallide complex and other Ligands.

Boron-Nitrogen Compounds

Here we shall discuss about the B–N compounds.

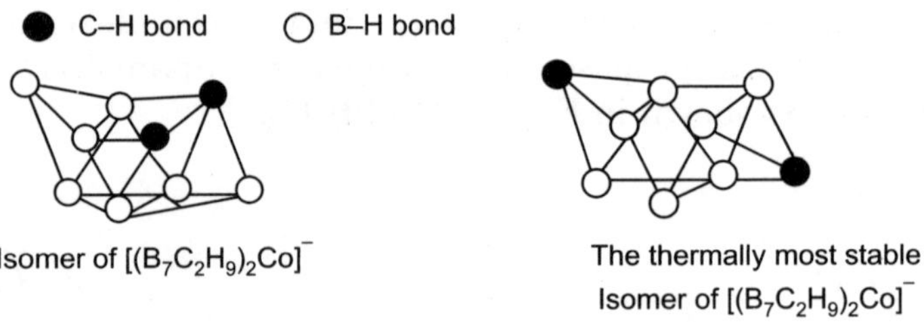

Fig. 25.13. The isomers of boranes.

B–N compounds resembles to C-C compounds but possesses some difference. Since the covalent bond-radius and the electronegativity of carbon are each intermediate between those of B and N it is to expect that C–C compounds will be as similar as B–N compounds. But, less polar than the less isoelectric B–N counter parts. We shall consider *amine-boranes* (similar to alkenes) and *borazines* (similar to benzenes).

Amine Borane

Amine boranes are Lewis Acid-base adducts containing a boron-nitrogen donor bond lengths found in simple alkenes are formed by symmetrical cleavage of diborane or by reaction of ammonium salts the equation following :

$$[H_3NR]Cl + LiBH_4 \longrightarrow RH_2N \longrightarrow BH_3 + LiCl + H_2$$

The B–N bond strength varies from adducts to another. The weakest B–N bonds are represented in a diagrammatic form below :

As in the structure (*a*) an arrow indicates a slight donor—acceptor interactions. More complete sharing of nitrogen electrons with boron is represented by structure (*b*), which is expected to be polar.

Steric hindrance can prevent the formation of some adducts, for example the 2, 6-dymethylpyridine with trymethylborane. Bridged diadducts can be obtained in the following reactions:

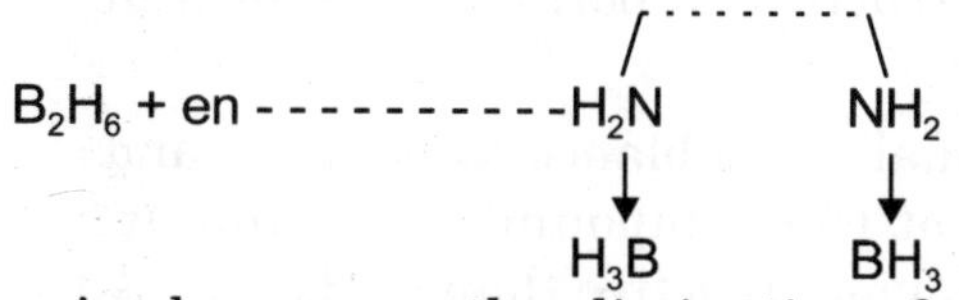

The chief reactions of amine borane are the elimination of either of AX or RH, to give aminoboranes.

Amino Borane

Amino boranes are B–N compounds that is similar to alkenes. Boron is trigonal in amino boranes and the three constituents at boron are planner, or very nearly so. Two resonance forms may by shown in structure (1) and (2).

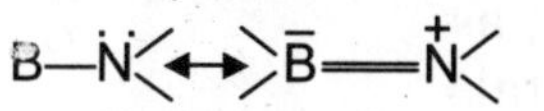

In most aminoboranes there is something less than a full double bond between a boron and nitrogen but in some cases high barriers rotate about the B–N bond indicate that the B–N bond order exists 1.0. The multiple bond aminoborane is formed by over lap of atomic p-orbitals as shown in Fig. 25.14. Average between the two nitrogen π-boron (see Fig. 25.14) donors for the empty p-orbital of boron. Rotational barriers are lower still for tri (amino) boranes.

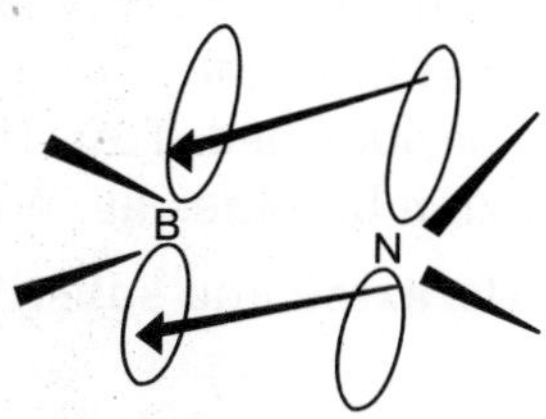

Fig. 25.14. The overlap of atomic p orbitals in multiple bond aminoborane.

The formation of a π-bonds in amino boranes as a result of donations of electrons form a filled p-orbital on nitrogen to an empty p-orbital on boron.

The sequence of reaction following serves as useful example:

$Me_2HN: + BCl_3 \longrightarrow Me_2HN \longrightarrow BCl_3$

$Me_2HN: \longrightarrow BCl_3 \longrightarrow Me_2N \longleftrightarrow BCl_2 + HCl$

$Me_2N^+ \longleftrightarrow BCl_2 + 2Et\ MgCl \longrightarrow Me_2N^+ \longleftrightarrow Bet_2 + 2MgCl_2$

The chief reaction (other than substitution) of aminoboranes is condensation to cyclic system. As in the formation of dimmer, structure is shown below:

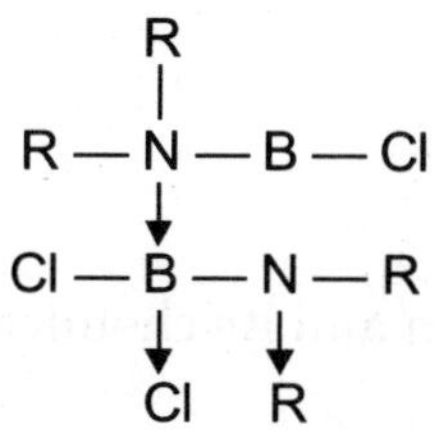

Condensation of aminoboranes to cyclic trimmers when accompanied by the elimination of either HX or RH leads to the borazine derivative.

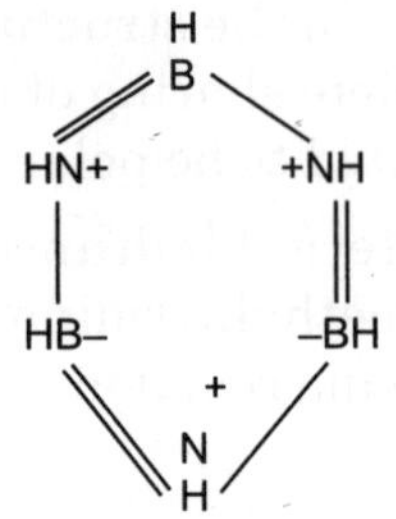

Fig. 25.15. The of Borazine $B_3N_3H_6$

Borazine

One of the most interesting B–N bond compounds is borazine, $B_3N_3H_6$ the structure of borazine is shown in Fig. 25.15.

It has an obvious formal resemblance to benzene and the physical properties of the compounds are similar. However borazine is much more reactive than C_6H_6 that is benzene and readily undergoes addition reactions, which is shown in the equation below:

$$B_3N_3H_6 + HX \longrightarrow (-H_2N-BOX)_3$$

Where X + Cl, OH, OR and so on

This does not happen in the case of benzene. Borazine also decomaposes slowly gets hydrolysed to ammonia and boric acid at higher temperature, as with the Benzene π complexes with transition metals may be obtained. Thus the structure of Hexamethyl-borazine becomes as in Fig. 25.16.

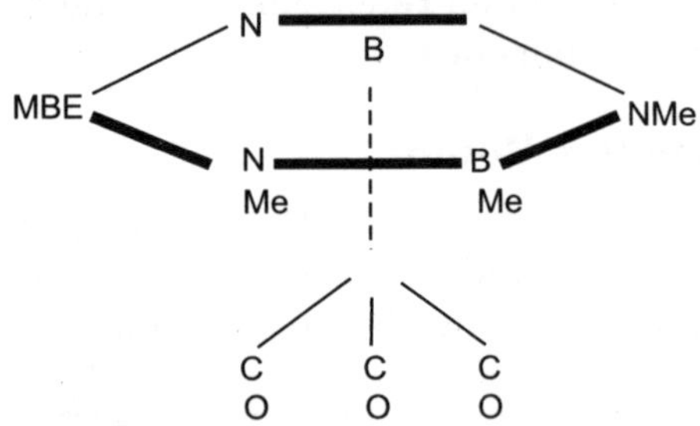

Fig. 25.16. The structure of Hexamethy-lborazine.

Borazine and substituted borazines may be synthesised by reactions such as :

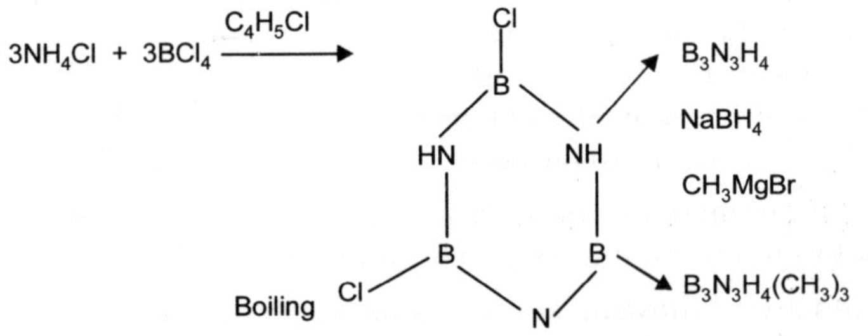

$$3Cl_3B.\ NH_2CH_3 + 6(CH_3)_3N \xrightarrow[C_6H_5Cl]{Boiling} Cl_3{\cdot}NH_2CH_3 \text{ (melting point 126 to 128°C)}$$

$$3Cl_3B.\ NH_2CH_3 + 6(CH_3)_3N \longrightarrow 6(CH_3)_3NHC_1 + Cl$$

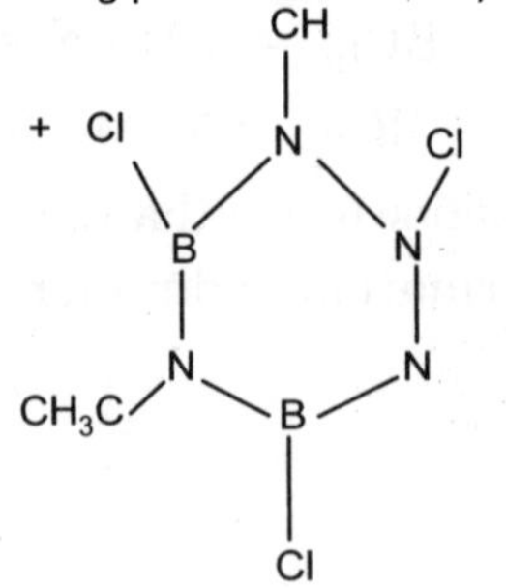

This ends the chemistry of boron and its chemical properties.

CHAPTER 26

Photochemistry

Introduction

Photochemistry is defined as the branch of chemistry, which deals with the reactions promoted by the action of light, either by absorption or by radiation of the visible or the ultraviolet region (200Å to 8000Å).

The reactions staking place due to the action of light are called photochemical reactions. A photon being the primary source of radiations, photochemical reactions involve the absorption of these photons.

What is Light?

The light is the form energy, considered as one of the family member of electromagnetic spectrum, with a wavelength between $10^4 - 10^{14}$ m.

Within these wide range the light is only about 10^{-7} to 10^{-2} m, wavelength. The total electromagnetic spectrum is shown in Fig. 26.1.

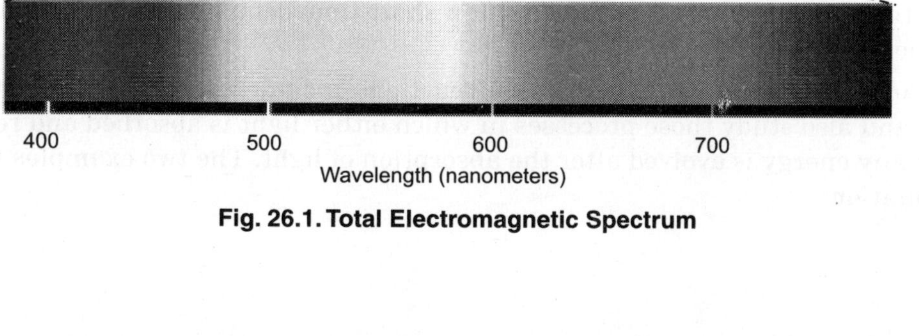

Fig. 26.1. Total Electromagnetic Spectrum

This light consists of particles called photons, which is called the primary unit of radiation. The photochemical reactions result in the adsorption of these photons. The reactants, which can be brought to the exposure of such radiations, include combination, oxidation, reduction and decomposition etc. Ordinary reactions are generally referred to as thermal reactions and are usually accompanied by decease in free energy. But it has been noted that certain photochemical reactions are accompanied by the increase of free energy. Photosynthesis of carbohydrates form carbon dioxide and water taking place in nature accompanying the increase of energy. In thermal reactions, molecules of the reactants undergo chemical reactions, if they acquire activation energy by collision. On the other hand photochemical reactions depend upon an outside source of energy reaction. The absorption of radiation is always quantised. Quantum energy which is absorbed in a photon and its value depends upon the frequency of radiations according to the equation

$E = h\nu$ Where E = energy

H = Plank's Constant

ν = Frequency of light in vacuum.

The energy absorbed can also be expressed in terms of wavelength of radiation or the wave number of radiation.

$$E = \frac{\lambda - c}{\lambda} [v = c/\lambda]$$

Where, λ = wavelength, c = velocity of light

Also $E = hc\bar{\nu}$ [$\bar{\nu} = 1/\lambda$ in cm].

Thus, energy absorbed is directly proportional to the wave number.

Photoprocess

Photorocesses are of two types, they are as follows :

(a) Photophysical process : These are the process in which light is absorbed by a substance without undergoing any chemical change depending upon the nature of the absorbed light and the manner in which it is emitted, process are of three types :

1. If the absorbed light is sufficiently energetic, the electrons jump to the other orbits (after getting excited) or even leave the atoms completely.
2. If the absorbed light is emitted instantaneously, the phenomenon is regarded as fluorescence.
3. If the absorbed light is emitted after a short time delay the phenomenon is called phosphorescence.

(b) Photochemical Process : These reactions include mainly the photochemical reactions and also study those processes in which either light is absorbed and reaction is promoted any energy is evolved after the absorption of light. The two examples will clear the explanation:

1. $H_2(g) + Cl_2(g) \xrightarrow{\text{light}} 2HCl(g)$
2. $6CO_2 + 6H_2O \xrightarrow{\text{light}} C_6H_{12}O_6 + 6O_2 + 674\text{Kcal}$

The table below shows the difference between photochemical reactions :

Photochemical Reactions	Thermochemical Reactions
1. These reactions involve the absorption of photons.	1. These reactions involve the evolution and absorption of heat.
2. The presence of light is an essential requirement for the reaction to take place.	2. The presence of light is not necessary for these reactions to take place; these reactions can be carried out in dark also.
3. Temperature has no significant role in the photochemical reactions. But the intensity of lights has an importance in these reactions.	3. Temperature has an important role in the thermochemical reactions. Intensity of the temperature increases the rate of thermochemical reactions.
4. The free energy change (ΔG) of a photochemical reaction is not negative always. For example in synthesis of carbohydrates by plants and decomposition of hydrogen and chlorine (ΔG) is positive but reactions are spontaneous.	4. The free energy (G) of the thermochemical reaction is always negative.

The light rays as they are the only promoters in the photochemical reactions, only that part of light is concerned which gets absorbed.

Light rays when emitted from a source, as we all know consists of photons, these photons travel at different velocities in same light beam, so they are called *non-coherent beams* and the example is ordinary light; they are also called multichromatic (having many colours) in nature. Some lights when emitted from a source consists photons having equal velocities are called coherent beams, and they are also monochromatic (single coloured) beams e.g. laser.

These lights effect different chemical reactions in photochemistry. Now we are going to discuss the different Laws, which govern photochemistry.

(i) Lambert's Law : It states that when a beam of monochromatic radiation passes through a homogeneous absorbing medium, the rate of decrease of intensity of radiations with thickness of absorbing medium. The rate of decrease of intensity of radiation with thickness of absorbing medium is proportional to the intensity of incident radiation.

Mathematically we can express the law:

$$\frac{-dI}{dx} \propto I$$

where dI = very small decrease in the intensity of radiation on passing through very small thickness dx of the medium.

or, $\frac{-dI}{dx} = kI$ or, $\frac{-dI}{I} kdx(i) \frac{-dI}{dx}$ Rate of decrease of intensity of Radiation with thickness of the absorbing medium.

K = constant, called absorption co-efficient of the medium of light of a particular wave length.

Let (1) I_0 be the intensity of radiation before entering the absorbing medium ($x = 0$) and (2) I be the intensity of radiation after passing through any thickness, say x of the medium.

Thus relation (1) can be integrated between the limits $I = I_0$ to 1 and $x = 0$ to x, we get

$$\int_{I_0}^{I} dI/I = \int_{I=0}^{k=x} kx$$

$$I_0 x = 0$$

or $$Ix\ I/I_0 = -kx \quad \text{...(II)}$$

or $$I/I_0 = e^{-kx}$$

or $$I = I_0 e^{-kx} \quad \text{...(III)}$$

This relation is called *Lambert's Law Equation*. It tells how the original intensity (II) after passing through a thickness x of the medium.

The intensity of radiation absorbed is given by

$$I_{absorption} = I_0 - I = I_0 - I_0 e^{-kx}$$
$$= I_0(1 - e^{-kx}) \quad \text{...(IV)}$$

Relation (I) can also be written as

$$\log I/I_0 = -kx$$
$$\log I/I_0 = kx/2.303 \text{ or , } \log I/I_0 = -k'x \quad \text{...(V)}$$

or $$I/I_0 = -10^{-kx} \text{ or } I = I_0 .\ 10^{-k'x}$$

Here, it is related to the absorption co-efficient k by the expression k = k/2.303.

We know that the relation

$$\text{Log } I/I_0 = -k'x$$

Or, $$K' = 1/x \log I/I_0$$
$$K' = 1/x \log I_0/I$$

If $$\log I_0/I = 1 \text{ , then } k' = 1/x$$

As $\log I_0 = 1$, we say that $I_0 = 10$ or $I = 1/10\ I_0$.

Extinction Coefficient

It is the reciprocal of that layers thickness (expressed in cm) at which the intensity of light falls to one tenth of the original value.

(i) Beer's Law (or Lambert's Beer's Law) : This law is applicable to solutions.

When a beam of monochromatic radiation is passed through a solution of an absorbing solution is proportional to the intensity of incident radiation as well as to concentration of the solution.

Mathematically

$$\frac{-dI}{dx} \propto I.C$$

or $$\frac{-dI}{dx} \;\varepsilon\; I.C \text{ [where C = concentration of the solution in moles litre}^{-1}] \quad ...(I)$$

ε = (epsilon) = constant called, molar absorption co-efficient and its value depends upon the nature of the absorbing substance.

Integrating the equation (I) between the same links (as in Lambert's law)

$$\int_{10}^{1} \frac{dI}{I} = -\int \varepsilon c dx$$

Or $$Ix\frac{I}{I_0} = -\varepsilon cx \quad ...(II)$$

Or $$I/I_0 = e^{-} \varepsilon cs$$

Or $$I = I/I_0\, e^{-}\, \varepsilon cs \quad ...(III)$$

The equation (III) tells how the original intensity I_0 of a monochromatic light falls to intensity, I on passing through a thickness x of a solution of concentration.

It is also clear that the solution is also given by the difference I_0 and I.

$$I_{absorption} = I_0 - I$$
$$= I_0 - I_0 e^{-} \varepsilon cx$$
$$= I_0 (1 - e^{-} \varepsilon cx) \quad ...(IV)$$

Also the equation (II) can be written as $2.303 \log I/I_0 = -\varepsilon cx$

Or, $$\log I/I_0 = -\varepsilon/2.303\, cx$$
$$\log I/I_0 = -\varepsilon' cx \quad ...(V)$$

Or, $$I/I_0 = -10^{-\varepsilon' cx}$$

Where ε' = molar extinction co-efficient of the absorbing solution.

Its relation with molar absorbing co-efficient (ε) of the absorbing solution is denoted by,

$$\varepsilon' = \frac{\varepsilon}{2.303}$$

Physical Significance of Molar Extinction Co-efficient

We know that from relation (v) $\log I/I_0 = -\varepsilon' cx$

Or $\varepsilon' = -1/cx \log I/I_0$ or $1/cx = \log I/I_0$

When c= 1m and $\log I_0/I = 1$ then $\varepsilon' = 1/x$

But we know that when $\log I_0/I = 1$ then $I_0/I = 10$

$\therefore I = 1/10 I_0$

Now we can define molar extinction co-efficient as the reciprocal of that thickness of the solution layer of 1 molar concentration which reduces the intensity of light passing through it to one tenth of its original value.

Laws of Photochemistry (Laws of Photochemical Reaction)

Two basic laws known as the Laws of Photochemistry govern all photochemical reactions.

1. ***Grotthus-Draper Principle of Photochemical Activation:*** It is also known as the first law of photochemistry. According to this principle, all the incident light is not effective in bringing about the chemical change. It is stated as when light falls in an object, a part of it is deflected and a part of it is transmitted and the rest is absorbed. It is only absorbed part of light, which is responsible for the photochemical change.

But this law does not clearly state whether the absorbed light will necessarily bring changes in the photochemical reactions. The absorbed light may bring about following changes, which are not chemical in nature.

(*i*) In several cases the absorbed light converted into the kinetic energy of the absorbing molecules and only heat effects are produced, for example, potassium permanganate solution strongly absorbs light but no chemical change is produced.

(*ii*) In many cases, the absorbed light may be re-emitted as Fluorescence and phosphoresce.

(*iii*) In many cases, the absorbed light is transferred from one substance to another when both are present together. As a result, some chemical changes are observed. Such a process is called *Photosensitisation*. The substance, which initially absorbs light then, transfers it to the reacting substance known as *sensitiser.*

In the photosynthesis of carbohydrates, chlorophyll acts as photosensitises.

2. ***Stark's Einstein's Law of Photochemical Equivalence :*** This is also known as second law of photochemistry. The law states that In a primary photochemical process, each molecule is activated by the adsorption of one quantum of radiation (one photon).

The absorption of one quantum of energy (photon) by a molecule results in the formation of an activated complex (primary step). It is not necessary that an activated complex must react. As activated complex can also cause the reaction of many molecules through a chain reaction mechanism. It may be noted that for a primary process, the number of molecules activated is equal to the number of quanta absorbed. Thus, the mass of the substance containing Avogadro's number (6.023×10^{23}) of molecules absorb (6.02×10^{23}) quanta of energy. This mole of quanta is called *Einstein.*

For a primary process, it is supposed that the law applies and the light is actually absorbed. Let ν be the frequency of the absorption of light. The corresponding quantum of energy absorbed per molecule will be $h\nu$.

Thus the quantum energy (E) absorbed per mole of the reducing substance is given by

$$E = N_0 h\nu$$

We know $N_0 = N_0 h\nu$

$$N_0 = 6.023 \times 10^{23}$$

$$H = 6.023 \times 10^{23} \text{ joules/sec}$$

$$E = 6.023 \times 10^{23} \times 6.62 \times 10^{-34}$$

$$= 3.39 \times 10^{11} c/\lambda$$

$$= \frac{3.39 \times 10^{-11} \times 3 \times 10^{8}}{\lambda} \text{ joules per mole}^{-1} \text{ (}\lambda\text{= wavelength in metres)}$$

If lambda is expressed in Å, then the above relation is multiplied by 10^{10}(1m = 10^{10}Å)

$$E = \frac{119.7 \times 10^{5}}{\lambda} \text{ kJ mol}^{-1}$$

It is seen from the above relations *that shorter the wavelength, greater will be energy absorbed*. Einstein is bigger in magnitude for the ultraviolet light absorbed as compared to the one where the infrared light is absorbed.

The shorter the wavelength the greater is the energy of the Einstein.

For example, different reactions has the following order of wavelengths

X-Rays < ultraviolet < visula light < Infrared

Therefore, the energy per Einstein, will be

X-Rays > ultraviolet > visula light > Infrared.

Quantum Efficiency or Quantum Yield

We know that according to Einstein's Law of photochemical equivalence, every reacting molecule absorbs one quantum of radiation and clearly the number of reaching molecules should be equal to the number of quanta absorbed. But it is not fully true and in many cases, it has been observed that :

(*i*) A small amount of light absorbed can bring a large amount of reaction.

(*ii*) A large amount of light absorbed brings only a small amount of reaction.

This peculiarity is explained on the basic of the fact that a photochemical reaction consists of two processes :

1. ***Primary Process :*** In such process, light radiation is absorbed by an atom or a molecule to produce an excited atom or excited molecule. The formation of excited atom or molecule does not necessarily mean that reaction will take place only one molecule will react.
2. ***Secondary Effect :*** The excited molecule during the primary process may under go one of the following changes to complete the reaction.
   (*i*) It may react to form the final product.
   (*ii*) It can be partly or fully deactivated and no reaction occurs.
   (*iii*) It may start a chain reaction.

It is very important to note that the secondary process may take place in the dark provided the products of primary reactions are available.

As the law of photochemical equivalence is applicable to primary process light radiations are essential only for these processes (primary) secondary reactions or some other complicated circumstances usually mark the operation of the law does not seem to be always obeyed.

Quantum yield in photochemical reactions is the ratio of the numbers of molecules reacting chemically to the number of quanta of radiation absorbed. It is denoted by ϕ.

$$\phi = \frac{\text{Number of molecules reacting in a given time}}{\text{Number of molecules reacting in the same time}}$$

or $$\phi = \frac{\text{Number of moles reacting carefully}}{\text{Number of Einstein of energy absorbed}}$$

In other words, the quantum yield or efficiency is the number of moles of substance under going photochemical change per absorption of one Einstein of radiation.

For a primary photochemical process, $\phi = 1$. When the quantum yield is greater or less than one, only one conclusion can be drawn regarding the mechanism of reaction.

To determine the quantum yield of any photochemical reactions, we have to measure.

(*i*) The quanta of light absorbed per second.

(*ii*) The number of moles of light absorbing substance that react in one second.

SOME PHOTOCHEMICAL REACTIONS AND QUANTUM YIELDS

Boudenstein studied quantum yield of a large number of photochemical reactions. The quantum yield of some important photochemical reactions along with effective wavelength is given in table.

Reaction	Effective Wavelength (nm)	Quantum Yield (ϕ)
1. Those for which quantum yield is small integer		
$SO_2 + Cl_2 \longrightarrow SO_2Cl_2$	425	1
$H_2S \longrightarrow H_2 + S$	208	1
$2HBr \longrightarrow H_2 + Br_2$	579	1
$2HI \longrightarrow H_2 + I_2$	207 - 282	2
$3O_2 \longrightarrow 2O_3$	170 - 190	3
2. Those for which the quantum yield is very large		
$H_2 + Cl_2 \longrightarrow 2HCl$	400 - 436	-10^4 to 10^3
$CO + Cl_2 \longrightarrow COCl_2$	400 - 436	10^3
3. These for which the quantum yield is very low		
$H_2 + Br_2 \longrightarrow 2HBr$	510 0.01	
$2NH_3 \longrightarrow N_2 + 3H_2$	310	0.05
$2NO_3 \longrightarrow 2NO + O_2$	405	0.7

From the above table, it is evident that law of photochemical equivalence is strictly valid for a very few reactions. On the basis of their quantum yields, the photochemical reactions can be divided into three categories:

(1) Those for which the quantum yields is small integer.

(2) Those for which the quantum yields is very large

(3) Those for which the quantum yields is very low.

In order to explain the variation in quantum yield, Boudenstein pointed out that photochemical reactions involve two distinct processes, i.e. a primary process and a secondary process.

Experimental Determination of Quantum Yield

The apparatus used for this determination of the number of quanta of light absorbed consists of the following sections :

- ***A Light Source :*** Which emits radiation of suitable intensity in the desired spectral range. A common source of light used is mercury vapour lamp. Iron and carbon arcs and in some cases metal filament lamps are used as source of light.
- ***A lens of suitable focal length***
- ***Monochromator :*** which cuts off all radiations except the radiations of desired wavelength (λ).
- ***A Reaction Cell :*** Which is made of glass or quartz. Glass is used when the wavelength of light is within the visible range. For light radiations having wavelength below 350 nm only the quartz vessel is used (see the apparatus set up in Fig. 26.2).

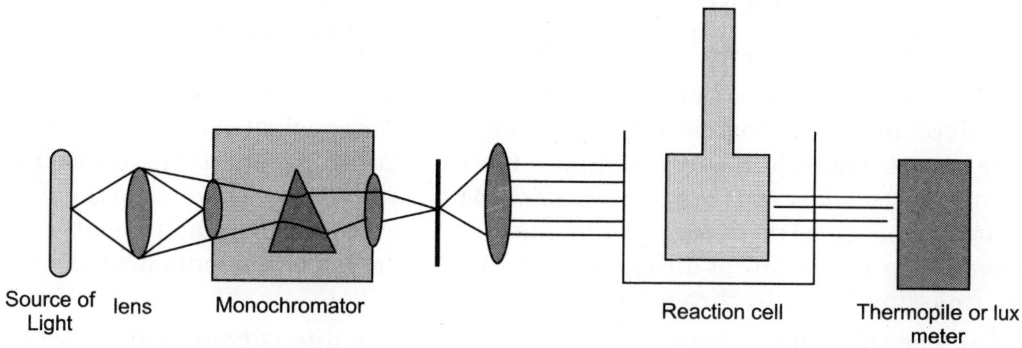

Fig. 26.2. The Experimental Set up for the Determination of Quantum Yield.

- ***Detector :*** Which is usually a lux meter used for the detection of the intensity of the light passing through the reaction cell, some times a thermopile is also used.
- ***Thermopile :*** It consists of thermocouples connected in series as shown in Fig. 26.3 it is silver-bismuth thermopile.

A thermocouple consist of two metals (here silver and Bismuth) joined together. One set of junctions of the thermocouple is blackened so as to absorb all the radiations. The other set is protected from radiations and maintained at constant temperature by placing the system in a box. The radiations falling on the blackened metal strip are almost completely absorbed by it. The heat radiations associated with light radiations absorbed by the blackened end, raise the temperature of this set of junctions. The temperature difference between the hot end and the cold end produces a current in the circuit. This thermoelectric current can be measured by connecting a miliampere meter to the thermocouple. The intensity of light radiations absorbed can be measured by calibrating the thermocouple with radiations of known intensity.

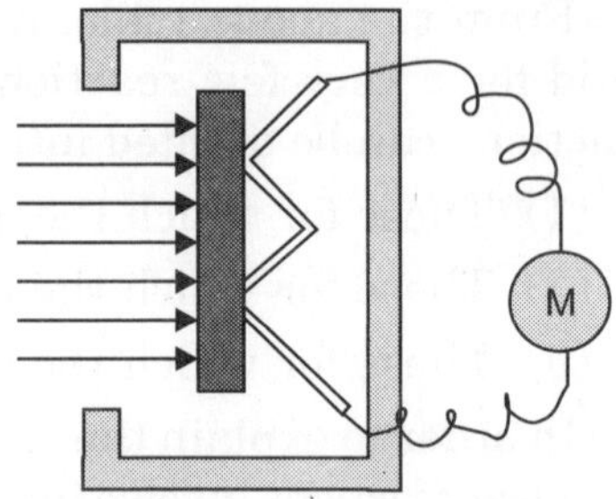

Fig. 26.3. A Thermopile.

From the intensity of light absorbed, the energy of incident radiations can be easily calculated.

The measurements are made before and after passing light through the cell. The difference of the two movements gives the energy of the radiation observed by the substance.

The energy associated with radiations absorbed can also be measured by using an *actinometer,* a device in which gas mixture or solutions sensitive to light are used. Actinomter is used to determine the definite amount of radiation absorbed by the substance, by which it can be, determined the nature of the chemical reaction.

One of the most commonly used actinometer is called *Uranyl Oxalate Actinometer*. It consists of 0.05 molar oxalic acid and uranyl sulphate (UO_2SO_4) solutions in water. When exposed to light the understated reaction takes place.

$$\begin{array}{c} COOH \\ | \\ COOH \end{array} (aq) \xrightarrow[UO_2SO_4]{h\nu} CO_2(g) + (CO)g + H_2O$$

Uranyl sulphate absorbs light and then it is allowed to pass thorough oxalic acid. The extent of decomposition of oxalic acid is determined at the conclusion of the experiment by titration against potassium permanganate. For calculating the energy associated with a radiation, *it is assumed that the amount of decomposition is proportional to the intensity of radiation of a given wavelength and the time of exposure*. The actinometer is first calibrated using radiations of different wavelength. The measurements are made before and after passing the light through the reacting cell.

The number of moles reacting in a given time and the intensity of light absorbed, the quantum yield of the reaction can be calculated by noting dôwn the change in some physical property.

Knowing the number of notes reacting in a given time and in intensity of the light absorbed the quantum yield of light can be calculated.

Primary Process in the Photochemical Reactions

The first step in a photochemical reaction is the absorption of a quantum of radiation by the molecule giving rise to the formation of an excited atom or an excited molecule as the care may be thus :

A	+	hν	$\longrightarrow$	A*
Atom or Molecule		One of Light		Excited atom or Excited molecule

The absorption of radiation by an atom or a molecule to form an excited atom or molecule is known as primary process in photochemistry. The excited molecule or atom formed by the absorption may be then behave in different ways. These are four distinct possibilities of excitation of the molecule:

1. In type I, the energy absorbed raises the molecule to an excited state which is suitable. The molecule breaks up into or radicals immediately on absorption of photon, i.e diminution of the molecule takes place. The fragment is obtained are associated with different kinetic energy and a continuous spectrum is obtained without any fine structure.
2. In type II, the electronic transition from a suitable ground state there will be no dissociation of the molecule. This is indicated by the corresponding spectrum consisting of discontinuous bonds with a fine structure.
3. In step III, the molecule is raised to a higher level. The energy acquired is more than the binding energy. Therefore, molecule would under go dissociation and the spectrum will be continuous through out.
4. In step IV, transition occurs from the lower level to a stable upper level. During the course of vibration, the molecule may shift from the stable state to the unstable state. When a shift of this type takes place, the molecule would dissociate producing atoms or radicals. This type of behaviour is referred to as *predissociation.* The spectrum would show fine structure in the region of predissociation. The rotational lines are absent and they have a diffuse appearance.

Secondary Photochemical Process

The activated molecules or the products of primary process may react with other molecules to form products or the activated molecules may emit the radiation of either the same or the different frequency. These processes are known as secondary process.

Some of the relevant chemical reactions support the above explanation.

1. Photochemical decomposition of hydrogen iodide in the gaseous phase (photosynthesis of hydrogen iodide) : On exposure to radiations having wavelength in the range 207 to 252 nm decomposition of hydrogen iodide takes place. The quantum yield of the reaction is 2. This has been explained with the mechanism shown below :

(*a*) Primary Process : $HI \xrightarrow{h\nu} H + I$ ...(*i*)

(*b*) Secondary Process : $H + HI \longrightarrow H_2 + I$ ...(*ii*)

$I + I \longrightarrow I_2$ ...(*iii*)

Possibility of other secondary processes such as

$$I + HI \longrightarrow I_2 + H \quad ...(iv)$$

$$H + H \longrightarrow H_2 \quad ...(v)$$

$$I + H \longrightarrow HI \quad ...(vi)$$

is cancelled out because the reaction (*iv*) is endothermic and takes place slowly and reactions (*v*) and (*vi*) are highly exothermic. The heated reaction produced results into the dissociation of the products of these reactions. The overall reaction is, therefore defined by adding reaction (*i*) (*ii*) and (*iii*).

$$2HI + h\nu \longrightarrow H_2 + I_2$$

Thus for every one quantum of light absorbed, two molecules of HI are decomposed. Hence, the quantum yield of reaction is 2. It may be noted that while adding the equation (*i*), (*ii*) and (*v*), no distinction has been made between the excited iodine atom and normal iodine atom.

The quantum yield for the photochemical decomposition of HBr is also 2 which suggests that the mechanism of this reaction is similar to that of photosynthesis of HI.

2. Photochemical Combination of Hydrogen and Bromine to form Hydrogen Bromide : When a mixture of hydrogen and bromine is exposed to radiations having wavelength (450-550 nm), they combine to form hydrobromic acid and reaction may be represented as :

$$H_2 + Br_2 \longrightarrow 2HBr$$

The quantum yield of the reactions is very low, i.e., 00.1. The low quantum yield may be explained by the mechanism given below :

(a) Primary Process : Bromine absorbs light in the given region of the spectrum (450-550 nm) and the molecule dissociates into atoms:

$$Br_2 + h\nu \longrightarrow 2HBr \quad ...(i)$$

(b) Secondary Process :

$$Br + H_2 \longrightarrow 2HBr + H \quad ...(ii)$$

$$H + Br_2 \longrightarrow HBr + Br \quad ...(iii)$$

$$H + HBr \longrightarrow H_2 + Br \quad ...(iv)$$

$$Br + Br \longrightarrow Br_2 \quad ...(v)$$

The bromine atoms formed in primary step (*i*) attack hydrogen molecule yielding hydrogen bromide and a hydrogen atom. The hydrogen atom then attack bromine forming another molecule of hydrogen bromide and another bromine atom. As a result, reactions (*ii*) and (*iii*) should repeat over and again i.e., a chain reaction should set up and the quantum yield of the reaction should be very high. However, in actual practice, the quantum yield is very low. This is explained as follows :

The step (*ii*) is highly endothermic and requires high energy of activation. This step is therefore, very slow at ordinary temperature. The step (*iii*), which depends upon step (*i*) it, is also slow. Step (*iv*) which is the reverse of (*ii*) becomes increasingly important and the rate of formation of HBr decreases. The only important reaction is (*v*) in which the bromine atoms recombine to form a molecule of bromine. It is just the reversal of the primary process. As a result, the quantum yield of the reaction is very low at ordinary temperatures.

The quantum yield of the photochemical combination of hydrogen and bromine increases with the increase in temperature. This is because the reaction (*ii*) is endothermic and requires high energy of activation become available in the form of heat. Hence, the reaction becomes faster at high temperatures.

The quantum yield of the photochemical combination of hydrogen and bromine increases with the increase in temperature. This is because the reaction (ii) is endothermic and requires high energy of activation become available in the form of heat. Hence, the reaction is faster at high temperatures.

3. Photosynthesis of Ammonia : The photochemical decomposition of ammonia has been found to take place by radiations of wavelength 210 nm. Its quantum yield is about 0.2 at ordinary temperature and pressures. This can be explained with help of mechanism given below:

(*a*) Primary process: -

(*i*) $$NH_3 + h\nu \longrightarrow NH_2 + H$$

(*b*) Secondary Process: - Since the final products of the photosynthesis of ammonia have been found to be nitrogen, hydrogen and hydrazine, the probable secondary process are represented by the following equations

(*i*) $$NH_2 + NH_2 \longrightarrow N_2 + 2H_2$$

(*ii*) $$NH_2 + NH_2 \longrightarrow N_2H_4 \text{ (Hydrazine)}$$

(*iii*) $$H + H \longrightarrow H_2$$

(*iv*) $$NH_2 + H \longrightarrow NH_3$$

The low observed quantum yield is due to the fact that product of the primary process i.e. NH_2 and H recombine to form ammonia.

4. The Hydrogen Chlorine Action : When a mixture of hydrogen and chlorine is exposed to ultra-violet light or visible of wavelength less then 480 nm, a first reaction takes place resulting in the formation of hydrogen chloride. The quantum yield of the reaction is exceptionally high, (10^5). The presence of oxygen slows down the rate of reaction and lowers the quantum yield also.

The extremely high quantum yield can be explained by the *Nernst* in 1918.

The different steps involved are as follows:

(a) Primary Process : According to Nernst's mechanism, the primary process is the dissociation of chlorine molecules into atoms as a result of adsorption of light. Thus,

(*i*) $$Cl_2 + h\nu \longrightarrow Cl + Cl \text{ [Chain initiating step]}$$

(b) Secondary Process

(*ii*) $Cl + H_2 \xrightarrow{\text{Exothermic}} HCl + H$

(*iii*) $H + Cl_2 \longrightarrow HCl + Cl$ {Chain propagating steps}

(*iv*) $Cl + Cl \longrightarrow Cl_2$ {Chain terminating step}

The step (*ii*) and (*iii*) get repeated giving rise to a chain reaction. According to the mechanism proposed above, there is possibility of the formation of an infinite number of molecules of HCl close to the absorption of a single quantum of radiation. The reaction stops when the chlorine atoms left off combine with each other to form Cl_2, according to the step (*iv*) this reaction takes place on the walls of the reaction vessel.

The process in the hydrogen-chlorine reaction is called chain initiating step. The reaction (*ii*) chain propagating step and the reaction (*iv*) is called chain terminating steps.

The reaction for the high quantum yield is that the step (*ii*) and (*iv*) are repeated by a large number of times before the chain termination. It is important to note that the quantum yield in the combination reaction of hydrogen and bromine is very low as compound to hydrogen and chlorine although the same mechanism was proposed for the combination reaction of hydrogen and bromine to form gaseous hydrogen bromide. The reason is the difference in the quantum efficiency is that while the reaction between chlorine atom and hydrogen molecule immediately following the primary process is exothermic and takes place readily, where as the corresponding reaction in hydrogen-bromine reaction it is highly endothermic and therefore, extremely slow at ordinary temperature.

As already mentioned the quantum yield of the reaction between hydrogen and chlorine decreases in presence of a small amount of oxygen.

This is because in presence of oxygen, The chain may be terminated by the reaction.

$$H_2 + O_2 \longrightarrow HOO \text{ [radical]}$$

Substance like oxygen which reduces the quantum yield of photochemical reaction are called photo inhibitors.

The chain quantum mechanism for the photochemical combination of hydrogen and chlorine is supported by the following facts:

(*i*) The combination of hydrogen and chlorine can be brought about even in absence of light by introducing either hydrogen or chlorine atoms in the reacting vessel.

(*ii*) The quantum yield of the reaction decreases continuously if the reaction is allowed to take place in ordinary tubes. This confirms the view that the chain is terminated at the walls of the containing vessel.

Photochemical Equilibrium or Photostationary State

In a certain photochemical reaction is followed by another photochemical or thermal reaction in the backward direction. The rate of photochemical reaction is proportional to the concentration of the photochemical product. Suppose a substance A changes to B by a photochemical process and then the B again changes to A by a Photochemical process. The situation may be represented as follows :

$$A \underset{\text{Light}}{\overset{\text{Light}}{\longleftrightarrow}} B$$

Now, if A changes to B by photochemical process and B changes to A by thermal process (dark reaction), then the situation may be represented as :

$$A \underset{\text{Thermal}}{\overset{\text{Light}}{\longleftrightarrow}} B$$

In either case the rate of forward reaction will become equal to the rate of backward reaction after sometime and the adsorption of light after this will produce no further chemical change. At this stage, the reaction is said to have attained a *photochemical equilibrium* or a *photo stationary state.* An example involving photochemical equilibrium is as following :

1. Photochemical decomposition of sulphur trioxide

$$2SO_3 \xrightarrow{\text{Light}} \text{Sulphur Dioxide + Oxygen}$$

Luminescence

We know that light in a bulb is produced by heating a metal filament. This is a case of heat energy changing electricity into light without the application of heat energy. *Luminescence is the phenomenon of the production of light by other agents besides heat.* Since temperature is not the main object, it is often called *cold light.* Some means other than heat can excite electrons, these excited electrons when return to the original state loose energy in the form of radiation, giving luminescence. The body emitting radiation under such condition is called luminescent.

Types of Luminescene

There are three types of luminescence.

Chemiluminescence : Emission of visible radiation during a chemical reaction is called chemiluminescence, examples are as follows :

(*i*) Green glows due to slow oxidation of phosphorous. It is assumed that an oxide of phosphorous in the excited state results. This is responsible for emission of lights an ordinary temperature.

(*ii*) The light of a firefly: It may be due to the oxidation of a protein called luciferin by atmospheric oxygen in the presence of enzyme, luciferase. The product of reaction in the excited state looses radiation of appropriate wavelength slowing the glow. It is actually *biochemiluminescence*.

Fluorescence : The adsorption of radiation simply excites some atoms or molecules of some gases, liquids of solids. These emitted atoms start emitting radiation of the same wavelength as of the absorbed radiation. This is called resonance radiation. In some cases, the excited molecules do not break but also start emitting radiations of usually larger wavelength than the absorbed radiation. *Such a process where the emitted radiation have* *different wavelengths is called fluorescence.* It may be noted that in the absence of the incident light, no fluorescence takes place. In fluorescence, the emitted radiation is caused immediately when the exciting source is removed. A fluorescent substance always gives of its characteristic fluorescence. Naturally occurring fluorescence substances are calcium fluoride solution with some dyestuff, vapours of sodium, mercury and iodine.

Some examples of artificial Fluorescence are :

(*i*) Certain organic dyes such as eosin, fluoroscein etc. When their solutions are placed in the light they show fluorescence from green to violet colour.

(*ii*) Chlorophyll

(*iii*) Uranyl Sulphate

(*iv*) Vapours of Mercury

(*v*) Ultramarine

(*vi*) Luminol, 5 amino 2, 3 Dihydro 1-4, pthalazinedione

Application of Fluorescence : The phenomenon of fluorescence has a number of applications as described below:

(*i*) *The intense green fluorescence* of aqueous fluorescence solution is used to delete links by adding it to water systems.

(*ii*) *Small quantities of optical brightness* are added to detergents and are retained by the fabric during washing. In sunlight, it fluoresces blue and adds brightness to the fabric after washing.

(*iii*) Fluorescent material dissolved in solution or in solid plastic bases is used to detect radioactive decay. This forms the basic of scintillation counters.

(*iv*) The three dimensional structure of proteins can be studied by measuring the proximity of known fluorescent groups within the protein.

(*v*) Fluorescent tubes are used for lighting purpose.

(*vi*) Fluorescent indicator luminol, is used for the detection of bloodstains, when applied on the blood, the reagent and the blood mixture glows in the dark, it is used for the detection of blood in the forensic science.

Phosphorescence : It has been found that in some substance, the emission of light continues, even when the source of radiation causing excitation is removed. *Such a phenomenon of delayed fluorescence is called phosphorescence.* It is to be noted that some fluorescent substance dissolving in boric acid or

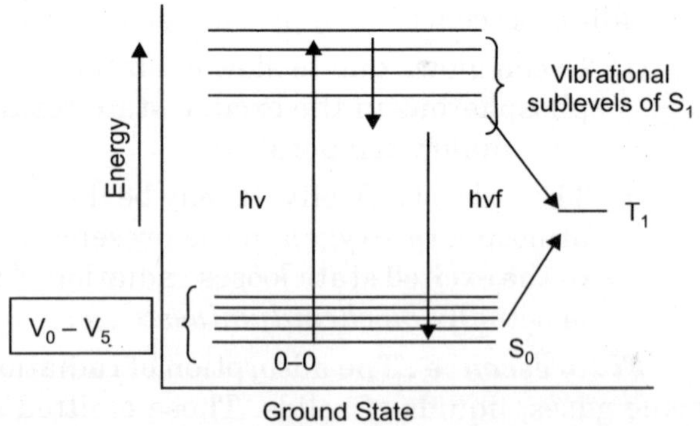

Fig. 26.4. The Phosphorescence Fluorescence Mechanism.

glycerol and then cooled becomes phosphorescent. The highly viscous substances or solids in which molecules have very slow movement are generally phosphorescent. Alkali metal sulphides containing some impurities are phosphorescent. (Fig. Fig. 26.4 showing the fluorescence phosphorescence mechanism).

Fluorescence	Phosphorescence
(*i*) Emission of radiation is rapid (decay time 10^{-6} to 10^{-8}).	(*i*) Emission of radiation takes place slowly (decay time 10^{-4} to 100s).
(*ii*) It is the radiation emitted when transition between two states of same .multiplicity takes place.	(*ii*) It is the radiation emitted when trans-mission between two states of different multiplicity takes place.
(*iii*) It can be observed in the solutions at room temperatures.	(*iii*) It cannot be observed in the solutions at room temperatures.

Photosynthesis and Its Mechanism

The formation of carbohydrates in nature occurs in green plants by the process, called photosynthesis. Plants contain the green pigment chlorophyll which catalyses the conversion of atmospheric carbon dioxide and water into carbohydrate. The reaction is thermodynamically unfavourable but proceeds because the necessary energy is supplied by sum-light. The reaction is very important since it produces us food and oxygen.

$$6CO_2 + 6H_2O \xrightarrow[\text{Chlorophyll}]{\text{Light}} C_6H_{12}O_6 + 6CO_2 + 283 \text{ kJ}$$

$$6nCO_2 + 6nH_2O \xrightarrow{\text{Light}} (C_6H_{10}O_5)$$

Carbohydrates form the major part of our diet. These functions as *biofuels* to produce energy for the functioning of living organisms. They slowly get oxidised in our body in the presence of enzymes to produce heat energy. This energy maintains the normal temperature of our body and supplies energy for doing work.

Mechanism

Carbon dioxide and water vapours present in the atmosphere are enable to absorb radiations emitted by the sun. However, chlorophyll (green colouring matter present in plant) absorbs the visible light energy form sunlight transfers it to carbon dioxide and water molecules, which then react photosynthesis, chlorophyll acts as a photosynthesiser.

$$\text{Chlorophyll} + h\nu \longrightarrow \underset{\text{(excited chlorophyll)}}{\text{Chlorophyll*}}$$

$$\text{Chlorophyll*} + 6nCO_2 + 6nH_2O \longrightarrow n(C_6H_{12}O_6) + 6nO_2 + \text{Chorophyll}$$

Thus, it involes the conversion of carbon dioxide and water into energy-rich carbohydrates. Photosyntheis is considered generally in two series of reactins i.e. *light reactions and dark reactions.* In light reactions chlorophyll absorbs solar energy which in term is grabbed by very efficient sub-cellular components of the plants called chloroplasts through a series of steps. The energy is utilized to synthsise energy rich ATP (Adenosine Triphosphate) molecules.

$$\text{Water} + \text{Carbon Dioxide} \xrightarrow{hv} \text{Organic Molecules} + \text{Heat} + \text{Oxygen} + \text{ATP}$$

In the reaction, the conversions of CO_2 into glucose and other carbohydrates occurs through a series of setps. The energy needed for the synthesis of one molecule of glucose, 18 molecules of ATP are used up.

Cell acquire chemical energy through photosynthesis and carry out the energy requering (endonergenic) process. Phototropic cells (plants algae and certain bacteria like halobacteria) syntheisise all nucleotides (nucleic acid) amino acids (protiens) sugar (carbohydrates) and the other nutrients are required for cell growth, maintanence and chemographs. These consume organic molecules (produced by phototrops) to get chemical energy needed for their cellular activities and break them to simple molecules.

Application of Photochemistry : The Photocells

Photoelectric cells generate electricity in response to the incidence of light or infrared or ultraviolet radiation. They serve to measure radiation intensifies and as light sensitive elementsin control devices.

Photoelectric cells are classified according to the basic of operation, into photoemmisive cells, photovoltaic cells and photoconductive cells.

(The operation of these cells will lead the syllabus of the book in a different track, so only gist is discussed)

Photoemissive Cells : They are also called phototubes consist of an envelop (partly or fully vacuumed) containing two electrodes one of which is photocathode, which is impregnated with a photosensitive substance, the envelop is transparent. The envelop being transparent, allows light to enter, which stimulates the electrons from the photocathode to come out and are collected by the anode, it is to be noted that the number of the electrons emitted is proportional to the radiation admitted (see Fig. 26.5).

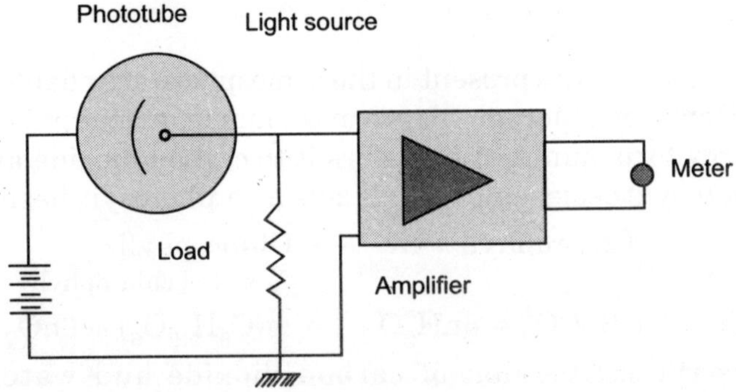

Fig. 26.5. The diagram of a photoemissive cell in a circuit.

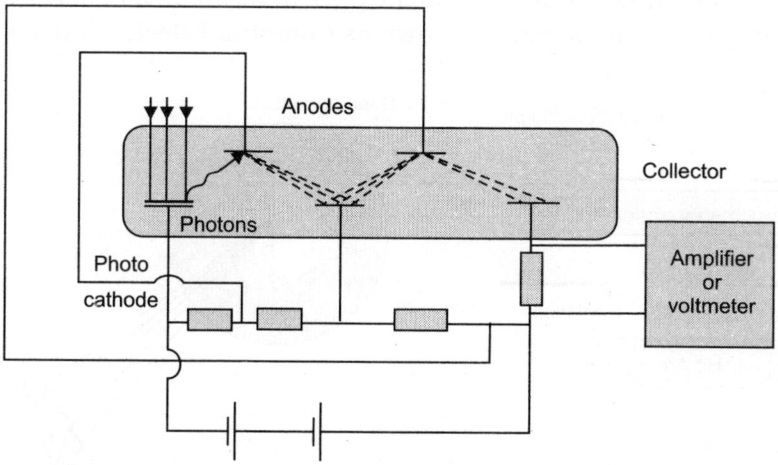

Fig. 26.6. The Photomultiplier Tube.

Photovoltaic Cell : Also known as barrier layer cells, photoelements, unlike photoemissive cells are generally employed without external voltage source. Light falling on the cell results in the flow of electrons resulting the potential difference between the two electrical conductors. If the terminals are connected by an electrical circuit, a current starts flowing in the circuit.

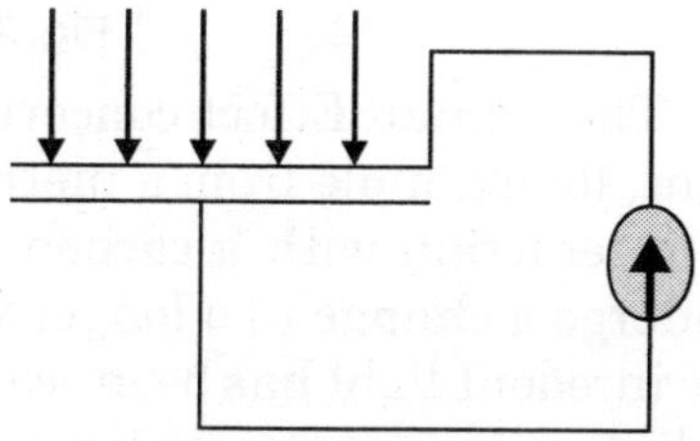

Fig. 26.7. The Diagram of a Photovoltaic Cell.

The most common form of the photovoltaic cell consists of a thin layer of semiconductor, such as selenium or cuprous oxide (see Fig. 26.7).

Photoconductive Cell : These cells contain layers of semi-conductor material whose conductance is increased by exposure of light.

Raman Effect

In March, 1928, *C.V. Raman* of University of Calcutta announced the discovery of a new phenomenon of light scattering, observed in various liquids and gases (9). In the same year, *Landsberg* and *Mandelstam* discovered the same effect in quartz. The possibility of the existance of this behaviour of light towards material substance had been predicted by *Smekal* in 1923, but theoritically.

It seems quite clear that Raman, who has studied 'light scattering" in general for may years, has the clearest grasp of the phenomenon and hence it carries its name, the Raman Effect. The Raman Effect is one of the powerful tool in physics and chemistry for the investigation of molecular structure. This type of information could therefore be obtained only from absorption and emission studies in the infrared region of the spectrum, but the discovery of Raman Effect makes it possible to carry out experiments in the vissible region.

This discovery compairs in scientific importance to the discovery of the non-classical x-ray scattering by A.H Compton, better known as Compton Effect, to effects are related in certain case.

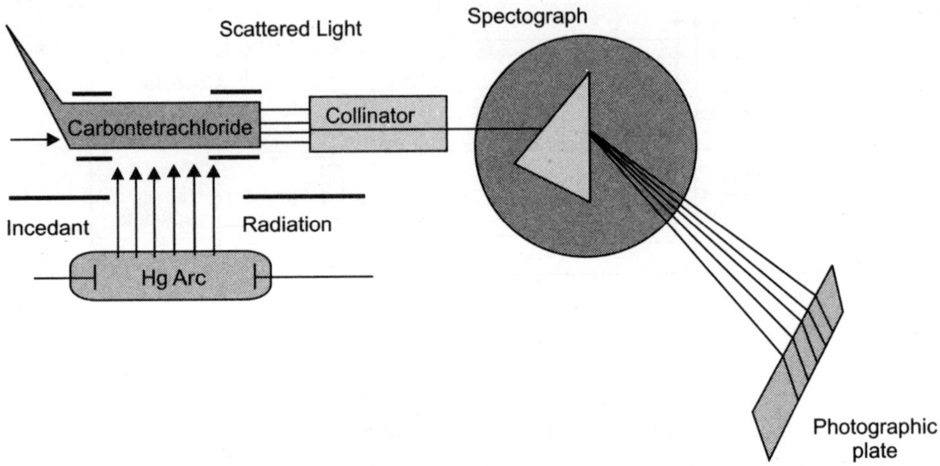

Fig. 26.8. The Experiment of Raman Effect.

The Raman Effect concerns the interaction of radiation and matter. Monochromatic light, for example from a mercury arc contains light of wave length 4358Å. It can happen by interaction with a carbon tetachloride molecule (as shown in the above experiment) undergo a change to a longer wavelength (toward the red of the spectrum). It is said that the incedent light has been scattered with a shift in wave length or frequency. The scattered radiation leaves the molecule in any direction in space. Usually it is observed at right angles to the direction.

So from this above expeiment Raman defined that:

The beam of light abosrbed in a longer wavelength, is scattred in a much shotter wavelength.

Here the white light having much longer wavelength, is absorbed and the red-light is of much shorter wavelegth is emitted. This also happens in the case of a clowdy sky, in the morning looks grey but the same sky looks red in the night. The nitrogen dioxide in the atmosphere absorbs the white light and thus looks black, and the same light is scattered in the night in the red region.

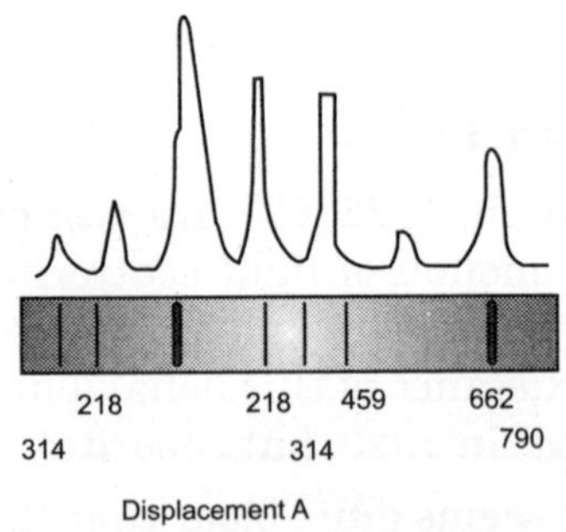

Fig. 26.9. The Raman spectrum of the Carbon tetra chloride.

Bose Einstein Condensate

From Quantum Mechanics we know that we have two types of particle one which obey Bose Einstein Statistics they are called *Bosons* and the other one is called *Fermions* which obey the Fermi Statistics.

Any particle has a fixed amount of energy may be different. Bosons can be dragged to the zero level energy, but Fermions cannot be brought to the zero energy level also more than one fermion cannot be placed in a single energy level if one electron is placed between two energy levels, there is no place to keep another electron. Therefore, in this case zero energy level can never be obtained. It is to be noted that quantum number of these particles have different quantum number their quantum energy is same. In this case the electron having coldest energy consist some energy. This is called zero point energy and it is very important.

This theory is only applicable to the fermions, but we can forcedly drag them to the zero energy level by producing a cold condensed state, where particles will take a very little place in the phase space.

According to Heisenberg's Uncertainty Principle which is $\Delta x \bullet \Delta y \geq h/4\pi$ where Δx is the uncertainty indetermination of position and the Δy is the determination of momentum.

Applying this principle we can determine the position of the quantum particles.

Here $\Delta x = 0$ $\therefore$ there is no uncertainty in the position of the quantum particle. We know that if a mathematical factor is multiplied with zero the product will be zero, here it is not possible because zero is getting multiplied to infinity because if the value of the position is zero [$\Delta x = 0$]. Then the momentum will be excessive and quantum number will be zero, which is again not possible, so it is clear that $\Delta x \neq 0$ but $\Delta x \rightarrow 0$.

The condensate conception has been developed in these theoretical expressions. According to Bose Einstein condensate when the gases are transferred to zero energy level, they produce a very cold gas at a temperature of 10^{-9} °K. In this state momentum of these particles are absolutely zero.

Now Momentum (ρ) = Energy of the Quantum particle /Velocity of the light in vacuum. Now if $\Delta p = 0$ then $\Delta x = \infty$. Which means that boson forms a cluster by overlapping on each other and making a honeybee comb. This cluster is so condensed that the bosons are not separable. This system is called Bose Einstein Condensate.

Through this system when light passes the velocity of the light slows down, because the refractive index of this medium is high. Now what is refractive index?

$$\text{Refractive index } (\mu) = \frac{\text{Velocity of Light in Vacuum}}{\text{Velocity of light in ordinary medium}}$$

$\therefore$ It is true that if the refractive index increases the velocity of the light decreases and if the refractive index decreases the velocity of the light increases.

Now we shall discuss about the decreasing velocity of the light, when this condensate was developed a laser beam was bombarded and the frequency was kept such that the laser beam did not came out from the condensate. Now it was seen that a few photons came from the condensate at a very slow speed. It was very surprising that what would be the cause. It was noticed that bosons in the condensate and the bosons present in the condensate did interact mechanically with each other and they did not get dissolved in the condensate. This can be easily understood by a simple example. Suppose in a fair some people get lost for a time but they usually meet after some times, this same thing happens to the photons.

It has been proved experimentally that after the development of Bose-Einstein Condensate it was possible to decrease the velocity of the light upto 32 miles per hour or 51 km/hr.

We shall now see what are the advantages of this condensate experiment; it will be possible to collect few photons, which will enable us to light up areas as much as we like. It will be possible to take very low light photographs, where there is a very less infra red radiation.

This Bose-Einstein Condensate was prepared by the mixture of sodium ion and photons. Though this condensate was developed a few years ago, but the phenomenon to decrease the velocity of light is entirely new invention.

CHAPTER

27

Eudiometry

Introduction

This involves the problems based on reactions taking place in gaseous state. These reactions are carried out in a graduated tube inverted over a trough of mercury and a spark plug of platinum is provided at the sealed end (see Fig. 27.1).

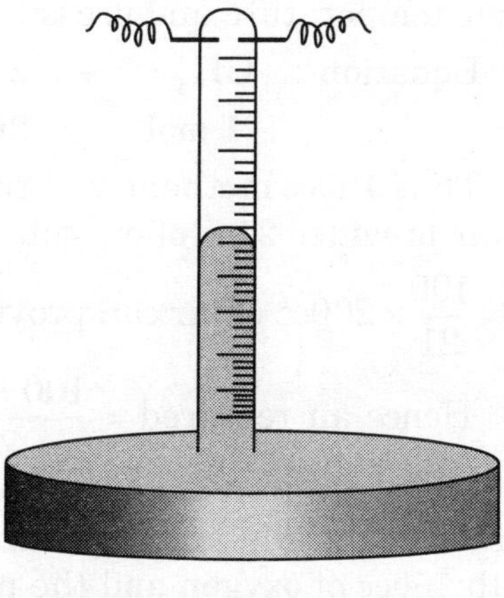

Fig. 27.1. The Eudiometry Apparatus.

Reaction is usually started by making electrical sparks in the substance filled in the tube, before sparking the initial level of the minuscus is noted. After sparking the tube is allowed to cool, and the level of the final mecury minuscus is noted. It may be recalled that :

1. Number of molecule of gases represent the number of volume (Avogadro's Law)

H_2	+	Cl_2	=	2HCl
One g molecule		One g molecule		2g molecules
i.e., 2 grams of		i.e 71 g (35.5 + 35.5)	or	
or		or		2 × 22.4 litres at N.T.P.
22.4 litres at N.T.P		2 × 22.4 litres at N.T.P. or same T and P or 1cc		or same T and P or 1cc

2. One gram molecule of any gas occupies is 22.4 litre at N.T.P.

Note: (*i*) When level of mercury in the eudiometry tube is same level inside, the pressure of closed gas is equal to the atmospheric pressure and the same can be read from the barometer.

(*ii*) A volume of a solid in liquid reactants or resultants is counted as zero. Thus ordinary volumes of steam condensed to form negligible volumes.

(*iii*) Volume of resultant gases may or may not be equal to the volume of the reacting gases.

(*a*) CH_4	+	$2O_2$	$\longrightarrow$	$CO_2 + 2H_2O$	
1 vol		2 vol	$\longrightarrow$	1 vol	2 vol [steam no change in vol]
1 vol		2vol		1 vol	water [molecules 2 vol]
(*b*) CO_2	+	C	$\longrightarrow$	2CO	
1 vol		solid	$\longrightarrow$	2 vol [expantion = 1 vol]	
(*c*) $2H_2S$	+	SO_2	$\longrightarrow$	$2H_2O + 3S$	
2 vols	+	1 vol	$\longrightarrow$	Water solid....contraction = sol 1	

PROBLEMS GASEOUS REACTIONS

Examples assuming that air contain 21% of oxygen by volume, what volume of air at 27°C and 755mm pressure will be required for complete combination of 100cc of methene at same temperature and pressure as air.

Equation : $CH_4 + 2O_2 \longrightarrow CO_2 + 2H_2O$

1 mol 2mol $\longrightarrow$ Avogrado's Law 1 vol to 2 vol

Thus 100cc methene will require 200cc of oxygen at 27°C and 755 mm pressure 100cc of air provides 21cc of oxygen.

$\frac{100}{21} \times 200$cc of air can provide 200cc of oxygen.

Hence air required = $\frac{100 \times 200}{21}$.

Example 2. A sample of coal gas contains 50% of hydrogen, 30% Marsh's gas, 14%. carbon monoxide and 60% ethylene, one hydred oxalic centimetre of the gas as mixed with 150cc of oxygen and the mixture exploded what will be the volume and composition of the resulting gas when cooled to the original temperature.

Volume of hydrogen in 100cc coal gas(50%) = 50cc

" " Marsh " " " (30%) = 30cc

" " Carbon dioxide " " " (14%) = 14cc

" " Ethyle " " " (60%) = 150cc

When such a gaseous mixture burns, reaction takes place which are as follows :

	$2H_2$	+	O_2		$2H_2O$
	2 moles	+	1mol		2 moles
Avogadros's Law	2 vols		1 vols		2 vols
	Or 2cc		1cc		water.

Thus 50cc of hydrogen will require 25cc of oxygen to yield a negligible amount of water. By similar arrangement from the eqations.

$$
\begin{array}{llllll}
CH_4 & + & 2O_2 & \longrightarrow & CO_2 & + & 2H_2O \\
30cc & & 60cc & \longrightarrow & 30cc & + & \text{water} \\
CO & + & 1O_2 & \longrightarrow & CO_2 & + & 2H_2O \\
14cc & & 7cc & \longrightarrow & 12cc & + & \text{water} \\
C_2H_4 & + & 3O_2 & \longrightarrow & 2CO_2 & + & 2H_2O \\
6cc & + & 18cc & \longrightarrow & 12cc & + & \text{water}
\end{array}
$$

∴ Total volume of oxygen used up = 25 + 60 + 7 + 18 = 110cc

The resulting mixture will contain, unused oxygen = 150 – 110 = 40cc

Carbon Dioxide = 30 + 14 + 12 = 56 cc

It will occupy volume = 40 + 56 = 96cc.

Problems of Mixture Analysis

Twenty volumes of a mixture of hydrogen gas and carbon dioxide was mixed with 13 volumes of oxygen and the mixture exploded. On cooling the resulting volume was found to be nine. Determine the composition of the mixture.

Let the mixture contain x cc of hydrogen and (20 - x)cc carbon monoxide

$$
\begin{array}{lllll}
\text{Equations :} & 2H_2 & + \; O_2 & \longrightarrow & 2H_2O \\
 & 2\text{ mol} & 1\text{mol} & & 2\text{mols} \\
\text{Avogadro's Law} & 2cc & + & & 1cc \text{ water} \\
 & xcc & + & & x/2 \text{ water}
\end{array}
$$

By same arrangement

$$
\begin{array}{llll}
2CO & + \; O_2 & \longrightarrow & 2CO_2 \\
2cc & + \; 1cc & \longrightarrow & 2cc
\end{array}
$$

or $$20 - xcc \; \frac{(20-x)}{2} cc \; (20-x)cc$$

Resulting volume (after cooling) will consist of CO_2 formed that is (20 – x) cc and

= 20 – x + 13 – x/2 – 10 + x/2 = 23 – x

Hence $23 - x = 9$ or $x = 14cc$

Thus original mixture contains H_2 = 14cc or , CO = 6cc.

Example: 20cc of a mixture of a carbon monoxide and ethylene were mixed with 50cc of oxygen and exploded. The volume after explosion was 45cc. After treatment with caustic soda solution only 15cc of oxygen where left behind. Calculate the volume of each gas in the mixture.

Let 20cc of the mixture contain xcc of carbon monoxide and 20 – xcc of ethylene

$$
\begin{array}{lllll}
\text{Equations} \therefore & 2CO & + \; O_2 & \longrightarrow & 2CO_2 \\
 & 2\text{ mols} & 1\text{mol} & & 2\text{mols} \\
\text{Avogadro's} & C_2H_4 & + \; 3CO_2 & \longrightarrow & 2CO_2 + H_2O \\
\text{Or} & 20 - x & 60 - 3x & & 40 - x
\end{array}
$$

Thus total oxygen used up $= (x/2 + 60 - 3x)\text{cc} = 50 - 15 = 35$ cc

Or, $x/2 + 60 - 3x = 35$

$x/2 - 3x + 60 = 35$

Or, $x - 3x/2 + 60 = 35$

Or, $5x/2 + 60 = 35$

Or, $5x/2 = 35 - 60$

Or, $5x/2 = 25 = 5x = 25 \times 2$ or $x = 10$

Example : How is ozone prepared? Describe its important chemical properties. When 100cc of a sample of ozonized oxygen where shaken with turpentine, the volume reduced to 84cc. When another 100cc from the sample was heated the volume increaed and cooled to original temperature the volume was 108cc. What conclusion do you draw about the composition of ozone from the above data?

Volume of ozone in 100cc of ozonized sample = volume absorbed by turpentines = 100 – 84 = 16cc.

On heating ozone decomposes to give oxygen, when 16cc ozone decomposes volume increases by 8cc that is 16c of ozone yield 24cc of oxygen.

16cc ozone yield 24cc of oxygen

or 2cc ozone yield 3 cc of oxygen

or 2 vol ozone yield 3 cc of oxygen

By Avogadro's Law

2 mol of ozone will yield 3 moles of oxygen.

Or, 2 moles of ozone contain 3 moles or 6 atoms oxygen, that is one molecule of ozone consists of three atoms of oxygen.

∴ The formula of ozone is O_3

Example: Hofmann's tube was fitted with chlorine and a few drops of strong ammonia where introduced rapidly from the upper top. After vigorous reaction subsided, the lower end is opened under the water. The residual gas nitrogen was found to occupy $\frac{1}{2}$ of the original volume. Use the above data to determine the empirical formula of ammonia.

Equation: - $2NH_3 + 3Cl_2 \longrightarrow 6HCl + N_2$

Since the hydrochloric acid is formed from equal volume of hydrogen and chlorine. It is evident that the amount of ammonia which yields.

(*i*) Hydrogen = volume of chlorine in the table also

(*ii*) Nitrogen = $\frac{1}{2}$ of the volume of chlorine

Thus ammonia gas is formed form hydrogen and nitrogen in the ratio 3:1 by volume

Or

$$\underset{1\text{vol}}{\text{Nitrogen}} + \underset{3\text{vol}}{\text{Hydrogen}} \longrightarrow \text{Ammonia}$$

By Avogadro's Law

$$1 \text{ mol} = 3 \text{ mol}$$
$$1 \text{ mol} = G \text{ mol}$$
$$1 \text{ atom} = 3 \text{ atoms}$$

Determination of the Formula of Gaseous Hydrocarbon

Procedure : A known volume of gaseous hydrocarbon is mixed with known excess of oxygen and sparked.

$$\text{Hydrocarbon + Oxygen} \longrightarrow \text{Carbon Dioxide + Steam}$$

While hot the resultant mixture contains CO_2 gas unused oxygen and steam. On cooling steam condenses to water, the residue is a mixture of CO_2 gas and gets absorbed leaving a residue of unused oxygen.

It is clear that :

1. (vol of hydrocarbon + vol of oxygen) – (vol of resulting gases while hot) = change in vol due to chemical reaction.
2. (vol of hydrocarbon + vol of oxygen) – (vol of resulting gases after cooling) = change in vol due to chemical reaction + Volume of steam that has condensed.
3. Construction on introducing alkali solution = Vol of carbon dioxide gas formed
   (*a*) Vol of gases before expansion (sparking) = vol of hydrocarbon + Vol of oxygen
   (*b*) Vol gases after expansion (hot) = volume of CO_2 + Vol of steam of volume of unused oxygen
   (*c*) Vol of gases after explosion (cooled) = vol of CO_2 + Vol of unused oxygen.

Example : 20cc of gaseous hydrocarbon were exploded with 100cc of oxygen. On cooling to the room temperature the volume of the gaseous mixture was found 70cc. Introduction of potassium hydroxide reduced the volume to 100cc the residual gas being only oxygen. Caculate the molecular formula of the hydrocarbon.

Let the formula by C_xH_y

Combustion of C_xH_y molecule reperesented by the equation

$$(C_xH_y) + (x + y/4)O_2 \longrightarrow xCO_2 + y/2O_2$$

$$\left\{\begin{array}{l} \text{Compare } C_3H_8 + 5O_2 \longrightarrow 3CO_2 + 4H_2O \\ \text{Or } C_3H_8 + (3 + H_2O) \longrightarrow 3CO_2 + RH_2O \end{array}\right\}$$

$$C_xH_y + (x + y)/4\ O_2 \longrightarrow xCO_2 + y/2\ H_2O$$
$$1 \text{ mol } x + y/4 \text{ mol} \longrightarrow x \text{ mols} + y/2 \text{ mols}$$

By Avogadro's Law

	1 vol	$x + y/4$ vols	x vols $y/2$ vols
or,	1cc	$x + y/4$cc	xcc $y/2$cc.

It is evident that if C_xH_y is the formula of the hydrocarbon it will require x + y/4cc of oxygen and will yield x cc Co_2 gas and y/2cc.

In the problem above

Volume of hydrocarbon = 2cc

Volume of oxygen taken = 100cc

Volume of unused oxygen = 10cc

Volume of used oxygen = 90cc

Volume of CO_2 formed + unused oxygen = vol after cooling = 70cc

Volume of CO_2 alone = construction = 70 – 10 = 60cc

Thus in this case 20cc of hydrocarbon needs 90cc oxygen and yields 60cc of CO_2 gas. Hence CO_2 formed 1cc hydrocarbon = 3cc = xcc.

Volume of oxygen used by 1cc hydrocarbon = 90/20 = $x + y/4$.

Or, 9/2 = 3 + $y/4$ or $y/4$ = G or y = 6

Hence hydrocarbon is C_3H_6.

Example : 10cc of hydrocarbon were named with 30cc of oxygen and exploded. There was a contraction of 20cc. After treatment with potassium hydroxide solution a further construction of 10cc took place. Determine the molecular formula of hydrocarbon.

Volume of mixture before explain = 10 + 30 = 40cc

Volume of unused oxygen = 20 – 10 = 10cc

Volume of used oxygen = 30 – 10 = 20cc.

Thus 10cc hydrogen needs 20cc oxygen and yields 10cc carbon dioxide or 1cc hydrocarbon will need 2cc oxygen and yield 1cc of CO_2 gas.

Let the formula of the hydrocarbon be C_xH_y

$$C_xH_y + (x + y)O_2 \longrightarrow xCO_2 + y/2\ H_2O$$

$$1\text{mol}\ (x + y/4)\ \text{mols} \longrightarrow x\ \text{moles}$$

By Avogadro's Law

1 vol $(x + y/4)$ vols $\quad$ x volumes

1cc $x + y/4$cc $\quad$ $x$cc

1cc 2cc $\quad$ 1cc

Evidently $x = 1$ $x + y/4 = 2$ or $y = 4$ $\therefore$ the Hydrocarbon is CH_4

Example : 20cc of a gaseous hydrocarbon are exploded with excess of oxygen. The contraction observed is 30cc. There is further contraction of 40cc on treatment with potassium hydroxide. What is the molecular formula of hydrocarbon?

Let C_xH_y be the formula of the hydrocarbon.

$$C_xH_y + (x + y/4)O_2 \longrightarrow xCO_2 + y/2\ H_2O$$

$$1\text{mol}\ x = y/4\text{cc} \longrightarrow 20\text{cc}\ 20y/2/2\text{cc}$$

Volume of CO_2 = Contraction on treatment with potassium hydroxide = 40cc or $20x = 40$

Or $x = 2$

Volume before expansion – volume after expansion on cooling = 30cc(given)

Or [20 + 20 $(x + y/4)$ + Surplus Oxygen –[$20x$ + Surplus oxygen] = 30

Or 20 + $2x$ + 5y –20 = 30 or 5y = 10 or y = 2 $\therefore$ the formula if the hydrocarbon is C_2H_2

EPITOME–1

Inert Gases and Uses

Introduction

The study of the annexure is to be done after the completion of reading of the original chapter the Inert gases and uses. In epitome we are going to discuss about the further advanced nature and properties of these inert gases.

Specific Heat of Inert Gases

In the case of inert gases we shall have to deal with two types of specific heat, one at C_v at a constant volume and the other one C_p constant pressure. The difference between the two corresponds to the internal work done when the gas expands from the Volume V_1 to volume V_2 at the pressure P.

$$C_p = C_v + \text{work done on expansion}$$
$$= C_v + P(V_2 - V_1)$$

If multiply the value of the specific heat (per gram) by the molecular weight of the substance, the product is known as the *molecular heat* and we shall represent them by C_p and C_v receptively. For a gram-molecule of a gas the expansion which takes place when the temperature is raised from 0°C to 1°C

$$\frac{22.4\times 1}{273} \text{ or } \frac{22400}{273}\text{cc}$$

Since the expansion has taken place at 1 atmospheric pressure (as then alone the volume 1g mol of the gas is 22.4 litres at 0°C), the work done is

$$P(V_2 - V_1) = \frac{76\times 13.6\times 981\times 22400}{273}\text{ergs}$$
$$= 8.31 \times 10^7 \text{ ergs or } 1.987 \text{ cal}$$

$$\therefore \quad C_v - C_p = 1.987 \text{ cal.}$$

Ordinarily when a gas is heated the energy taken up is utilized (*i*) to increase the kinetic energy of translation of the molecules (*ii*) to increase the rotational velocity of the molecules and (*iii*) to increase the motion of the atoms within the molecule.

In the case of monatomic gases, it is assumed that all heat up is converted into kinetic energy of transnational motion.

Since $P_v = m\theta^2 = \rho.K.E = RT.$

$$\text{K.E. of translation} = R.T$$

$\therefore$ C_v = the K.E of translation per 1°C for monoatomic gases it is $= R = \rho \times 1.98 = 2.98$ cal

Since $C_p = C_v = R$ and $C_v = 2.98$

$$C_v = 2.98 + 1.98 = 4.96 \text{ cal}$$

$\therefore$ The ratio is of $C_p/C_v = 4.96/2.98 = 1.667$.

This ratio is the value for a monoatomic gas.

But if the gas is diatomic, we shall have to consider the two degrees of freedom of rotational motion also, besides the three degrees of freedom of translation. For each degree of freedom, the energy shared is R for each degree rise of temperature.

$$\therefore \quad C_v = \text{K.E for translation + Rotational energy.}$$

$$= \text{3 degrees of freedom + 2 degrees of freedom}$$

$$= 5/2R = 5/2 \times 1.98 = 4.95$$

Since $C_p - C_v = 1.98$ and $C_v = 9.95$

$$\therefore \quad C_p = 1.98 + 4.95 = 6.93$$

$\therefore$ For diatomic gas, $C_p/C_v = \dfrac{6.93}{4.95} = 1.39$

Similarly, it has been found that the value of C_p/C_v for a triatomic gas is 1.33, and is much less for higher polyatomic molecules.

The ratio $C_p/C_v = \gamma$ is determined by measuring the velocity of sound in the gas. We know that the velocity, u of sound in cms is given by.

$$u = \sqrt{\gamma P/D}$$

Where P = Pressure is dynes per cm^2 and D is the density in grams per cc. For this, an experiment is set in the Kundt's tube. The following table shows some of the ratios of the gases.

Gas	$C_p/C_v = \gamma$
Monoatomic	
Helium	1.666
Argon	1.65

Gas	$C_P/C_v = \gamma$
Diatomic	
Oxygen	1.396
Nitrogen	1.405
Hydrogen	1.408
Carbon Monoxide	1.405
Triatomic	
Carbon dioxide	1.302
Nitrous Oxide	1.300
Polyatomic	
Ammonia	1.31
Ethylene	1.23

Some of the parameters of the inert gases have been given in the next table.

Liquid Helium		Liquid Argon		Solid Argon	
Temperature	Vapour Pressure	Temperature	Vapour Pressure	Temperature	Vapour Pressure
–268.71°C	760 mm	–195.7°C	215.0 mm	–207.51°C	21.97 mm
–269.03°C	565 mm	–190°C	985.8 mm	–202.51°C	59.26 mm
–270.66°C	51 mm	–187.5°C	633 .4 mm	–193.99°C	255.6 mm
–271.63°C	3 mm	–185.2°C	821.6 mm	–189.07°C	512.17 mm
		–150.9°C	10313 mm		
		–136.2°C	20700 mm		
		–117°C	40200 mm		

The constants a and b in the Van der Waal equation of state ($P + a/V^2$) (v – b) = RT, for the rare gases are.

	Helium	Argon	Krypton	Xenon	Radon
a	0.046898	0.002671	0.009633	0.008149	0.012191
b	0.001063	0.001435	0.001778	0.002277	0.002768
Compressibility					
	+0.001	0.000841	-0.00218	-0.00592	-0.09815

Two Forms of Liquid Helium, Helium I and Helium II

In the year 1927 Keesom and Woolfake made a very important discovery about liquid helium. They write :

"When measuring the dielectric constant of the liquid helium between the boiling points and 1.9°K, we observed that a temperature almost corresponding with that.

Kammerlingh Onnes has found a maximum in the density curve; the di-electric constant shows a sudden jump, or at least a jump made in a small temperature region. The thought suggested itself that, at that temperature the liquid helium transforms itself into another phase, which is liquid as well. If we call the liquid which is stable at the higher temperature, as liquid helium I and the liquid helium which is stable at the lower temperature as liquid helium (*ii*), then the dielectric constant of the liquid helium (*i*) should be greater than helium (*ii*)."

Further they say, "the liquid helium ii compared with liquid helium (*i*) has (*a*) smaller density (*b*) a greater heat of vapourization (*c*) and a smaller Surface tension.

The Transformation of liquid helium II to liquid helium I takes place within absorption of heat, of which the amount can be valued at 0.13 cal/g. The temperature at which the transformation takes place is 2.186°K. This is known as *lambda point.*

If one watches the liquid helium as the vapour is pumped away, ordinary ebullition is obtained until the λ-point is reached. Then as a complete evaporation is continued, there is a complete cessation of the appearance of boiling and the liquid becomes perfectly quiescent, although, the rapid evaporation and the consequent lowering of the temperature continue quite regularly. No bubbles are formed in the body of the liquid and the surface cannot be seen unless it is illuminated obliquely, so that it reflects light.

Physical Properties

At the translation point (the λ point) certain classical characteristics show charges discontinuity; for example, specific heat, viscosity thermal conductivity, compressibility and velocity of sound and velocity of sound and surface tension.

(*i*) λ-point is marked with a very steep maximum in the specific heat curve.

(*ii*) The viscosity particularities are very significant. Wilhelm, Misener, did the first measurements in these connections and Clark (1935) A metal cylinder suspended by a wire was immersed in the liquid helium in the cryostat and set into torsional vibration. Values of the viscosity were divided from the rate of decay of oscillations. This experiment indicated a certain change in the viscosity at the λ-point, the value for helium II being only about one tenth of that for helium I. Just below the λ-point. The viscosity was found to be in the order of 10^{-5} cgs units.

Temperature		Viscosity (Keesom and Macwood)
He I	4.0°K	29×10^{6}
	3.0°K	25
	2.3°K	19
He II	2.16°K	16
	1.72°K	3.8
	1.32°K	1.9

In 1838, Allen Misener and Kapitza independently measured the rate of flow of liquid helium II through very fine aperture and made the extraordinary discovery that the gently flowing was particularly independent of the pressure and corresponded to an estimated viscosity of the order of 10^{-9} cgs units. In addition, three other extraordinary properties of helium II were discovered.

(*a*) The fact that a film of the liquid will creep over any solid surface with which it is in contact.

(*b*) The extraordinary conduction of heat.

(*c*) The "fountain effect", a material flow of the liquid induced by a temperature gradient. To demonstrate, Daunt and Mendelssohn took a small tube closed at the bottom (see Fig. 1 (A)) and dipped the closed end into the liquid helium II. It soon filled to the same level inside and out, when the partly filled tube was raised it emptied again. When the tube was raised completely out of the helium drops could be seen gathering on the bottom and falling off (see Fig. 1 (B)).

(*iii*) The first measurements of the thermal conductivity of liquid helium was made by Keesom (1935), who found the extraordinary value of 10 cgs units in helium II a little below the lambda point. That is, liquid helium II appeared to conduct over a hundred times more conducting material of heat (at ordinary temperature). The abnormal conduction disappears at about 0.5°K. Interesting experiments have been desired on the basis of the peculiar heat conduction. For example, Allen and Jones used a tube (Fig. 1 (C)) with the fine tube at the bottom filled with emery powder. When light was focussed on this tube, it was observed that a jet of liquid helium came out of the tube taking a height of several centimetres of height, just like a fountain. Emery powder in fact absorbed the energy, which was abnormally conducted by liquid helium thus expanding the liquid enormously.

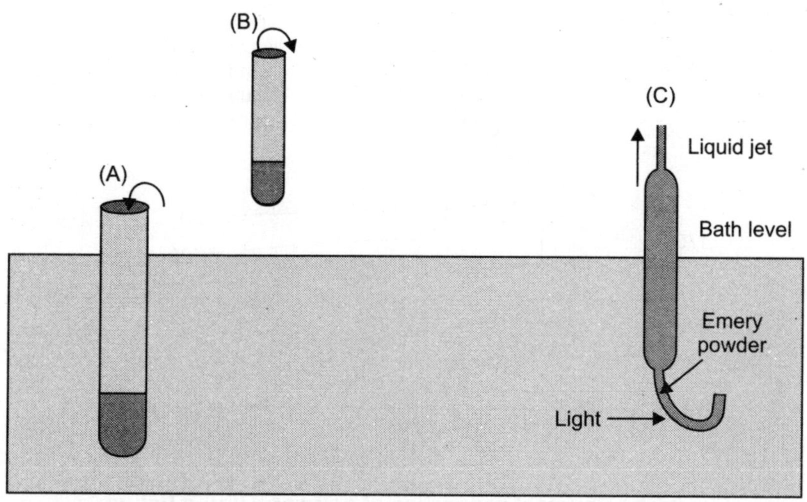

Fig. 1. The Transparent Phenomenon of Liquid Helium II

(*iv*) The Isothermal compressibility of liquid and solid helium has been studied by Keesom and Clusius.

$$K_T = 1/V\ (dV/dP)_T$$

For the solid, the value of K = 0.0015 per atmosphere at 3.7°K and 115 atm. The compressibility of a liquid can also be deduced from measurements of the velocity of sound by the relation

$$V = 2\pi\sqrt{K/P}$$

The ultrasonic waves were used and the velocity measurement thus gave the adiabatic compressibility.

$$K_3 = 1/V\ (dV/dP)_S$$

It must be noted that $K_T/K_S = \gamma$, where, γ is the ratio of specific heats. At the λ-points the velocity of sound shows a distinct discontinuity.

(*v*) The surface tension of liquid helium is extremely small so small that there is no visible curvature when the liquid meets the wall of any container. Urk, Keesom and Onnes measured the surface tension of the liquid helium over a wide range of temperature. At λ-point there was a slight discontinuity.

At lambda point there is a distinct change in temperature co-efficient of such physical properties as vapour pressure, density (see Fig. 2) internal energy. Dielectric constant refractive index and latent heat of fusion and vapourization. However neither the discontinuity or any change at the lambda point is observable in electrical polarization, molecular refractivity, and electrical conductivity, scattering of light and X -Ray structure.

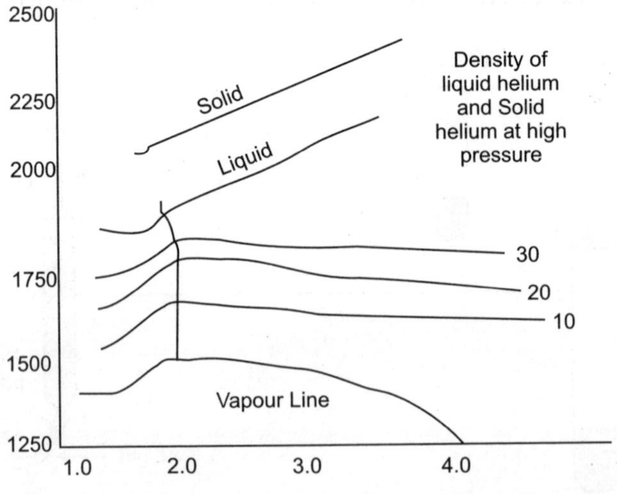

Fig. 2. The Physical properties of solid and liquid helium.

The Phase Diagram

Ones cooled liquid helium down to a very low temperature, 0.71°K, but still there was no evidence of its solidification. It was thus clear that simply lowering the temperature to any extent couldn't solidify liquid helium. In 1926, Keesom tried the experiment of applying a pressure to liquid helium. When a small bomb containing helium at a pressure of 250 atm was cooled in the helium cryostat, the connections at it were blocked at all temperatures below 4.2°K. Examinations soon proved that elusive solid state of the helium has been discovered and the pressure required solidifying liquid helium at different temperatures were measured. What ever may be the lowest value of the liquid helium it cannot be solidified below 25-atm pressure.

As the pressure increases, the melting point of solid helium also increases.

Pressure	Melting Point of Solid Helium (°K)
25 atm	1.2
50 atm	2.3
75 atm	2.95
100 atm	3.3
125 atm	3.9
600 kg/cm^2	10
1800 k/cm^2	20
3250 kg/cm^2	30
5200 kg/cm^2	40
5800 kg/cm^2	42

The table above shows that the helium gas can be solidified even at the temperature of 40°K at which it does not exist as liquid, provided we apply an enormous pressure more than 5000kg/cm^2. It then corresponds to the sublimation of a solid (like iodine) which vapourizes without passing to any liquid phase. The phase diagram is shown in Fig. 3.

The diagram (Fig. 3) shows the change of vapour pressure with temperature for liquid helium II (CA in graph and for liquid helium I (AD in graph). A is transition point or lambda point in the graph, for the two types of liquid helium. The curve AB indicates the influence pressure on lambda point. The curve EBF shows the pressure at which liquid helium can be solidified at different temperature. Helium II is depicted as extending towards the absolute zero. Direct equilibrium between the solid and vapour phases occur only at temperature above the critical temperature.

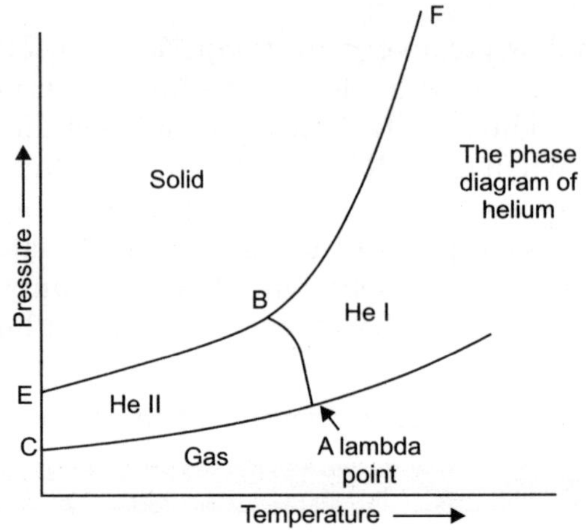

Fig. 3. The Phase Diagram of Helium.

Isotopes of Helium

Four Isotopes of helium are known they are: · ^4He, ^3He, and ^5He.

The natural helium is almost wholly ^4He. ^3He occurs in the atmosphere to an extent of 10^{-7} and is artificially produced when deuterons bombards deuterium.

$$^2D_1 + {}^2D_1 \longrightarrow {}^3He_2 + {}^1n_0$$

^5He is rather unstable and is produced when heavy paraffin is bombarded with alpha particle.

$$^2D_1 + {}^4He_2 \longrightarrow {}^5He_2 + {}^1H_1$$

and also from the bombardment of lithium with swift deuterons.

$$^7Li_3 + {}^2D_1 \longrightarrow {}^4He_2 + {}^5He_2$$

Compounds of Helium

(i) Absorption in Water : Ramsay (1895) gave 0.0073 for the co-efficient of absorption of helium in water at 18.2°C and 0.0394 for argon at 12°C. Lord Rayleigh found that 100cc of water dissolves 4.05 cc of argon at 13.9°C. The solubilities of rare gases in water are given below in the table.

Temperature	Helium	Neon	Argon	Krypton	Xenon
0°C	0.0134	0.0114	0.0561	0.1027	0.2189
10°C	0.01	0.0118	0.0438	0.0921	0.1500
20°C	0.0138	0.0147	0.0379	0.0729	0.1109
30°C	0.0161	0.0158	0.0348	0.0679	0.0990
40°C	0.0191	0.0203	0.0338	0.0650	0.0812
50°C	0.0226	0.317	0.0343	0.0716	0.0878

From the above table it would be seen that minima occur at 10°C for helium 0°C for neon, 40°C for argon, 35°C for krypton and 40°C for xenon. Villard (1896) considered that an *Argon hydrate* existed with a dissociation pressure of 150 atm at 0°C and 210 atm at 8°C. R de Forcrand in 1902 gave A.xH_2O for the formula of argon hydrate at 8°C. Where *n* has the value of 5.14 – 5.40. He also reported about krypton hydrate, Kr 5.08 H_2O (vapour pressure 1.45 atm at 0°C). The dissociation pressure of xenon hydrate in 1.45 atm at 0°C. No neon hydrate was formed at 260 atm pressure and 0°C. The dissociation pressure of the hydrates of Kr 5H_2O and A.5(or 6) H_2O were dissociated by Tammann and Kridge in 1925 also.

Crystals of the hydrate of radon, the existence of which was indirectly proved by Nikitin (1936) can be obtained by passing radon through water at 0°C dissociation pressure being less than atmosphere at 0°C.

Similar to hydrates of radon, some deuterohydrates, G. y D_2O, are also known, where y is near about 6, and krypton deuterohydrate has a dissociation pressure of 1 atmosphere at –25.1°C and neon deuterohydrate 1 atm at –3.2°C. Since it is difficult to polarize small molecules of neon and helium it has not yet been possible to prepare to the hydrate or deuterohydrate from them.

The hydrate of inert gas are said to be formed by introducing water vapour into an atmosphere of the gas at low temperatures The solid crystalline bodies are formed, which under atmospheric pressure decomposes according to the list below:

Argon Hydrate	–29.8°C
Krypton Hydrate	– 39.2°C
Xenon Hydrate	0°C

Xenon hydrate is stable at +23.5°C under 23.5 atm pressure.

(ii) Absorption in other Substance: Liquid oxygen at –205°C dissolves one part of helium in 80 parts. Neon is less soluble than argon in oxygen. Helium is insoluble in benzene and alcohol (Ramsay). Argon is not absorbed by compounds of the fatty series but the members of the aromatic series absorb the gas, benzene absorbs 8% by volume and aniline 1%. Krypton and xenon are slightly soluble in pentane and acetic acid. One litre of blood absorbs 25.3 cc of argon.

The inert gas form definite solid addition compounds with a number of organic substances as quinol $C_6H_4(OH)_2$. When quinol is crystallized from solution in the presence of an inert gas (argon, krypton or xenon under high pressure), the small molecules are incorporated in a cage like structure. With argon, the compound obtained has a formula A, 3$C_6H_6O_2$, but may loose argon, if not all the available crystallographic positions are occupied. The compound is a solid with no significant dissociation pressure of argon. When the crystals are heated, the cage structure is broken down and the gas is dissolved. The same thing happens when the crystals are dissolved in alcohol.

The corresponding krypton and neon compounds (also known as caltherite compounds, where one component is trapped in the crystal lattice.) contain 15.8% and 26% of the inert gas by weight. The quinol caltherite have the general formula $3C_6H_6O_2$ X (X = any rare gas).

(iii) Occlusion by Substance : There has been a considerable dissociation on the occlusion by the glass of the electrical discharge tube. Plüker noted that when a discharge tube has been running for some time, the gas pressure becomes progressively lower with the continuations of the discharge. This may be partly due to the absorption of helium by the cathode metal also (Brodetsky and Hodgson). The chemical action between glass and the gas may also be probable. Ramsay found helium, a trace of neon and hydrogen in the glass of an old X-Ray tube. Goldsmith found that high speed molecules of helium and hydrogen could penetrate a mica sheet 0.001-0.006 mm, thick, under conditions where air argon or carbon dioxide cannot penetrate. Claude noted that neon, gets occluded in the metal spluttered* from the metal cathode of a discharge tube. Argon is not absorbed by platinum black.

According to some reports, argon hydrogen helium and neon are able to pass through glass, but they are contradictory.

(iv) Hydrides of Helium : Ramsay and Callie (1896) postulated that if all the combination between the rare gasses and other elements were probable, the reactions will be endothermal; but no method known for the preparation which would be successful for the endothermal preparation of rare gases.

It has been emphasized that the chemical affinity of the rare gases is not necessarily zero under all conditions. Helium has two electrons in the K-shell and when bombarded by cathode rays, one of these might be knocked off (when a new electron darts back, it might not get into the K-Shell at once, but be less firmly bound in some metastable position farther out). Under these conditions, helium might appear to act like an atom of hydrogen. J.J Thomson observed particles of mass 5 in the gas, which has been in contact with radium chloride for 13 years. Aston observed particles of mass 5 in his examination of helium. It was supposed that these particles are molecules of helium hydride HeH.

Hydrogen compounds, HeH^+ and HeH_2^+, can be detected by the mass spectrographs, but their life period is only of the order of 10^{-8} seconds. The diatomic helium molecule He_2 can be compared with the halides, it existence shown by the band spectrum of helium.

(v) Mercury Halides : It is said that the vapour density of zinc in argon is 12% higher than in nitrogen. It was concluded from that zinc and argon have a distinct tendency to form a compound. This tendency is found to exist in the case of cadmium and helium, mercury and argon, mercury and helium, selenium and argon (slight) selenium and helium (slight).

Manely reported that in presence of an electric glow discharge, mercury and helium could be made to combine, forming mercury decahelide $HgHe_{10}$ as a stable substance, which decomposes at a bright red heat. The formation of the helide is a companied by a

---

*Spluttered: - Making a noise and leave.

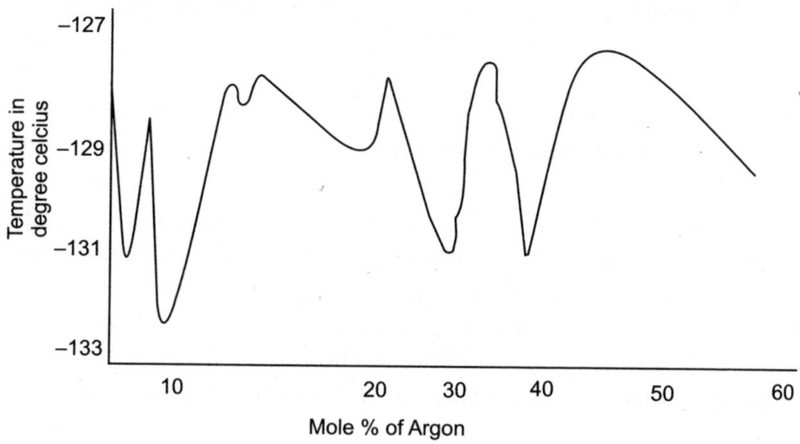

Fig. 4. The Freezing points in the system of Argon-Boron Trifluoride.

(*f*) Helium arc lights give an intense red and yellow light. In Geissler Tube helium works as the light in the spectrophotometer.

(*g*) Helium may also be used in metallurgy when inert atmosphere is necessary. Since it is insoluble in molten metals, this is an additional advantage.

(*h*) Helium thermometers are used to read very low temperatures.

(ii) Neon : Its main use in the lighting devices, especially those used for decorative and advertising purposes. The tube used for such purposes is filled with helium, neon or argon or mixture of these gases. When an electrical discharge passes through the tube, electrons are given off and a glow is obtained. The glass tubes are also sometimes coloured. The glass tubes are also sometimes coloured. The neon sign also depends upon the type of the electrode used; usually then electrodes are of copper and various alloys (containing thallium 82% and cadmium 18% for example). The signs may operate as hot or cold cathode lamps. Various colours are obtained in these signs by the use of different gas mixtures.

Gas	Glass	Pressure	Colour
Helium	Clear	3 - mm	White
Helium	Noviol	3 - 4 mm	Yellow
Neon	Clear	10 - 18 mm	Red
Neon	Soft red	19 - 18 mm	Deep-red

Usually a fluorescent material is also introduced in the tube to make the signs more attractive. The intense red light of neon lamp is of great value as beacon lights in aviation and marine services, especially in stormy and fogy weather. Neon lamps are also used in green houses. The light is effective in the formation of chlorophyll.

(iii) Argon : Its best use is in the filling of incandescent bulbs, because of its low thermal conductivity, its inertness and its density which holds in check the vapourization of the tungsten filament and thus prolongs its life. Sometimes, argon is mixed with 10-25% nitrogen for this purpose. Along with other rare gases, it may be used in advertising tubes for the blue and green display. It has been used in tungsten rectifier, which uses tungsten filament in argon bulb. Argon is also used for sometimes with oxygen as an atmosphere for arc welding, because its stability allows the use of a lower voltage in the arc.

(iv) Krypton : The very little availability of this gas imposes restriction on its wide use. To some extent krypton mixed with other gases may be used for display tube lighting. The mixture of krypton and neon has been tried in incandescent lamps. When such lamps are subjected to over voltage they could be run for several hours and developed inappreciable heating effects only. Under similar conditions, the lamps filled with nitrogen and argon lasted for few minutes only. It is said that 25W lamp filled with krypton or neon gave a 33% greater yield.

In the cosmic ray measurements, a krypton ionization chamber has been found to be 40 times as efficient as when air is used.

(v) Xenon : Xenon is not much commercially used except in gas mixtures for tube lighting as mentioned in the use of krypton.

It's higher boiling points, low ionization voltage and its high electrical conductivity these physical characteristics could have been usefully availed, has been a good supply of gas at a low cost. In research laboratories, where cost is immaterial it finds its little use.

Estimation of Rare Gases

Helium : *(i) Thermal Conductivity Method :* It is estimated as a stream of gas flowing through as automatic device developed by the U.S. Bureau of Standards. The thermal conductivity of the gas stream during its flow is regularly and continuously recorded. This gives fluctuations if there is any impurity in the gas. If the mixture contains only two gases the estimations are reliable. The thermal conductivity method can measure helium up to even 10^{-15} to 10^{-8} cc.

(ii) Gas Flow Method through an Aperture : A 20-30% quantity of helium in a mixture can be determined by staying the flow through a minute hole in a piece of platinum foil. The instrument is calibrated with respect to a pure sample of nitrogen. This experiment takes only 2 to 3 minutes. If the mixture contains only one other gas, the accuracy is 1 to 2%.

(iii) Panneth's Method : The apparatus contains a heated wire manometer and a compensating manometer as two arms of a whetstone bridge. This method determines the composition of He-Ne mixture at pressures down to 6×10^{-6} mm with a accuracy of 2%.

Argon : This gas is quantitatively estimated by its di-electric cohesion, since 1% of any diatomic gas increases the constant by more than two hundred percent.

Neon mixtures, if quite simple, may be assayed by the dielectric cohesion. The differences in the specific gravities of helium and neon may also provide a method for estimating neon in a mixture with helium.

Ionization Energy

The energy expanded in transferring an element from the normal state (the electron in the intermost available orbits) is known as the ionization energy of the atom. The ionization may proceed through a single stage or multiple stages.

$$He \longrightarrow He^{+} + e \text{ (first stage)}$$

$$He^{+} \longrightarrow He^{++} + e \text{ (second stage)}$$

Let us take helium gas in a tube and introduce into it electrons of low energy obtained from an incandescent wire, moving under a definite potential difference with a defined velocity. If e is the charge of electron and v the accelerating potential, m the mass of the electron and v the velocity of electron, the energy $mv^2 = eV$. It is measured in electron volts (eV).

$$1eV = 1.602 \times 10^{-12} \text{ erg} = 3.825 \times 10^{-20} \text{ cal}$$

Let us gradually increase the applied voltage. We shall find that the current first gradually uses, but suddenly at a potential of 19.75 volts, its practically fall to zero. We know that the current strength depends on (i) the number of electrons and (ii) their velocity so long as the temperature of the filament is constant. The number of electrons emitted is constant and so the initial increase in the current is due to the increase in the velocity of electrons as the potential is used. As soon as the field at the 19,75 volts is reached, the electrons completely loose their energy they give up their entire energy to the helium atom with which they collide. If the potential is increased, the current once more rises, corresponding to the excess of voltage over the critical value 19.75 volts. A second break now appears at 10.55 volts. Two more breaks are observed in the case of helium at 21.2 and 22.9 volts.

We must remember that so long as the energy of the colliding electron is insufficient to lift one of the electrons of the atom into a higher orbit, the electron is reflected according to the laws of electric collision. The voltage, which is necessary in order to excite an atom in the manner indicated, so that a certain spectral line is emitted, is called the existing potential of the line in question. The potential, which is required in order to impart to the colliding electron so great a velocity that it completely removes an electron from the atom and so ionizes it is called the ionization potential. The first ionization potential of helium is 24.48 volts and the second is 54.14 volts. The total energy for the $He \longrightarrow He^{2+}$ conversion is $24.48 + 4.14 = 78.62$ electron volts. From one mole of helium gas to one-gram ion of helium, one would need

$$78.62 \times 3.8285 \times 10^{-20} \times 6.0238 \times 10^{23} \text{ cal} = 1.813.3 \text{ kilo cal of energy.}$$

Ionization potential of elements from hydrogen to argon

Atomic number	Symbol	Valency Configuration	Ionization potential, eV							
			I	II	III	IV	V	VI	VII	VIII
1	H	$1s^2$	13.95							
2	He	$1s^2$	24.48	54.14						
3	Li	$2s^2$	5.390	75.6193	122.420					
4	Be	$2s^2$	9.320	18.206	153.850	217.657				
5	B	$2s^2.2p^1$	8.296	25.149	37.920	259.298	340.127			
6	C	$2s^2.2p^2$	11.264	24.376	47.864	64.476	391.986	489.84		
7	N	$2s^2.2p^3$	14.54	29.605	47.426	77.450	97.863	551.925	666.83	
8	O	$2s^2.2p^4$	13.614	35.146	54.934	77.394	113.873	138.080	739.114	871.12
9	F	$2s^2.2p^5$	17.42	34.98	62.646	87.23	114.214	157.117	185.139	953.60
10	Ne	$2s^2.2p^6$	21.559	41.07	64	97.16	126.4	157.91		
11	Na	$3s^1$	5.138	47.29	71.65	98.88	138.60	172.36	208.44	264.155
12	Mg	$3s^2$	7.644	15.03	80.12	109.29	141.23	186.86	225.31	265.97
13	Al	$3s^2.3p^1$	5.984	18.823	28.44	119.96	153.77	190.42	241.93	285.13
14	Si	$3s^2.3p^2$	8.149	16.34	33.46	45.13	166.73	205.11	246.41	303.87
15	P	$3s^2.3p^3$	11.0	19.65	30.156	51.354	65.007	220.414	263.31	309.26
16	S	$3s^2.3p^4$	10.357	23.4	35.0	47.29	72.5	88.029	280.99	328.80
17	Cl	$3s^2.3p^5$	13.01	23.80	39.90	53.5	67.80	96.7	114.27	348.3
18	A	$3s^2.3p^6$	15.755	27.62	40.90	59.79	75.0	91.3	124.0	143.346

EPITOME–2

Radioactive Chemistry

Introduction

In this annexure we are now going to discuss more about the trans-uranium elements, the Actinides and the Lanthanides.

La^{57} | Lanthanide series 58-71 | Ha^{72}

The electronic configuration of the lanthanum 57 is

$1s^2.2s^2.2p^6.3s^2.3p^6.3d^{10}.4s^2.4p^6.4d^{10}.5s^2.5p^6.5d.6s^2$

After lanthanum, we have what are known as the 14 more rare earth element, which have also the same configuration with only difference is that 4f shell is successively built up. The rare earths following lanthanum are known to constitute the lanthanide series, which has 4 elements, thus placed between La^{57} and Hf^{72}

Actinides : These elements are found between the 90 and 98 and are called actinides. The configuration of actinium (89) is also similar to that of lanthanum.

Ac (89) - $1s^2.2s^2.2p^6.3s^2.3p^6.3d^{10}.4s^2.4p^6.4d^{10}.5s^2.5p^6.5d.6s^2$

$6s^2.6p^6\ 6d\ 7s^2$

After actinium, the building of 5f shell states that (of to 4f shell in rare earths) and it would be complete only when 14 electrons have been added to this shell. Thus we shall have after actinium, a series, homologous to the Lanthanides which we shall cal as actinide series.

Thorium (90) is the first member d this series and protactinium (91), uranium (92) and the other so named as the transuranium elements really belong to this actinide series.

Ac^{89} | Actinide Series 90-103 | Ku^{104}

All the elements of this series are now almost known. The artificial methods have given us element up to nobelium 102.

There is a resemblance between the properties of the elements of the rare earth series (lanthanide series) and the trans-uranium elements (actinide series).

The electronic configurations are fairly comparable

Lanthanides	Actinides
$4f^5$ Promethium (61)	$5f^4$ Neptunium (93)
$4f^8$ Samarium (62)	$5f^5$ 6d Plutonium (94)
$4f^7$ Europium (63)	$5f^6$ Americium (95)
$4f^7$.5d Gadolinium (64)	$5f^7$ 6d Curium (96)
$4f^8$5d Terbium (65)	$5f^8$6d Berkelium (97)
$4f^9$5d Dysprosium (66)	$5f^9$6d Californium (98)
$4f^{10}$5d Holmium (67)	$5f^{10}$6d Einsteinium (99)
$4f^{11}$5d Erbium (68)	$5f^{11}$ 6d Fermium (100)
$4f^{12}$5d Thulium (69)	$5f^{12}$ 6d Mendelevium (101)
$4f^{13}$5d Ytterbium	$5f^{13}$6d Nobelium (102)

We may compare for the trivalent ion (not always releasable). The number *n* of electrons in the penultimate quantum group, that is in the fourth shell of the lanthanides and the fifth shell of the actinides, assuming that there are 8 electrons in the outer most shell in the series.

Lanthanides Actinides Number *n*	57 La 89 Ac 18	58 Ge 91 Th 19	59 Pr 91 Pa 20	60 Nd 92 U 21	61 Pm 93 N p22
Lanthanides Actinides Number *n*	62 Sa 94 Pu 23	63 Eu 95 Am 24	64 Gd 93 Cm 25	65 Tb 97 Bk 26	66 Dy 98 Cr 27
Lanthanides Actinides Number *n*	67 Ho 99 Es 28	68 Eb 100 Fm 29	69 Tm 101 Md 30	70 Yb 102 No 31	71 Lu 103 Ku 32

Just as the number of the rare earths have a close similarity amongst each other, similarly there is the similarity amongst the members of the actinide series also. We shall group out the striking features as follows :

In reference to especially 5 elements U Np Pu Am and Cm

All the five elements exist in the trivalent state (this we shall call (III) oxidation state) and in this respect are similar to the rare earths. As the atomic number increases, the relative stability of this (III) oxidation state with reference to the other states increases. Thus in Uranium U^{+++} ion in solution is oxidised by air, but the Cm^{+++} ion is very stable. Infact, Curium, compounds have been obtained in the trivalent state only.

In the actinide series, the trihalides of MX_3 form (M= metal and X = halogen) are all isomorphous with one another also isomorphous with the halides of actinium.

Uranium, neptunium, plutonium and americium form compounds of the (iv) state (i.e tetravalent) and in the case of neptunium and plutonium, these tetravalent compounds are the most stable. The tetravalent halides MX_4, are isomorphous with the halides of actinium, it is also isomorphous with thorium tetrachloride.

The (v) oxidation state has been observed in solution of compounds in this state is unstable, only neptunium (v) state compounds in this state most stable.

Uranium, neptunium and plutonium from salts in (vi) state also uranyl UO_2^{++}, neptunyl NpO_2^{++} and plutonyl PuO_2^{++} of these uranyl salts are more stable. These compounds having is general the ions MO_2^{++} having the same crystalline structure.

The Table given below shows the relative stabilities of the various oxidation states (stability proportional to the number of stars).

Valency	Uranium	Neptunium	Plutonium	Americium	Curium	Berkelium
III (M_2O_3)	*	**	***	***	****	*****
(IV) (MO_2)	***	****	****	*		**
(v) (M_2O_5)	**	*	*	*		
(vi) (MO_3)	****	**	**			

In 1923 Bohr postulated the existence of an actinide series homologous to the rare earths in the light of his atomic structure and periodicity. G.T Healberg, revived this term (the names lanthanide and actinide were covered in the fact by V.M Goldschmidt in 1925). The parallelism between the two series is quite marked especially as regard the (III) oxidation state. The similarities exist among (a) atomic spectra (b) magnetic properties and (c) *"actinide contraction"* in the trivalent (III) ions as we go up the series with increasing atomic number.

In this contraction, it will be interesting to observe that just as gadolinium in the lanthanide series shows an absence of absorption light in the visible region, so does curium in the actinide series also.

In the following table is given the actinide contraction that is as the atomic number increases in the actinide series, radii of the trivalent (IV) ions successively decrease. Ordinarily in such a series, the ionic radii should increase with the atomic behaviour is due to the smaller screening effect of the electrons in the f-sub group. It has been shown by X-Ray measurements that this occurs when the radii of the trivalent and the tetravalent ions, as well as with the A O distances in the hexavalent $[AO_2]^{++}$ ions.

Element	Ionic radii		A—O distance in hexavalent $[AO_2]^{++}$
	III state	IV state	
89. Ac	1.11Å	–Å	—
90. Th	—	0.95	—
91. Pa	—	—	—
92. U	1.04	0.89	1.91
93. Np	1.02	0.88	1.90
94. Pu	1.01	0.86	1.89
95. Am	1.00	0.85	—

Paramagnetism : We shall be comparing the trivalent ions of the members of the actinide series with those of the lanthanide with respect to their magnetic moments or magnetic susceptibilities. The total number of electrons in the system of lanthanum atom is 57 and therefore, the trivalent ion, La^{III}, would have the number 54 and similarly in actinium trivalent ion, Ac^{III}, will have 86 electrons. The distribution along different shell is :

La^{III}	2, 8, 18, 18, 8	total 54
Ac^{III}	2, 8, 18, 32, 18, 8	total 86

2, 8, 18 and 32 all these figures denote the saturation numbers and therefore, there is a type of perfection in all levels in the atomic systems of the trivalent ions of lanthanum and actinium. They must be therefore, diamagnetic. Scandium and yttrium trivalent ions are also diamagnetic for the same reason.

Se^{III}	2, 8, 8	Total 18
Y^{III}	2, 8, 18, 8	Total 36

After lanthanum, in the lanthanide series is cerium, the trivalent ion of which would slow imperfection since one more electrons has been added. It would, therefore, be paramagnetic, the tetravalent cerium ion would, however, be still diamagnetic just as trivalent lanthanum ion.

Ce^{iii}	2, 8, 18, 18 + 1, 8	Total 55 (paramagnetic)
Ce^{iv}	2, 8, 18, 18, 8	Total 54 (diamagnetic)

So long as in the fourth shell, the number 32, in not reached, the imperfection of the shell will give rise to paramagnetism.

Similar is the case with members of the actinide series. The trivalent actinium ion shows diamagnetism, where as the trivalent thorium ion, Th^{III} if any, would have shown paramagnetism, but the tetravalent thorium ion, would be diamagnetic, hexavalent uranium ion U^{vi}, would also be diamagnetic (the total number of electrons in U is 92 - 6 = 86 giving perfect shells).

Ac^{iii}	2, 8, 18, 32, 18, 8	Total 86 (diamagnetic)
Th^{iv}	2, 8, 18, 32, 18, 8	Total 86 (diamagnetic)
U^{vi}	2, 8, 18, 32, 18, 8	Total 86 (diamagnetic)

All other ions of the member of the actinide series would be paramagnetic. In the following table are given the actinide and lanthanide series.

Number of electrons in penultimate quantum group	Lanthanide		Actinide	
	Ions	Paramagnetic susceptibility $\times 10^3$	Ions	Paramagnetic susceptibility $\times 10^3$
18	La^{III}	Diamagnetic	Th^{IV}, U^{VI}	Diamagnetic
19	Ce^{III}	2.35	Np^{VI}	2.45
20	Pr^{III}	5.05	U^{IV},Np^{V}, Pu^{VI}	4.0
21	Nd^{III}	5.05	Np^{IV}	3.9
22	—	—	Pu^{IV},	1.65
23	Sa^{III}	1.0	Pu^{III}	0.25
24	Eu^{III}	3.6	Am^{III}	0.70
25	Gd^{III}	24.6	Cm^{III}	2.65

If these susceptibilities are plotted against the atomic numbers of the elements, two humps will be observed in the case of lanthanide series, but only one in the case of actinide series. In both the cases, the first hump is, however, at the identical place.

Absorption Spectra : The absorption spectra of the salts of the members of the actinide series are very much similar to these of the lanthanide. Both are compound of very narrow bands, many of them are only 5Å to 10 Å broad at low temperatures. The coloured ions of the other series give in solutions broad absorption band, 500 to 1000 Å. Thus the absorption spectra in the case of lanthanide and actinide salts is an evidence of the fact that in both areas the orbits concerned in the absorption are far less compared to the fields of neighbouring molecules. This appears to be a very significant confirmation of the conception that in the case of rare earth salts and the salts of the actinide series the absorption has to be ascribed to an electronic group that is relatively deep down in the atom. Instead of being as it is for example, in the cupric ion 2, 8, 17 on the surface.

False Transuranium Elements

The discovery of neutron by Chadwick in 1932 provided a new instrument for the synthesis and transmutation of elements. Since it possesses no charge, it easily brings about elastic and non-elastic collision with nuclei of atoms. It was further noted that if the neutrons are passed through water or paraffin, they become much more effective in the radioactive

Densities Atomic Volumes and Atomic Radii and ionic radii of some of the actinides are given below

Actinide	Thorium 90	Uranium 92	Neptunium 93	Plutonium 94	Americium 95
Density	11.71	19.0	19.5	—	11.7
Atomic Volume	19.8	12.5	12.15	—	20.6
Atomic Radius	1.82	1.48	1.32	—	—
Radius of					
(i) Trivalent Ion	—	1.04	1.02	1.01	1.00Å
(ii) Tetravalent Ion	1.10	1.03	1.02	1.00	0.99Å
(iii) Hexavalent Ion	—	0.58	0.57	0.56	—

capture reactions. In 1934, Fermi and his collaborators studied these reaction details. These (x, γ) reaction yielded some important result such as :

$$_1H^1 + {}_0n^1 \longrightarrow {}_1H^2 + \gamma$$

$$_{45}Rh^{103} + {}_0n^1 \longrightarrow {}_{45}Rh^{104} + \gamma$$

$$_{49}In^{115} + {}_0n^1 \longrightarrow {}_{49}In^{116} + \gamma$$

In analogy to these reactions, Fermi and collaborators led to think that when uranium is bombarded with neutrons, the (n, γ) reactions is as follows:

$$_{92}U^{238} + {}_0n^1 \longrightarrow {}_{92}U^{239} + \gamma$$

It was also shown that this U^{239} is a radio active Beta-emitter with a half-life of 23 minutes. The decay is represented this.

$$_{92}U^{239} \longrightarrow {}_{93}X^{239} + {}_{-1}e^0$$

The new daughter element x is the element with atomic number 93. This element with atomic number 93. This element does not occur in nature, and it further decays to yield another element 94 :

$$_{93}X^{239} \longrightarrow {}_{94}Y^{239} + {}_{-1}e^0$$

These elements, having a place beyond uranium in the Periodic table came to be known as *Transuranium elements*. It was however, shown by Hahn and Strassmann in 1939, that in the neutron reactions Fermi and his collaborators did not actually have trans uranium elements, but nearly the fission products of the nucleus barium or lanthanum.

$$U^{238} + {}_0n^1 \longrightarrow Ba^{138} + Ma^{101}$$

$$U^{235} + {}_0n^1 \longrightarrow Mo^{95} + La^{239} + 2{}_0n^1$$

NEPTUNIUM

Melting Point : 644°C
Boiling Point : 3902°C

Atomic Number : 93
Atomic Weight : 237.048

History

It was only during the year 1939 and 1940, that we had the conclusive evidence of the identification of the element 93 which has now been named as neptunium (It has been named as neptunium after the planet Neptune next to Uranus). Among the several beta-particle decay periods which had been observed following by the intersection of uranium with neutrons was on of 23 minutes. E. M Mcmillan and P. H Abelson showed in 1940 that Uranium 239 also generated another product with 2-3 days as half-life period. It was significant to observe that the rate of the formation of the 2-3 days product and the rate of decay of the 23 minutes half-life product were the same. Hence, the 2-3 days period is due to the element 93, mass number 239, formed from the beta decay of the uranium 239.

$$_{92}U^{239} \longrightarrow {}_{93}Np^{239} + {}_{-1}e^0$$

The chemistry of neptunium was supposed to be very much similar to that of rhenium, but Mcmillan and Abelson have shown that this is not case.

A large number of isotopes of neptunium are now known of these isotopes, neptunium 237, is the longest-lived so far known. This isotope is an alpha-emitter, with a half-life of 2.2×10^6 years. Wahl and Seaberg discovered it in 1942. It is obtained as a beta decay product of Uranium 237 (which itself is obtained from U^{238} by the action of fast neutrons).

$$_{92}U^{238} + {}_0n^1 \longrightarrow {}_{92}U^{237} + 2{}_0n^1 \text{ (6 days)}$$

$$_{92}U^{237} \longrightarrow {}_{93}Np^{237} + {}_{-1}e^0 \text{ (} 2.2 \times 10^6 \text{ years)}$$

The other isotopes are 231, 234, 235, 236

Np^{231} from U^{233} (d, 4x)

Np^{234} from Pa^{231} (α, x)

Np^{235} from U^{235} (α, p3x)

Np^{236} from U^{235} (d, x)

Isolation of Neptunium

When uranium is bombarded by fast neutrons, we get neptunium-237 in addition to (*i*) Unchanged uranium, (*ii*) Some plutonium and (*iii*) Some fusion products. Np 237 on account of its low radioactivity (half-life 2.2×10^6 years) is difficult to trace and consequently a small quantity of 2.3 day Np^{237} isotope is added.

(A) The whole material is now observed in nitric acid.

(B) The nitric acid solution is then reduced by sulphur dioxide. To this solution is now added hydrofluoric acid and a small amount of lanthanum is also added as carrier. Thus fluoride, of Np, Pu and La are precipitated where as uranium remains in solution as Uranyl nitrates.

(C) The fluorides are then dissolved in sulphuric acid and oxidized by means of bromate at room temperatures. Neptunium is then oxidized to (VI) nitrate (NpO_3 type).

(D) Hydrofluoric acid is again added to the solution and this time, lanthanum and plutonium, where as neptunium remains in solution and thus separated.

(E) The neptunium solution is (VI) state is now reduced by sulphur dioxide to (IV) state (NpO_3 type) and it yields a precipitate of hydroxide solution. This precipitate, when heated to redness, yields the oxide NpO_2.

Nature of Neptunium

Neptunium differs from rhenium, unlike rhenium (*a*) neptunium does precipitate with hydrogen sulphide in acid solution (*b*) it is not reduced to the metallic state by zinc and (*c*) its oxide is not volatile at red heat.

Moreover in some respects neptunium is similar to rare earths. In reducing solution (reduced by SO_2) it can be precipitated with cerium (VII) as fluoride (CeF_3;NpF_3, NpF_3;NpF_4 either or both) and with thorium (IV) as iodate $Th(IO_3)_4$ $Np(IO_3)_4$

Neptunium resembles uranium in this respect that it also exists in many oxidation states III, IV, V and VI. However, whereas uranium in state VI is not easily reduced, neptunium is salts of state (IV) as NpO_2 and NPF_4.

Compounds of Neptunium

It has been already stated that, neptunium salts exist in the states of III, IV V and VI, corresponding to oxides Np_2O_3 NpO_2 NpO_5 and NpO_3 the last one giving salts of neptunyl series corresponding to uranyl. The salts of IV and VI salts are most stable. The salts in the state IV are oxidized to the state VI are reduced to the state V by sulphur dioxide.

$$Np \xrightarrow[HBrO_3]{SO_2} NpO_2$$

From the salts of NpO_2 series [as NpF_4, $Np(SO_4)_2$], one gets the precipitate of $Np(OH)_4$ on the addition of ammonia and this on ignition gives NpO_2. The salts have been obtained in the micro gram scale.

Neptunium fluorides have been prepared in the III, IV states. When the dioxide is heated in a mixture of hydrogen and hydrogen fluoride at 500°C, the trifluoride is formed.

Neptunium Dioxide	+	Hydrogen	+	Hydrogen Fluoride	⟶	Neptunium Trifluoride	+	Water
$2NpO_2$	+	H_2	+	6HF	⟶	$2NpF_3$	+	$4H_2O$

When this trifluoride is heated in a current of oxygen and hydrogen fluoride at 500°C tetrafluoride is produced.

Neptunium Trifluoride	+	Oxygen	+	Hydrogen Fluoride	⟶	Neptunium Tetrafluoride	+	Water
$4NpF_3$	+	O_2	+	4HF	⟶	$4NpF_4$	+	$2H_2O$

The tetrafluride is similarly obtained from the dioxide also :

Neptunium Dioxide	+	Hydrogen Fluoride	+	(oxygen)	$\longrightarrow$	Neptunium Tetrafluoride	+	Water (+ oxygen)
NpO_2	+	$4HF(O_2)$			$\longrightarrow$	NpF_4	+	$2H_2O(+ O_2)$

If the trifluoride is heated in fluorine, a hexafluoride, NpF_6 is obtained which is very much similar to uranium hexafluoride, UF_6 Neptunium tetrafluoride is obtained by heating the dioxide in carbon tetrachloride vapour.

Neptunium Dioxide	+	Carbon Tetrachloride	$\longrightarrow$	Neptunium Tetrachloride	+	Carbonyl Chloride
NpO_2	+	CCl_4	$\longrightarrow$	$NpCl_4$	+	$2COCl_2$

If will be interesting to see that uranium tetrachloride is also prepared in the same manner.

Uranium Dioxide	+	Carbon Tetrachloride	$\longrightarrow$	Uranium Tetrachloride	+	Carbonyl Chloride
UO_2	+	CCl_4	$\longrightarrow$	UCl_4	+	$2COCl_2$

When this tetrachloride is heated in current hydrogen at 450°C, the trichloride, $NpCl_3$ is formed.

Neptunium Tetrachloride	+	Hydrogen	$\longrightarrow$	Neptunium Tetrachloride	+	Hydrogen Chloride
$2NpCl_4$	+	H_2	$\longrightarrow$	$2NpCl_3$	+	$2HCl$

When a mixture of neptunium dioxide and excess aluminium is expressed to the action of bromine vapours and the unreacted aluminium bromide is removed by sublimation at 250°C. The residue finally heated to 800°C, *Neptunium Trichloride, $NpBr_3$* gets volatilized and deposits in the cooler portions of the tube.

If in the above experiment, aluminium is not taken in excess, a tetrabromide, $NpBr_4$ is obtained which sublimes at 500°C.

Similarly, if the mixture of neptunium dioxide and aluminium is exposed to iodine vapour and finally the residue is heated to 800°C tri-iodide, NpI_3 volatilizes.

It has been clearly shown by the X-Ray diffraction patterns that the halides of neptunium possess the same crystalline structure as those of uranium.

When the hydride of neptunium is dissolved in sulphuric acid and treated with bromate solution, it is oxidised to (VI) state. Now if to the mixture a concentrated solution of sodium nitrate and sodium acetate is added, we get *sodium neptunyl acetate $NaNpO_2(CH_3COO)_3$*.

Metallic Neptunium

When neptunium trifluoride is heated with barium vapour at 12,00 it is reduced to the metallic state ($2NpF_3 + 3Ba \rightarrow 2Np + 3BaF_2$) This reduction is carried out in beryllium crucible. The tetrafluoride is convenient for the experiment since it gets soaked into the crucible at the high temperature.

The tetrafluoride is convenient for the experiment, since it gets soaked into the crucible at high temperature.

Neptunium is a silvery metal, which does not tarnish in air.

Isotopes of Neptunium

Isotope	Half-life	Mode of Formation
Np 231	53 mins	U -238 (d. 4n)
234	4.4 days	U 235 (p, 2n) or (d, 3n)
235	240 days	U 235 (d, 2n) or (α, p, 3n)
236	20 hrs	U 235 (d, n) Np 237 (n, 2n)
237	2.25×10^6 hours	U 237 (β); am 241 (α)
238	2.0 days	U 238 (d, 2n); U-235 (α, p)
239	2.31 days	U 239 (β); U 238 (d, n)

PLUTONIUM

More About Plutonium

Some of the neptunium isotopes show beta activity and have after exactly a beta particle is converted to an element with atomic number 94.

$$_{93}Np^{239} \longrightarrow {}_{94}Pu^{239} + {}_{-1}e^0$$

Mcmillan and Abelson in their study of neptunium 239 were unable to detect any new activity, when this nuclide decayed and hence they concluded that the new product of atomic number 94 was an alpha emitter of long life. This fact gets confirmation in March 1991, by Kennely, Seaborg, Segre and Wahl. This plutonium 239, was actually an alpha emitter with a half-life of 24, 000 years. This isotope of plutonium was found to undergo fission by slow neutrons and hence, immerse importance was attached to its production in appreciable amounts for the purposes of atom bomb. Does this plutonium 239 isotope occur in nature, since it has such a half-life period? The results obtained so far indicate that the mineral pitchblende and carnolite contain if at all $1/10^4$ of plutonium.

The first isotopes of plutonium to be definitely identified were with a mass number-238. Since it has short half life and consequently, a relatively higher activity Seaborg, Mcmillan and Kennely and Wahl discovered the 20 day neptunium 238 in the winter of 1940-41 and they also noted that this substance developed an alpha activity which could be separated chemically both from uranium and neptunium. This new activity the presence of an element of atomic number 94, which was subsequently named as plutonium by the authors, after Pluto, this planet beyond Neptune in the solar system. Plutonium 238 is an alpha-emitter with years of half-life period.

$$_{93}Np^{238} \longrightarrow {}_{94}Pu^{238} + {}_{-1}e^0$$

Isotope	Half-Life	Mode of Formation
Pu 232	22 mins	U 235 (α, 7n)
234	8.5 hrs	U 233 (α, 2n)
236	2.7 yrs	Np 236 (β; U 235 (α, 3n)
237	40 days	U 235 (α, 2n), U 238 (α, 5n)
238	50 yrs	Np 238 (β)
239	2.48×10^4 yrs	Np 239 (β), U 238 (α, 3n)
240	6.6×10^3 yrs	U 238 (d, 2n)
241	13.0 yrs	U 238 (a, n)
242	5×10^5 yrs	Pu 241(n, γ)
243	5.0 hrs	Pu 241 (n, γ)

Toxicity of Plutonium

Plutonium as well as the entire alpha emitting fission products of sufficient half-life is extremely injurious to tissues. They form insoluble phosphates and get incorporated in growing bone cells. They enter the blood stream, trough and injured tissue. Plutonium (IV) phosphate is so insoluble that if once precipitated, its elimination from the system is very slow. A speak of plutonium compounds is sufficient to cause injury through its alpha emissions and produce wound in the bone marrow cells which generate red blood corpuscles. Absolute precautions are necessary for field of radioactivity since the effects are cumulative.

Nature of Plutonium Compounds

Plutonic compounds exist in four states of valency III, IV, V and VI like neptunium it resembles uranium in many of its chemical properties. Where as plutonium is most stable in (IV) state, uranium, however, exhibits its maximum stability in (VI) state.

Oxides : Oxides of plutonium known are Plutonium Monoxide, PuO, Plutonium Sesquioxide, Pu_2O_3 and Plutonium Pyrooxide Pu_4O_7 and Plutonium Dioxide PuO_2. PuO is iso-structural with Sm_2O_3 (as might be expected).

Plutonium dioxide, PuO_2 is the most stable amongst oxides. This oxide, however can be reduced by carbon or barium at 1000°C to the sesquioxide, Pu_2O_3(most probably the sesquioxide of uranium and neptunium do not exit or are very unstable).

Plutonium Peroxide, PuO_4 : It is obtained by treating plutonium (IV) salt with hydrogen peroxide. This reagent, in small quantities at first produces a blood red solution but in high quantities it gives an intense green precipitate of plutonium peroxide. It is isomorphous with ThO_4, This compound is also precipitated from Pu(VI) salts in solutions although slowly.

Plutonium Sulphides : One sulphide P_2S_2 the monosulphide and an oxysulphide Pu_2SO_2 are known. When $PuCl_3$ is heated in a current of hydrogen sulphide the sesquioxide is formed, it is isolated with La_2S_3 Ac_2S_3 and Am_2S_3 (cubic structure). The oxysulphide is obtained when PuO_2 is heated to 250°C in a current of hydrogen sulphide. In its hexagonal structure, it is similar to La_2O_2S and Ce_2O_2S.

Halides : Plutonium hexafluoride is not so stable with the formula of PuF_6, but the uranium hexafluoride is much more stable than this compound. Plutonium hexafluoride is obtained by heating plutonium dioxide in the mixture of hydrogen and hydrogen fluoride, at about 500°C, its trifluoride, PuF_3 is obtained.

Plutonium Dioxide	+	Hydrogen +	Hydrogen Fluoride	⟶	Plutinium Trifluoride	+ Water
$2PuO_2$	+	H_2	+ 6HF	⟶	$2PuF_3$	+ $4H_2O$

Plutonium tetrafluoride can be precipitated from plutonium salts in (IV) state from solutions or addition of alkali fluorides. This fluoride is quite stable, although it can be reduced to the trifluoride. The tetrafluoride forms a soluble complex, when sodium oxalate is added to its precipitate.

Plutonium tetrachloride and tetrabromide have not been obtained so far, at least in the solid state. If a mixture of plutonium dioxide and excess aluminium is expressed to the action of bromine vapour and the unreacted tribromide is removed by sublimation at 250°C. If the residue is heated to about 800°C, *plutonium tribromide* $PuBr_3$ volatilizes and condenses at the cooler portions of the tube. Plutonium tri-iodide PuI_3, is also prepared in an analogous manner from the dioxide, aluminium and iodine vapour. The iodide also volatilizes at about 800°C.

Neptunium Halides			Plutonium Halides		
halides	colour	density	halides	colour	density
NpF_3	Purple	9.12	PuF_3	Violet	9.32
NpF_4	Light green	6.8	PuF_4	Light green	7.0
NpF_6	White	—	$PuCl_3$	Emerald green	5.70
$NpCl_4$	Red brown	4.92	PuOCl	Blue green	8.81
$NpBr_3$	Green	7	$PuBr_3$	Green	6.69
$NpBr_4$	Red brown	—	PuOBr	Deep green	9.07
NpI_3	Brown	6.82	PuI_3	Bright green	6.9
			PuIO	Green	8.46

If to a solution of salts in (III) state, ammonium hydroxide is added, we get a precipitate of *plutonium hydroxide* $Pu(OH)_3$.

In respects to these precipitates platinum shows as analogy to the rare earth elements.

Oxyhalides : The dioxide with hydroiodic acid gives the oxyiodide, PuOI, a green substance. The other hydralogen acids (in gaseous state) in presence of water vapours produces other oxy halides as PuOCl and PuOBr.

Carbides, Nitrides and Silicides : These compounds of plutonium are refractory in character. Plutonium carbide, PuC is a semi-metallic black substance, stable up to zero. The nitride, PuN is brown in colour. The silicide, $PuSi_2$ is silvery metallic and can be heated in vacuum to 1000°C.

Other Compounds

The aqueous solution of the positive plutonium (III) Ion P^{+++} are oxidized in air to the plutonium (IV) state, although not as are solutions of uranium ion (III) state. The compounds of plutonium are most stable in (IV) state. The fluoride PuF_4 the iodate $Pu(IO_3)_4$ and the hydroxide $Pu(OH)_4$ are insoluble and can be precipitated by the usual double decomposition reactions. The chloride, nitrate, perchloride and sulphate, $Pu(SO_4)_2$ are soluble in water. Numerous complexes, both organic and of the plutonium (IV) state, have been prepared; a soluble oxalate complex for example makes it possible to dissolve PuF_4 and other insoluble Pu(IV) compound in water.

A large number of "Chelete" complexes of plutonium have also been prepared. These complexes are soluble in organic liquids.

Plutonium (IV) compounds can be reduced to the plutonium (III) state be means of worm sulphur dioxide, by hydroxylamine, by zinc amalgam an also by the iodide ion. Similarly, by the use of mild oxidizing agent plutonium (III) compounds can be oxidized the (IV) state.

From Trivalent Plutonium

Iodate, $Pu(IO_3)_4$: Bright brown, sparingly soluble.

Trichloroacetate, $Pu(Cl_3COO)_3$: Pale red, slightly soluble in water fully soluble in acetone.

Perchlorate, $Pu(ClO_4)_3$: Very soluble

Oxalate, $Pu_2(C_2O_4)_3$: Violet; fairly soluble

Sulphate, $Pu(SO_4)_3$: Fairly soluble.

Its half-life has been found to be about 500 years.

Isotopes of Americium

Isotope	Half life	Mode of formation
Am 238	1.3 hrs	Pu 237 (d, n)
239	12 hrs	Pu 239 (p, n), Np 237 (α, 2n)
240	2.1days	Pu 239, (d, n), Np -237 (α,2n)
241	470 hrs	Pu 241 (β), Bk 241 (α)
242	18 years	Am 241 (n,γ)
243	10^4 years	Am 242 (n,γ); Pu 243(β)
244	26 years	Am 243 (n,γ)

Separation of Americium from Rare Earths

From earlier experiments, it has been found that its compounds in (III) oxidations state are very stable and in this respect, this element resembles the rare earth elements. The separation of americium from the rare earth is difficult and there are two process involved in this job.

(*i*) Americium fluoride is an insoluble substance like rare earth fluorides, but still americium fluoride has little more solubility than rare earth. On the basis of this observation if fractional precipitation is carried out carefully with a control of fluoride ion concentration, the separation may be effected.

(*ii*) Americium may also be separated from the rare earth, by making the use of ion exchange procedure. The mixture of the solution of the salts of americium and the rare earth in the dilute hydrochloric acid is passed through a column packed with particles of small resin. As a result the rare earth as well as the americium ions are almost separated from the solution, being taken up by the top layers of resin. But the adsorption characteristics of these ions differ from each other considerably. Now a solution containing a substance like ammonium citrate, with an adjusted pH, is allowed to flow slowly down the column of resin. Ammonium citrate forms complexes of varying stability with the rare earth carrier ions and americium ions. Therefore, two do not come down through the column simultaneously, but in different portions and hence their separations may be effected. In respect to this type of separation americium lies cerium and element 61, the promethium

Nature of Americium Compounds

Americium is undoubtedly most stable in (III) oxidation state, its compounds however, in the lower oxidation state (II) and higher oxidation states (IV) and (V) have also been prepared.

Americium exists in acidic solutions in the (III), (V) and (VI) oxidation states with the ionic species correspondingly to Am^{+3}, AmO_2^+ and AmO_2^{++}.

When a solution of americium (III) hydroxide, $Am(OH)_3$ in concentrated potassium carbonate solution is oxidised by heating with sodium hypochlorite, a dark precipitate is formed which may contain americium in state (IV) and (V). When americium hydroxide $Am(OH)_3$ or nitrate, $Am(NO_3)_3$ is heated in air to almost 1000°C americium dioxide AmO_2 is produced which has americium in (IV) and (V) state.

Since europium, which is the rare earth homologue of americium gives compounds in the state (II), it is very likely that americium would also give some compounds in this lower state. When compounds of trivalent americium in aqueous solution are treated with sodium amalgam, americium is reduced to divalent state. (This divalent americium can be precipitated by using barium chloride or europium sulphate as carrier) and in fact an oxide AmO has actually been isolated.

Compounds

All such methods, which yield tetravalent compounds of neptunium (as NpF_4, $NpCl_4$) when applied to americium compounds, give only trivalent salts. For example, when the hydroxide of americium $Am(OH)_3$ is heated in a current of oxygen and hydrogen fluoride, only the trifluoride AmF_3 is obtained. Similarly the dioxide, when heated with carbon tetrachloride, yields only the trichloride, $AmCl_3$

Americium Hydroxide + Hydrogen Fluoride + Oxygen ⟶ Americium Trifluoride + Water + Oxygen

$$Am(OH)_3 + 3HF + O_2 \longrightarrow AmF_3 + H_2O + O_2$$

Carbon Tetrachloride + Oxygen ⟶ Carbonyl Chloride + Chlorine

$$2CCl_4 + O_2 \longrightarrow 2COCl_2 + 2Cl_2$$

Americium Dioxide + Chlorine ⟶ Americium Trichloride + Oxygen

$$2AmO_2 + 3Cl_2 \longrightarrow 2AmCl_3 + 2O_2$$

When the oxide, Am_2O_3, is heated with aluminium in vapour of bromine or iodine, *americium tribromide, $AmBr_3$* or the *triiodide, AmI_3* is obtained. These halides are isomorphous with those of uranium.

No tetrahalides of americium has yet been obtained, only the fluoride is likely to exist. Even if it were prepared, it will be less stable than the tetrafluoride of plutonium. Fried in 1951 has reported the preparation of AmF_4.

Americium Trioxide + Aluminium + Iodine ⟶ Americium Triiodide + Aluminium Trioxide

$$Am_2O_3 + Al + 3I_2 \longrightarrow 2AmI_3 + Al_2O_3$$

Americium Sesquioxide + Aluminium + Bromine ⟶ Americium Tribromide + Aluminium Trioxide

$$Am_2O_3 + 2Al + 3Br_2 \longrightarrow 2AmBr_3 + Al_2O_3$$

Among the trivalent state of the americium salts the fluoride and the oxalate are insoluble and the phosphate and iodate only moderately soluble in acid solution, americium nitrate, halides, sulphate and perchlorate are soluble. The hexavalent (VI state) americium can be precipitated with sodium acetate, giving crystals isostructural with sodium uranyl acetate, $NaNO_2\ (CH_3COO)_3xH_2O$.

Metal

The metal of barium vapour in the trifluoride of americium at over 1000°C yields americium metal.

Americium Trifluoride + Barium ⟶ Americium + Barium Difluoride

$$2AmF_3 + 3Ba \longrightarrow 2Am + 3BaF_2$$

This metal has a silvery lusture and density 10.5, it is more malleable than the uranium and neptunium. It slowly tarnishes when left in dry air at ordinary temperature.

Other Isotopes : The most common isotope is Am^{241}. When this isotope is bombarded with neutrons from a nuclear reactor, the isotope Am^{242} is obtained in the (n,γ) process.

$$_{95}Am^{241} + {}_0n^1 \longrightarrow {}_{95}Am^{242} + \gamma$$

It has been absorbed that isotope 242 emits in two isometric states (I) one is a beta-emitter (half-life is 16 hrs) and (II) the other which decays both by α and β emissions (half-life 400 yrs). Other isotopes reported are 238, 239 and 240 some of them are obtained from plutonium 239 by reactions (d,3n); (d,2n) and (d,2n) respectively.

$$_{94}Pu^{239} + {}_1D^2 \longrightarrow {}_{95}Am^{238} + 3{}_0n^1 \text{ and so on}$$

Summary of Americium Compounds

Oxides : AmO, AmO_2 and $Am(OH)_3$ [AmO has NaCl structure and AmO_2 has fluorite structure]

Hydride : It is obtained by heating gently the metal in hydrogen

Fluoride : AmF_3 formed from AmO_2 + HF + O_2 (light red)

Chloride : $AmCl_3$ is formed from AmO_2 + CCl_4 (colourless)

Bromide : $AmBr_3$ from AmO_2 + Al + Br_2 Vapour (colourless)

Iodide : AmI_3 from AmO_2 + Al + Br_2 Vapour (yellow)

Oxychloride : AmOCl from black AmO_2 + dil HCl (light red)

Sulphide : Am_2S_3 from AmO_2 + CS_2 + H_2S at 1400°C

Ameryl compounds : (*i*) Sodium Ameryl acetate, $Na[AmO_2](C_2H_3O_2)$ from Am(VI) salt solutions and sodium acetate phenyl and uranyl acetate (ii) americyl nitrate $[AmO_2](NO_3)_2$ (uranyl nitrate) can be extracted with ether from aqueous solution, it forms complexes similar to the uranyl nitrate complexes.

CURIUM

Melting Point : 1340°C	Atomic Number : 96
Boiling Point :	Atomic Weight : 247.00

History

This element was discovered earlier than americium by G.T. Seaborg, R. A James and A. Ghiorso in 1944 and has been named so in the honour of Marie and Pierre Curie. It is analogous to gadolinium rare earth which has been named after J. Gladolin a Finnish chemist an early worker on rare earths. When plutonium 239 is bombarded with alpha particle from cyclotron, this new transuranium element is formed.

$$_{94}Pu^{239} + {}_2He^4 \longrightarrow {}_{96}Cm^{242} + {}_0n^1$$

The same nuclide is also formed in beta-decay of 16 hours, of americium 242.

$$_{95}Am^{242} \longrightarrow {}_{96}Cm^{242} + {}_{-1}e^{0}$$

It is also an alpha emitter with a half-life of 150 days. It was actually that curium in the form of its sesquioxide; Cm_2O_3 was first obtained by L. B Werner and I Perlman as a pure compound in 1947 in ultramicrochemical amounts free from carrier.

Isotopes

The five six other isotopes of curium with mass numbers 238, 240 244 and 245 have also been reported, form such reactions from Pu^{239}, Cm^{238} by (α, 5n) process and Cm^{240} by (α,3n) process.

$$_{94}Pu^{239} + {}_{2}He^{4} \longrightarrow {}_{96}Cm^{238} + 5{}_{0}n^{1} \text{ and so on } (Cm^{238} \text{ by } (\alpha, 5n) \text{ process})$$

$$_{94}Pu^{239} + {}_{2}He^{4} \longrightarrow {}_{96}Cm^{240} + 3{}_{0}n^{1} \text{ and so on } (Cm^{240} \text{ by } (\alpha, 3n) \text{ process})$$

Curium 240 is an alpha emitter and its half-life is 27 days.

ISOTOPE OF CURIUM

Isotope	Half life	Mode of Formation
Cm 240	27 days	Pu 239, (α,3n); Cf 244 (α)
242	162 days	Pu 239 (α, n), Am 242 (β) Cf 246(α)
243	100 yrs	Cm 242(n) Bk 243 (K)
244	10 yrs	Am 241 (α,p) Am 244(β)
245	2×10^4 yrs	Bk 245 (K)

For chemical investigation, the isotope 244, will be very useful some it is not only a long life isotope, it also has a lower specific alpha activity. The bombardment of Am^{241} with neutrons yields Pu^{242}.

$$_{95}Am^{241} + {}_{0}n^{1} \longrightarrow {}_{95}Am^{242} + \gamma$$

$$_{95}Am^{242} \longrightarrow {}_{94}Pu^{242} + {}_{1}p^{0}$$

The Pu^{242} is separated by chemical process and it can be transmuted to Am^{243} via the reactions given below:

$$_{95}Am^{243} + {}_{0}n^{1} \longrightarrow {}_{95}Am^{244}$$

$$_{95}Am^{244} + \longrightarrow {}_{96}Cm^{244} + {}_{-1}e^{0}$$

Separation of Curium from Americium

Curium is extremely stable in the (III) oxidation state and so far only trivalent compounds have been prepared. Separated of curium from the rare earths and also from americium (which may also be simultaneously present) is achieved by functional precipitation of fluorides.

It is not easy to separate curium from americium since both of them are very stable in their trivalent state. The separation of the two, in fact, is difficult as the separation of two neighboring rare earth elements.

The separation of the two has been achieved by partial precipitation of americium in its higher oxidation state by means of hypochlorite in potassium carbonate solution; then by absorption and elution with ammonium citrate from a column of ion exchange resin (for details see americium).

Analogy with Gadolinium : Most probably the arrangement of orbital electrons in curium is analogous to that in gadolinium and hence the solutions of the salts of curium though colourless exhibit strong adsorption in ultraviolet region. It would also exhibit peculiar magnetic susceptibility characteristics, as shown be gadolinium salts.

Alpha-activity : Since curium 242 isotope has a short life, it has an intense alpha activity, which makes, which makes the experimentation with this element a bit difficult. Not only the study of absorption is hindered, the alpha particle cause decomposition of the solution also with the evolution of gas and heating of water in which the curium compound is dissolved. Being injurious to heath, the experiments must be carried out in a closed system.

Chemical Properties

As we know that this element probably exhibits extremely on the (III) oxidation state in aqueous solution. This is also expected on the basic of the stable configuration of the half-filled 5f electron shell, that is the $5f^7$ configuration. With tracer concentrations of curium, a variety of strong oxidizing agents (dichromate, permanganate, peroxydisulphate and argentic ions) give no indication of oxidation. Curium is moreover, carried quantitatively by the rare earth fluorides in precipitation reactions and is very similar the tripositive rare earth elements in its so far known chemical characteristics.

Curium is not oxidized in alkaline solution by sodium hypochlorite under conditions where americium is oxidized to pentapositive state.

Like other tripositive actinide elements, curium fluoride and oxalate are insoluble in acid solution; while the nitrate, halide sulphide perchlorate and sulphate are soluble.

Solid curium trifluoride has been prepared by drying the fluoride, which precipitates from dilute nitric acid upon the addition of hydrofluoric acid. Curium trifluoride can be reduced to the metal by heating at 1276°C in a beryllia crucible with barium vapour.

Curium metal is silvery in appearance, and is similar to the other metals of the actinide series.

Like other transuranium elements, curium is determined analytically through its radioactive radiation.

BERKELIUM

Melting Point : 986°C
Boiling Point :

Atomic Number : 97
Atomic Weight : 247.00

History

During the end of the year 1949, S. G Thompson, A. Ghiorso and G.T Seaborg announced the discovery of berkelium; the name came from the Berkeley, the city in which it was discovered.

Berkelium has been obtained by bombarding americium 241 with alpha particle accelerated to about 35 MeV energy by means of cyclotron. The nuclear reaction may be represented in two ways.

(*i*) $_{95}Am^{241} + {}_{2}He^{4} \longrightarrow {}_{97}Bk^{244} + {}_{0}n^{1}$

(*ii*) $_{95}Am^{241} + {}_{2}He^{4} \longrightarrow {}_{97}Bk^{243} + 2{}_{0}n^{1}$

It is more likely that the isotope 243 has been obtained rather than isotope 244.

Isotope	Half life	Mode of Formation
Bk 243	4.6 hrs	Am 241 (α,2n)
245	5.0 days	Cm 244 (d,n)
249	1 yrs	Cm 249 (β)
250	3.13 hrs	Bk 249 (n,γ)

Berkelium shows its radioactive decay by way of orbital-electron captive and by way of alpha emission to a small extent (half-life 4.6). The ion-exchange method uses a resin column and subsequent elution with ammonium citrate, led to separation of breakdown from ammonium and curium.

Berkelium shows probably the maximum stability in a (III) oxidation state, but there is an indication of its forming compounds in a (IV) oxidation state also. In this respect, it must be similar to its rare earth homologue terbium.

The ionic species of berkelium (III) berkelium (IV) couple is about -1.6Volt on the scale where that for the hydrogen-hydrogen ion couple is taken on zero.

Just as in the case of other lanthanides and actinides oxalate and fluoride of tripositive americium are insoluble in acid solution and iodate and phosphate of tripositive americium are also insoluble in acid medium. Nitrate sulphate, halides and sulphides of both oxidation state (III) and (IV) apparently soluble.

Berkelium is best separated in the tripositive form by the ion exchange adsorption elution method and from the tripositive rare earth elements by the additional use of elution with concentrated hydrochloric acid in which actinide elements form complex ions under conditions where the lanthanides do to separate from the actinides. Lanthanides may also be affected by making use of tetrapositive state of berkelium.

CALIFORNIUM

Melting Point : 900°C | Atomic Number : 98
Boiling Point : | Atomic Weight : 252

History

S.G Thompson, A. Ghiorso and G.T Seaborg announced it in 1950 that they could get another transuranium element 98 named as Californium (Cf) by the action of alpha particle accelerated to about 35 MeV energy by means of a cyclotron in impinging on curium 242.

$$_{96}Cm^{242} + _{2}He^{4} \longrightarrow _{98}Cf^{246} + \gamma$$

or $$_{96}Cm^{242} + _{2}He^{4} \longrightarrow _{98}Cf^{244} + 2_{0}n^{1}$$

This new element is an alpha emitter with half-life of 45 minutes. It will be interesting to see that the amount obtained in the initial work was extremely small, 2×10^{-18} g (about 5000 atoms)

It is only through the use of tracer quantities, that it has been possible to have some idea of the chemical properties of californium. Americium is a homologue of dysprosium and as such, the oxidation state (III) is the most stable one in aqueous solution. It has neither been possible till now to reduce californium or oxidize it. As in other cases such compounds as the fluorides and oxalate appear to be insoluble in acid solution, where as the nitrate halide sulphate perchlorate and sulphide all appear soluble.

Isotopes of Californium

Isotope	Half life	Mode of formation
Cf 249	45 mins	Cm 242 (α,2n); U-238 (C12,6n) U238 (N14,p7n)
246	35.7 hours	Cm 244 (α,2n) U 238 (C 12 4n)
248	225 days	U 238 (C12, 4n)
249	400 days	Bk 249 (β)
253	20 days	Cf 252 (n, γ)

Oxidation Reduction Scheme for Actinides

The scheme outlined below is one the scale in which potential for $H_2 \longrightarrow H^+$ is taken as zero.

Uranium : Uranium ions in 1 molecular hydrochloric acid

For $U \longrightarrow U^{+3}$ 1.80 volt

For $U^{+3} \longrightarrow U^{+4}$ 0.61 volt

For $U^{+4} \longrightarrow UO_2^{4+}$, –0.62 volt

For $UO_2^{+} \longrightarrow UO_2^{++}$ –0.05 volt

For $U^{+4} \longrightarrow UO_2^{++}$ –0.334 volt

Neptunium : This scheme is one molar hydrochloric acid.

$$Np \xrightarrow{+1.8} Np^{+3} \xrightarrow{-0.14} Np^{+4} \xrightarrow{-0.79} NpO_2 \xrightarrow{-1.13} NpO_2^{++}$$

$$Np^{+3} \xrightarrow{-0.67} NpO_2^{++}$$

$$Np^{+3} \xrightarrow{-0.44} NpO_2^{+}$$

$$Np^{+4} \longrightarrow NpO_2^{++}$$

Plutonium : For one molar perchloric acid solution.

$$Pu \xrightarrow{+2.06} Pu^{+3} \xrightarrow{-0.982} Np^{+4} \xrightarrow{-1.15} PuO_2^{+} \xrightarrow{-0.93} PuO_2^{++}$$

$$Pu^{+3} \xrightarrow{-1.022} PuO_2^{++}$$

$$Pu^{+4} \xrightarrow{-1.042} PuO_2^{++}$$

Americium : In acidic aqueous solution

$$Am^{+2} \xrightarrow{>0.09} Am^{+3} \xrightarrow{2.32} Am^{+4} \xrightarrow{-1.2} AmO_2 \xrightarrow{-1.64} AmO_2^{++}$$

$$Am^{+3} \xrightarrow{-1.8} AmO_2^{+}$$

$$Am \xrightarrow{2.21} Am^{+3}$$

EINSTEINIUM FERMIUM AND MENDELEVIUM

History

Einsteinium and Fermium were discovered about simultaneously during end of 1952 by a group of workers in the California Radiation Laboratory, Argonne National Laboratory and Los Almos Scientific Laboratory. These discoveries were made, while a band of enthusiast workers were engaged for the study of transuranium products of uranium, which had been subjected to a very high instantaneous neutron influx in the thermonuclear explosion. The elements 99 and 100 were named after two distinguished physicists Albert Einstein and Enrico Fermi. Whilst these two elements are found also californium, Berkelium and Americium are also produced. Separation from these elements was achieved with ammonium citrate solution.

In this way, their cations are more eluted than those of elements Cf, Bk Cm and Am. The elution is, in fact in the following order:

$${}_{100}Fm^{3+} \quad {}_{99}Es^{3+} \quad {}_{98}Cf^{3+} \quad {}_{97}Bk^{3+} \quad {}_{96}Cm^{3+} \quad {}_{95}Am^{3+}$$

It will be interesting to compare this series with elution in the case of lanthanides under similar condition.

$${}_{68}Eb^{2+} \quad {}_{67}Ho^{2+} \quad {}_{66}Dy^{3+} \quad {}_{65}Tb^{3+} \quad {}_{64}Gd^{3+} \quad {}_{63}Eu^{3+}$$

Having studied their characteristics half lives and distinguished energies along with the elution order, it become easy to associate the newly discovered element with its proper atomic number. Einsteinium and Fermium were both α-emitters with half-lives of 20 days and 15 hours each and their mass numbers were found to be 253 and 255, thus $_{99}Es^{253}$ and $_{100}Fm^{255}$. The other isotopes of fermium and einsteinium are given in the table below.

Isotopes of Einsteinium and Fermium

Isotopes	Half-life	Mode of Formation
Es 247	7.3 mins	U-238(N,14,5n)
253	20 days	Cf 253 (b)
234	30 hrs	Es 253 (N, γ)
255	30 days	Es 254(n, γ)
Fm 250	30 mins	U 238 (O,16, 4n)
254	3.2 hrs	Es 254 (β)
255	15 hrs	Es 255(β)

In 1954 Melander and co-workers in Stockholm discovered a fermium isotope with number 250 (half 30 minutes). This isotope was obtained by irradiation of uranium with highly accelerated $^{16}O^{6+}$ particles. Similarly when U 239 was irradiated $^{14}N^{6+}$ particles of high acceleration; Ghiorso and his co-workers could obtain two short-lived isotopes of Einsteinium, Es 246 and Es 247. The Einsteinium isotope with the longest half-life is E 255 and it is beta emitter.

In 1955 Ghiorso obtained a new element 101, which was named in the honour of Russian chemist Mendeleef. Thus Mendelevium was obtained by irradiation of Es 253, with particles accelerated to a high degree of energy in cyclotron.

$$_{99}Es^{253} + {}_2He^4 \longrightarrow {}_{101}Mv^{256} + {}_0n^1$$

This new element has a small half-life of few hours and is an alpha emitter.

The success in obtaining artificial elements beyond californium in the periodic table would depend on how we have able to conduct our attempts towards the introduction of a large number of neutrons into the nuclei successively. This is usually accomplished in two ways.

(i) *By a succession of (n,γ) reactions and β-decay process.* The occurrence of x consecutive neutron capture process depends upon the cross-reactions at which each stage and upon the xth power of neutron concentration. This would necessitate the use of a most intense neutron flux.

(ii) *One may utilize the* highly accelerated nuclei of carbon $^{12}C^6$ or of nitrogen $^{14}N^7$ etc. in bombarding heavy elements in the cyclotron. In such ways one may succeed in transforming the nucleus considerably.

$$_{92}U^{238} + _{7}N^{14} \longrightarrow _{97}Bk^{243} + _{2}He^{4} + 5_{0}n^{1}$$

$$_{92}U^{238} + _{7}N^{14} \longrightarrow _{99}Es^{247} + 5_{0}n^{1}$$

Isotopes of Einsteinium Es(99), Fermium(100) and Mendelevium Mv(101) have been obtained in this way by both these methods. In the very intense neutron flux of the USAEC materials Testing Reactor, the californium isotope Cf 25 is built up from U238, or Pu 239, this gives in succession.

F 20 days 36 hrs 3.2 hrs 12 yrs

Cf 232 (n,γ) → Cf 253 (β) → Es253 (n,γ) → Es 254(β) → Fm 254 (α) → Cf250(α) → Cm 256

Chemical Properties

They have been obained at extreme dilutions on "weightless " samples. Co-precipitation experiment indicates that einsteinium formation and mendelevium exist in solutions as trivalent ions. On the basis of universal trend of oxidation potential in the actinide series one does not expect any higher valencies of their elements at least, in the state of an aqueous solution. Co precipitations of fluorides like EsF_3, FmF_3 and MvF_3 takes place with LaF_2.

On cation exchange columns, using ammonium citrate or lactate as eluent agent elements diminishes with the increase in the atomic number so that the separations are less effective using as eluent on cation exchange resins.

NOBELIUM

Melting Point : 827°C
Boiling Point :

Atomic Number : 102
Atomic Weight : 254

History

This element was known during 1957 by a group of workers in the Nobel Institute of Physics, Stockholm. They bombarded curium 244 with cyclotron produced $^{13}C^{+}$ ions and saw it was emitting α-particles. This element nobelium was named so far in he honour of the institute. The discovery of this new element could not be confirmed by Seaborg, Ghiorso and others at that time. These workers, however announced that the production of their element in April 1958, when they bombarded curium 246 with $^{12}C^{4+}$ ions and obtained the isotope No 258. They represented the reaction as follows:

$$_{96}Cm^{246} + _{6}C^{12} \longrightarrow _{102}No^{254} + 4_{0}n^{1}$$

This isotope of nobelium is also an alpha emitter with a half-life of 3 seconds. Seaborg and co-workers detected this element 102 by the chemical identification of its daughter Fm 250. This has been separated from the parent element (nobelium) by taking advantage of the recoil due to its alpha decay. (These workers may like to change the name but nothing has been announced so far.

Decay Series of Transuranium Elements

This type of decay series are available :

(*i*) Disintigration of transuranium nuclides with mass number of the M = 4n(n is an integer), developed after a nuclear decay such nuclei which ar of the 4n form (similar tot he thorium series).

(*a*) M-256 $\xrightarrow{\alpha}$ Es-252 $\xrightarrow{\alpha}$ Cf-252 $\xrightarrow{\alpha}$ Cm-248 $\xrightarrow{\alpha}$ Pu-244 $\xrightarrow{\alpha}$ U-240

(*b*) Cf-248 $\xrightarrow{\alpha}$ Cm-244 $\xrightarrow{\alpha}$ Pu-210 $\xrightarrow{\alpha}$ U-236 $\xrightarrow{\alpha}$ Th-232 $\xrightarrow{\alpha}$ MsTh-228

(*c*) Pu-232 $\xrightarrow{\alpha}$ U-228 $\xrightarrow{\alpha}$ Th-224 $\xrightarrow{\alpha}$ Ra-220 $\xrightarrow{\alpha}$ Em-216 $\xrightarrow{\alpha}$ ThC

(*ii*) Disintegration of nucleids with mass numbers of the M = 4*n* + 1 form :

(*a*) Cf-249 $\xrightarrow{\alpha}$ Cm-245 $\xrightarrow{\alpha}$ Pu-241 $\xrightarrow{\beta}$ Am-241 $\xrightarrow{\alpha}$ Np-237 $\xrightarrow{\alpha}$ Pa-233

(*b*) Bk-245 $\xrightarrow{\alpha}$ Am-241 $\xrightarrow{\alpha}$ Np-237 $\xrightarrow{\alpha}$ Pa-233 $\xrightarrow{\beta}$ U-233 etc.

(*iii*) Disintegration of nucleids with mass numbers of the M = 4*n* + 2 form :

(*a*) Fm-254 $\xrightarrow{\alpha}$ Cf-250 $\xrightarrow{\alpha}$ Cm-246 $\xrightarrow{\alpha}$ Pu-242 $\xrightarrow{\alpha}$ UI-238 $\xrightarrow{\alpha}$ Ux-234

(*b*) Fm-250 $\xrightarrow{\alpha}$ Cf-246 $\xrightarrow{\alpha}$ Cm-246 $\xrightarrow{\alpha}$ Pu-238 $\xrightarrow{\alpha}$ UII-234

(*iv*) Disintegration of nucleids with mass numbers of the M = 4*n* + 3 form :

Fm-255 $\xrightarrow{\alpha}$ Cf-251 $\xrightarrow{\alpha}$ Cm-247 $\xrightarrow{\alpha}$ Pu-243 $\xrightarrow{\beta}$ Am-243 $\xrightarrow{\alpha}$ Np-239

EPITOME–3

Heavy Metals

Extraction of Titanium

(i) ***Rose's Process :*** The mineral is heated to redness in a stream of hydrogen sulphide and iron sulphide is removed by extraction with concentrated hydrochloric acid and the hydrated Titania is well washed. The process is repeated to remove all iron.

(ii) ***Berthier's Process :*** The mineral is fused with a mixture of sodium carbonate and sulphur added in small portions at a time to hot crucible. The fused mass is dissolved in cold dilute sulphuric acid. The solution is treated with sodium carbonate, the ferrous ion remains in solutions and the titanic acid is precipitated.

(iii) ***Wöhler's Process :*** The mineral is fused with twice its weight of calcium fluoride, CaF_2 and the cold mass is digested with sulphuric acid the liquid is filtered from calcium sulphate, reduced with sulphur dioxide and boiled to precipitate the titanic acid

(iv) ***Friedel and Guerin's Process :*** In this process, iron is volatilized, as chloride by heating the mixed precipitate is a current of hydrogen chloride and chlorine.

(v) ***Wöhler's Process :*** The finely powdered mineral is fused with three times of its weight of potassium carbonate and the cake thus obtained is powdered and digested with dilute hydrofluoric acid. The sparingly soluble potassium fluorotitanate is boiled with sufficient water to dissolve all the salt; the solution is filtered while hot and cooling. The potassium fluorotitanate is obtained as lustrous scaly crystals. The salt is washed with cold water and purified by recrystallization from boiling water. The hot solution can be treated with ammonia and the washed precipitate is ignited as to produce highly pure titanium dioxide.

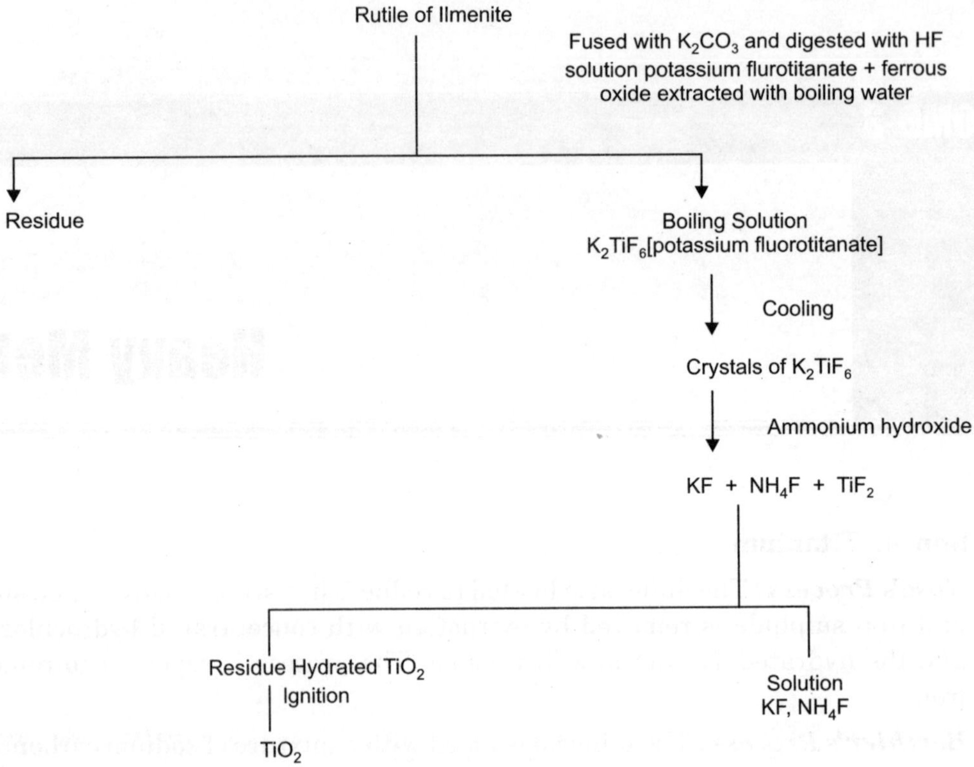

Purification of Titanium

1. By reducing potassium fluorotitanate, K_2TiF_6 with sodium or potassium:

Potassium Fluorotitanate	+	Sodium	⟶	Potassium Fluoride	+	Sodium Fluoride	+	Titanium
K_2TiF_6	+	4Na	⟶	2KF	+	4NaF	+	Ti

In vacuum if the reaction is conducted it produces 73% metal. It is usually contaminated with titanious nitrate, which also shines like metal. Reduction with sodium may be carried out in an atmosphere of hydrogen to yield the metal of higher purity.

2. Kern's Process : The vapour of titanium tetrachloride is passed over (*a*) molten sodium or (*b*) over sodium hydride at 400°C; then heated in a stream of carbon dioxide, it is again heated at 800°C to decompose all the titanium hydride.

(*a*) $TiCl_4 + 4Na \longrightarrow Ti + 4NaCl$

(*b*) $TiCl_4 + 4NaOH \longrightarrow 4NaCl + TiH_4$

$TiH_4 \longrightarrow Ti + 2H_2$ (800°C).

All these methods lead so much specimens of titanium, which invariably contain much oxide an impurity. M. A Hunter, however, heated, a mixture of highly purified titanium tetrachloride and sodium metal in a steel dome capable of with standing an internal pressure of 40,000 kg. The temperature was raised to dull red heat so as to start the reaction, which was over in few minutes. On leaching the product with water, a grey powder, mixed with small rounded phases of the fused metal was obtained. The fused metal contained 99.9% titanium.

3. Weiss Method : Weiss and Kaiser (1910) purified the metal as obtained by reducing the fluorotitanate with sodium or aluminium by fusing the mixture into pencils and using the pencils as electrodes in a vacuum arc furnace.

The arc furnace used by Newmann is illustrated in Fig. 1. The mixture of potassium fluorotitanate and aluminium was squeezed into rods or pencils (6.4 cm × 1.0 cm × 0.5 cm). These were used as electrodes in the arc furnace, which enabled them to be heated in vacuum or in an atmosphere of any desired gas under reduced pressure.

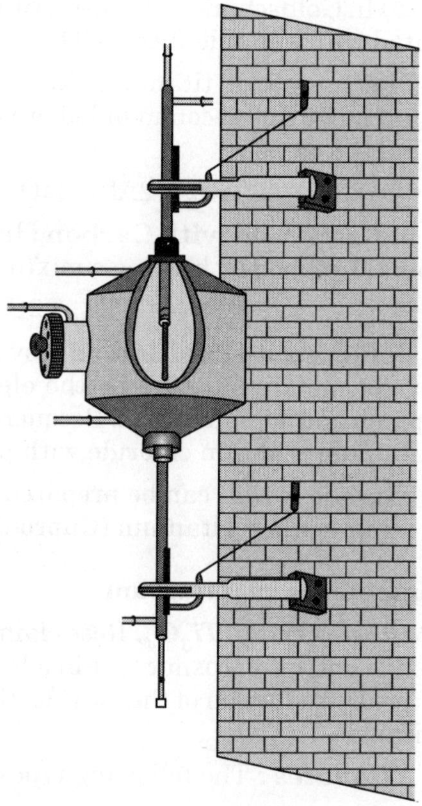

Fig. 1. The Vacuum Arc Furnace for Titanium Extraction.

The electrode holders, as well as the walls of the electric furnace is cooled by a current of water. The distance of the electrodes is regulated by a rack and pinion arrangement. There is a window in the furnace that enables one to have a view during the experiment. The lower negative electrode is insulated from the upper vessel for a distance of 2 - 3 mm between the electrodes. The electrical current used for the electrodes is about 25 volt @ of 60 - 70 ampere. Globules of titanium are formed at the lower electrode. The melting point of the metal is so high that the chief impurity aluminium is vapourized.

The apparatus is finally cooled and the titanium globules are detached.

The reaction for the equation is

$$\underset{\text{Fluorotitanate}}{\text{Potassium}} + \text{Aluminium} \longrightarrow \underset{\text{Fluoride}}{\text{Potassium}} + \underset{\text{Trifluoride}}{\text{Aluminium}} + \text{Titanium}$$

$$3K_2TiF_6 + 4Al \longrightarrow 6KF + 4AlF_3 + 3Ti$$

Reduction of Halogen Salts or the Oxide with Calcium Magnesium or Aluminium : (*a*) Titania is heated with calcium in a porcelain tube. Evacuated at about 600 and then heated at a temperature of 1,050°C, the titanium is isolated by washing it with a solution of ammonium chloride and drying at 100°C ($TiO_2 + 2Ca = 2\,CaO + Ti$).

(*b*) In Goldschmidt's Process, titanic oxide is mixed with aluminium and the mixture is ignited with suitable fuse $3TiO_2 + 4al = 3Ti + 2Al_2O_3$.

The sample of titanium thus obtained is allowed with about 5% of aluminium as impurity. Kütine recommended mixing barium or sodium peroxide along with aluminium for a quick action :

$$TiO_2 + 2Al + BaO_2 \longrightarrow BaO + Al_2O_3 + Ti$$

4. Reduction with Carbon : In 1895, Moissan obtained a sample contaminated with about 5% carbon by heating a mixture of titanic oxide and carbon in an electric arc furnace.

$$TiO_2 + 2C \longrightarrow Ti + 2CO$$

5. Electrode Position : Bussy claimed to have obtained silvery white titanium with a bright metallic lusture by the electrolysis of a solution of potassium titanate, sodium sulphate and sulphuric acid. Becquerel electrolyzed a concentrated hydrochloric acid solution of titanium and iron chloride with platinum electrodes. He got iron titanium alloy.

The alloys that can be prepared are (*i*) Silico-titanium Si 25-75% and titanium 5-70% (*iii*) Cupro-silico Titanium (Cupro-titanium and (*iv*) manganese titanium.

Compounds of Titanium

***Titanosic Oxide, Ti_3O_4*:** Rose claims to have obtained it by melting a mixture of titanium dioxide and microcosmic salt in a loosely covered crucible. Puccini and Marino prepared it by the dry reduction of the dioxide; they regard it as titanic magnetite $TiOTi_2O_3$ or titanium dititanate $Ti(TiO_2)_2$.

Titanates : The following types of titanates are important:

Metatitanate	R_4TiO_4	Na_4TiO_4
Paratitanate	$R_2Ti_3O_7$	$Na_2Ti_3O_7$
Mesotitanate	$R_4Ti_3O_8$	$Na_4Ti_3O_8$
Mesodititanate	$R_2Ti_2O_5$	$Na_2Ti_2O_5$
Orthodititanate	$R_6Ti_2O_7$	$2Zn_3Ti_2O_7$

1. *Potassium Metatitanate, K_2TiO_3 :* It is obtained by fusing titanium dioxide with potassium carbonate. Its hydride is prepared by precipitation of titanium dioxide and potassium hydroxide.
2. *Sodium Metatitanate, Na_2TiO_3 :* Rose has prepared this, he fused a mixture of titanium dioxide, with 5-12 time of weight of sodium carbonate. It comes down in the lower layer of the fused mass.
3. *Barium Metatitanate, $BaTiO_3$, and BaO TiO_2 :* Piccini has prepared it as an amorphous mass by the ignition of barium oxyfluorotitanate.
4. *Zinc Metatitanate, $CoTiO_3$:* This compound is obtained by heating to redness a mixture of titanic oxide, 2 parts zinc sulphate 8 parts, and potassium sulphate 3 parts. It is pale violet substance.

5. *Cobalt Metatitanate, $CoTiO_2$:* This compound is obtained by melting cobalt chloride with titanium dioxide.
6. *Sodium Orthotitanate, $NaTiO_4$ or $2Na_2O.TiO_2$:* Mallard prepared this compound of titanium by fusing sodium carbonate and titanium dioxide.
7. *Zinc Orthotitanate, Zn_2TiO_4 or $2ZnOTiO_2$:* This compound was prepared by Leavy, he heated to bright redness a mixture of 2 parts of titanium dioxide, 15 parts of zinc and potassium sulphate.
8. *Manganese Orthotitanate, $MnTiO_4$ or $2MnO.TiO_2$:* Hautefeville prepares this compound, by fusing a mixture of titanium dioxide, manganese fluoride or chloride and sodium chloride. Similarly he prepared ferrous orthotitanate Fe_2TiO_4 by fusion of a mixture of titanium dioxide a ferrous salt and sodium chloride.
9. *Potassium Pertitanate, K_2Ti_3 $2H_2O$ or K_23TiO_2 $2H_2O$:* This compound as prepared by Demely, he prepared it by adding potassium carbonate to a solution of titanium tetrachloride. It is insoluble in water.
10. *Sodium Metatitanate, $Na_4Ti_3O_8$ or $2Na_23TiO_2$:* This compound can be prepared by heating for 12 hours, 5 parts of sodium tungstate and a previously prepared, fused mixture of 2 parts of sodium carbonate and 1 part of titanium dioxide. It forms milk white short prismatic crystals.
11. *Sodium Mesotitanate, $Na_2Ti_2O_5$ or Na_2O $2TiO_2$:* This has been prepared by heating 20 parts of normal sodium tungstate and 5 parts of anhydride with a previously fused and powdered mixture of 4 parts of titanium dioxide and 8 parts of sodium carbonate. The crystals are insoluble in water.
12. *Barium Mesotrititanate, $Ba_2Ti_3O_8$ or $2BaO.3TiO_2$:* This compound is obtained by heating equivalent quantities of barium carbonate and titanium dioxide using an excess of barium chloride as a flux. The cold mass is extracted with dilute hydrochloric acid. It produces pale yellow-brown crystalline residue.
13. *Zinc Orthodititanate, $Zn_3Ti_2O_7$ or $3ZnO.2TiO_2$:* This was prepared by Leavy, he heated to redness a mixture of one part of titanic oxide, 5 parts of zinc sulphate and 1.2 parts of potassium sulphate and washing the product with dilute sulphuric acid. It forms a yellow crystalline powder.
14. *Zinc Mesopentatitanate, $Zn_4Ti_5O_{14}$ or $4ZnO,5TiO_2$:* This compound is formed at red heat from a mixture of titanium dioxide, 3 parts, zinc sulphate, 12 parts, potassium sulphate 1 part; the product being washed with dilute sulphuric acid. It forms small pale brown crystalline leaflets.

Titanium Trioxide, Pertitanic Acid and Pertitanate : Schömn and Walter showed it in 1870 that if hydrogen peroxide were added to a solution of titanium sulphate, an orange yellow colour is developed, and the reaction can be used fro the colourimetric analysis of titanium. According to Walter, if freshly precipitated titanium hydroxide is treated with neutral hydrogen peroxide or if a solution of titanium sulphide were treated with hydrogen peroxide in the presence of ammonium hydroxide, a yellow product is obtained which when dried over concentrated sulphuric acid produces a yellow heavy mass. The ratio

$TiO_2 : O = 1 : 0.8$ in the substances and when heated it gives off oxygen and water. Therefore, it is more or less decomposed hydrated titanic trioxide or Pertitanic acid TiO_3 nH_2O. It decomposes in acids forming a reddish yellow solution; sulphuric acid solution is freshly stable; the hydrochloric acid solution gives off chlorine; ether is without action and the reducing agent decolorize the liquid.

Classen (1888) prepared the substance by adding titanium tetrachloride to alcohol and treating the dilute solution with hydrogen peroxide in excess. When ammonia was added, a yellow liquid was obtained, which yielded a yellow precipitate. The formula proposed for the compounds are TiO_3; $TiO_2.H_2O_2$ or $Ti_2O_5.H_2O_2$. Faber (1906), Melikoff and Pissarjewsky (1898) suggested that during the action of hydrogen peroxide, the tetravalent titanium becomes hexavalent, but this is not symmetrical with the position of titanium in the Periodic Table. Perhaps in the hydrated titanium trioxide, the titanium is quadravalent.

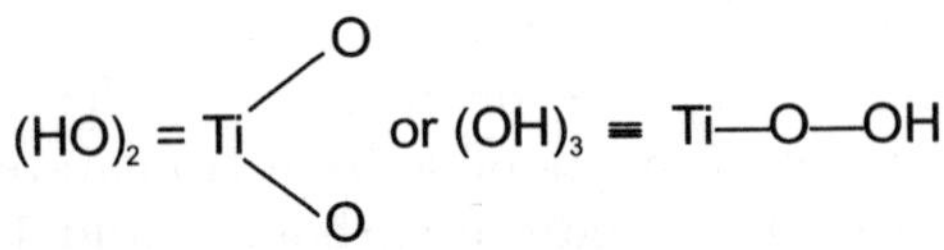

$TiO_2.H_2O$ $TiO(OH)_4$ or $TiO_3 2H_2O$

Pertitanic acid gives two series of salts (*i*) yellow and (*ii*) colourless.

Yellow Salts : They are prepared by adding well-cooled hydrogen peroxide to hydrated titanium trioxide and then sufficient well-cooled alkali type to dissolve all the solid. Now alcohol is added, which precipitate out of the salt, which is then washed with alcohol and ether. It is then dried on a porous plate. It is fairly stabled in the dried stated. Water prepared the sodium salt by treating titanium dioxide with sodium peroxide. The sodium salt Na_2O_2 TiO_3 $3H_2O$ has the following structure:

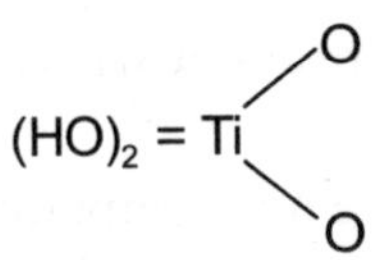

Pertitanic Acid

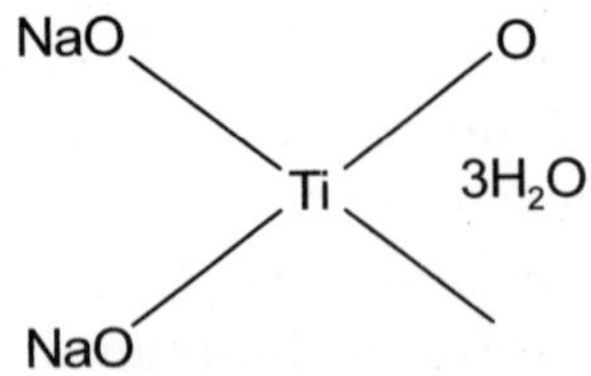

Sodium Peroxy Pertitanate

The ammonium peroxy pertitanate is $(NH_4)_2O_2TiO_3$ H_2O_2 or $(NH_4O_2)_2$ TiO_2H_2O.

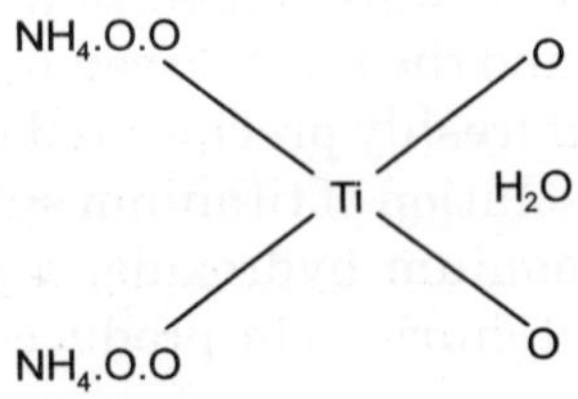

Colourless Salts : These salts are prepared by adding an extra amount of alkali and hydrogen peroxide otherwise the method is the same as used in the case of yellow precipitates. The alkaline lye is added until the yellow salts become colourless. Colourless sodium peroxyhydrate is $(Na_2O_2)_4\ Ti_2O_7\ 10H_2O$. The potassium salt $K_2O_4K_2O_2\ TiO_3\ 10H_2O$.

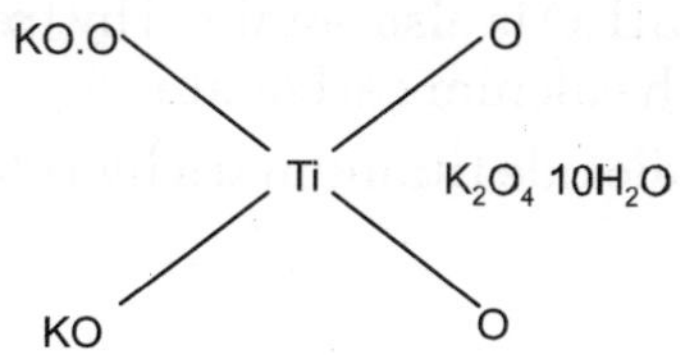

There is no much of evidence in favour of the structures propsed.

Potassium Salt

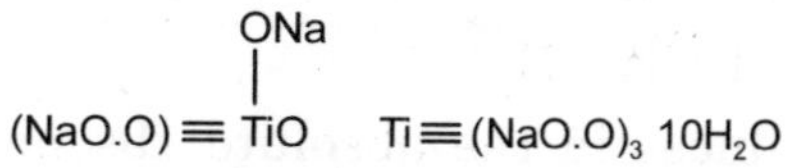

Fluorides : The fluorides of titanium are TiF_3 and TiF_4.

Titanium Trifluoride, TiF_3: Hautefelle has prepared it, he did it by heating potassium fluorotitanate, K_2TiF_6 in a stream of hydrogen mixed with a little hydrogen chloride. It produces dark violet prismatic crystals. Rammelsbery also obtained it by reducing a solution of potassium fluorotitanate with zinc and hydrochloric acid.

The complex salts of this trifluoride are known as fluorotitanates. Flurotitanic acid $HTiF_4$ cannot be isolated, when a strong solution of ammonium fluoride is added to titanium trifluoride a violet compound called ammonium pentaflurotitanate, $(NH_4)_2TiF_5$. Piccini obtained potassium pentaflurotitanate K_2TiF_5 as a deep violet precipitate by adding potassium hydrofluoride to an aqueous solution of titanium trifluoride.

Titanium Tetrafluoride, TiF_4: It has been prepared as yellow oily drops (i) by the distillation of titanium dioxide in a leaden retort, with flurospar sulphuric acid (ii) by the direct combination of titanium and fluorine (iii) by treating titanium tetrachloride with hydrogen sulphide.

$$TiCl_4 + 4HF \xrightarrow{100°C} TiF_4 + 4HCl$$

And also (iv) by the reaction of titanium tetrachloride with antimony tetrafluoride. The titanium tetrafluoride is a white solid at ordinary temperature; it has a boiling point of 284°C. It forms a clear solution with water and crystallizes from this solution as $TiF_4 \cdot 2H_2O$. It absorbs ammonia, forming TiF_34NH_3; at 120°C , TiF_42NH_3. Similarly, it unites with pyridines, forming a solid $TiF_4.C_5H_5N$. With phosphorous oxychloride, it reacts thus :

$$3TiF_4 + 4POCl_3 \longrightarrow 3TiCl_4 + 4POF_3$$

Hydroflurotitanic Acid, H_2TiF_6: As Berzelius suggested, the solution of titanium dioxide in hydrofluoric acid gives stable fluoride R_2TiF_6. This acid is also obtained by dissolving titanium in a worm mixture of hydrofluoric acid and nitric acid.

Ammonium Fluorotitanate, $(NH_4)_2TiF_6$: This compound is obtained by adding ammonia to a solution of titanium tetrafluoride so long as the precipitate first formed redissolves and evaporating the solution to crystallization. Sodium fluorotitanate Na_2TiF_6 is obtained by treating the tetrafluoride with caustic soda. Potassium fluorotitanate is prepared by adding potassium hydrofluoride to a solution of titanium dioxide in excess of hydrofluoric acid. Calcium fluorotitanate, $CaTiF_6\ 3H_2O$ is also obtained by treating solution of titanium tetrafluoride in hydrochloric acid with calcium carbonate.

Chloride : The chlorides known are titanium dioxide titanium trichloride and titanium tetrachloride.

Titanium Dichloride, $TiCl_2$: It is obtained :

(*i*) When titanium is dissolved in hydrochloric acid.

$$Ti + 2HCl \longrightarrow TiCl_2 + H_2$$

(*ii*) By heating the trichloride to redness in the absolute absence of moisture and atmosphere.

(*iii*) By reducing the tetrachloride of titanium with hydrogen sulphide.

$$TiCl_4 + H_2S \longrightarrow TiCl_2 + 2HCl + S$$

(*iv*) by the thermal decomposition of titanium trichloride below 600°C in a carbon tubes. It is a blackish powder, unstable in air and in presence of moisture. Thrown into water, it hisses like red-hot iron and dissolves with a large evolution of hydrogen, the black colour changes to white. When heated in air it catches fire, with dry ammonia in red heat it forms a nitride.

$$3TiCl_2 + 4NH_3 \longrightarrow TiN_4 + 6HCl + 3H_2$$

with bromine it forms a liquid $TiCl_2.Br_2$. Nitric acid oxidizes it first to the tetravalent (white) state. It reduces nitrous oxide to ammonia.

Titanious Trichloride, $TiCl_3$ or Titanious Trichloride : It is prepared

(*i*) by passing dissolved hydrogen through worm liquid of titanium chloride.

(*ii*) By leading the mixture of titanium tetrachloride and hydrogen through a red hot tube, the trichloride collects as a mass of dark violet scales in the tube just beyond the heated zone.

(*iii*) by the above reaction. Using a porcelain tube heated electrically by a wire spiral inside the tube and surrounded by a glass tube cooled externally by a rapid current cold water, the mixed gases transverse the space between the tubes at 785°C the yield is 21.7% and at 1200°C It is 94.3%.

$$TiCl_4 + Ag \longrightarrow TiCl_3 + AgCl$$

Or with mercury

$$TiCl_4 + Hg \longrightarrow HgCl + TiCl_3$$

(*v*) by heating a mixture of titanium tetrachloride and mercury diethyl ether at 100°C in an atmosphere of carbon dioxide.

$$2TiCl_4 + Hg(C_2H_5)_2 \longrightarrow AgCl_2 + C_4H_{10} + 2TiCl_3.$$

The trichloride, if amorphous from dark violet scaly crystals isomorphous with these off ferric chloride; it is delequescent, non-volatile, but decompose at 400°C.

$$2TiCl_3 \longrightarrow TiCl_2 + TiCl_4$$

When dissolved in water, it forms a violet-red solution. It is obtained, when titanium dissolves in hydrochloric acid, or by reducing a solution of the tetrachloride with tin and hydrochloric acid (removing the hydrogen sulphide). It can also be obtained by reducing a cold or boiling solution of chlorotitanate. The solution yields the crystals of titanium trichloride ($6H_2O$). The violet-red solution of trichloride is decolourized on exposure to air due to the formation of titanic acid. The aqueous solution on evaporation to dryness produces a blue oxychloride; nitric acid decolourizes the solution with the evolution of nitric oxide. Titanium trichloride solution with gold chloride produces colloidal gold. Selenates produce red selenium; telluride produces tellurium when heated with it.

Titanium Tetrachloride, $TiCl_4$: This compound is obtained

(*i*) by passing dried chlorine over titanium heated in a glass tube

(*ii*) by reacting titanium dioxide with chlorine at a very high temperature.

$$TiO_2 + 2Cl_2 \longrightarrow TiCl_4 + O_2$$

(*iii*) By passing chlorine over a red hot mixture of titanium dioxide:

(*a*) $$TiO_2 + 2Cl_2 + 2CO \longrightarrow 2CO_2 + TiCl_4$$

(*b*) $$TiO_2 + 2CHCl_3 \longrightarrow TiCl_4 + CO + 2HCl.$$

The tetrachloride is a transparent colourless mobile liquid, with a boiling point 135°C at 765 mm of pressure. It emits dense white fumes in air; it is reduced by hydrogen above 600°C forming the di- and trichloride with water, it forms a series of oxychloride $TiCl_3OH$, $TiCl_2(OH)_2$; $TiCl(OH)_3$ and with an excess of water $Ti(OH)_4$ is produced. Its hydrates are $TiCl_45H_2O$ and $TiCl_42H_2O$ etc.

With hydrogen fluoride or silver fluoride it forms citron-yellow $TiClF_3$, titanium trifluorochloride. Hydrogen sulphide reduces it to the dichloride, with alcohol, it gives crystals of Ti (OC_2H_5)Cl HCl.

Rose found that dry ammonia is rapidly absorbed by titanium tetrachloride, forming $TiCl_4.4NH_3$; and $TiCl_4.6NH_3$ and $TiCl_4.8NH_3$. Weber passed the dried vapours of ammonia form aqua regia over titanium tetrachloride and obtained a compound titanium dinitrosyl hexachloride.

$TiCl_4$. 2NOCl. with phosphoryl chloride, ir forms $TiCl_4.POCl_3$.

Titanium tetrachloride dissolves in fuming hydrochloric acid; the conductivity indicates the presence of hydrochlorotitanic acid, H_2TiCl_6. Its salt is ammonium chlorotitanate $(NH_4)_2$ $TiCl_6$ is obtained in yellow crystals by shaking a mixture of equivalent quantities of ammonia chloride and hydrochloric acid solution of titanium tetrachloride.

Iodides : The iodides known are $TiCl_2$, TiI_3 TiI_4.

Titanium Iodide, TiI_2 : It is obtained by the reduction of tetraiodide by silver and mercury. Stahler prepared hexahydrated titanium tri-iodide or titanious iodide. TiI_36H_2O

by the electrolysis of a solution of hydrated titanium dioxide in hydrochloric acid. It produces violet crystalline mass.

Weber made titanium tetraiodide, TiI_4, by passing iodine vapour over titanium and then removing the extra iodine by heating the product in on of the legs of sealed ∧- tube; the iodine and some tetraiodide dissolves in forming hydroiodic acid; probably hydroiodotitanic acid, H_2TiI_6 is formed.

Sulphides : These sulphides known TiS, Ti_2S_3 and TiS_2

Titanium Monosulphide, TiS : This is obtained by heating one of the higher sulphides in a platinum tube at a temperature above the softening temperature of glass (696°C); or by heating the sesquioxide in a current of hydrogen. It is reddish solid with bismuth like metallic appearance.

Titanium Sesquisulphide, Ti_2S_3 : This can be prepared

(*i*) by the action of carbon disulphide on red hot titanium dioxide or

(*ii*) by strongly heating powdered titanium dioxide in a mixture of hydrogen sulphide and the vapour of carbon disulphide and also

(*iii*) by heating the disulphide the in a current of nitrogen or hydrogen.

$$2TiS_2 \longrightarrow Ti_2S_3 + S$$

It is greenish black compound, crystalline in shape.

Rose prepared a dark-green mass, supposed to be titanium disulphide, TiS_2 by the action of carbon disulphide and titanium dioxide. It has also been prepared (i) by strongly heating a mixture of rutile, sodium carbonate and sulphur under a layer of charcoal or (ii) by the action of hydrogen sulphide vapours on titanium tetrafluoride in a hot tube.

$$TiCl_4 + 2H_2S \longrightarrow TiS_2 + 4HCl$$

It produces yellow titanious scales resembling mosaic gold.

Sulphates : The sulphates known are $TiSO_4$, $Ti_2(SO_4)_3$ and $Ti(SO_4)_2$. Titanium monosulphate, $TiSO_4$ is formed by evaporation of a solution of titanium in sulphuric acid, or by heating titanium trichloride in sulphuric acid and evaporating the solution in vacuum over line.

Stahler and Wirthwein prepared the sesqisulphate (rather the hydrosulphate) 3 $Ti_2(SO_4)_3H_2SO_4.25H_2O$, by repeatedly evaporating a concentrated solution of the chloride with dilute sulphuric acid in vacuum at 60°C. It forms double salts as sodium titanious sulphate $Na_2SO_4(SO_4)_3\ 5H_2O$; rubidium titanious sulphate, $Rb_3SO_43Ti_2(SO_4)_32H_2O$ etc. Trihydrated titanium disulphate, $Ti(SO_4)_{2.}3H_2O$ is a yellow amorphous mass obtained by evaporating a solution of the sesqisulphate oxidized with nitric acid and also by dissolving titanium dioxide in sulphuric acid. The basic sulphates as $Ti_2O(SO_4)_33H_2O$ or $TiSO_4$ are also known. Wöhler prepared the double salt as $K_2SO_4Ti(SO_4)_2$ by dissolving titanium dioxide in molten potassium hydrogen sulphate.

Carbonates : No carbonates is known, when ammonium carbonate is treated with titanic acid, ammonium titanic carbonate is produced.

Nitrates : When titanium dioxide is treated with nitric acid a solution containing titanium nitrate is formed. On evaporation, it produces the crystals of a basic salt $5TiO_2N_2O_56H_2O$.

Phosphates : Titanium dioxide dissolves in molten sodium metaphosphate, giving probably sodium titanium phosphate Na_2O $4TiO_23P_2O_5$. Various other basic double phosphates have also been obtained.

Complex Formation : In the group, titanium, zirconium, hafnium and thorium the tendency to form complex falls off as a whole from titanium to thorium the basicity increases. But since the simple salts of lighter elements are hydrolyzed by water, the number of double salt is much greater with thorium.

(i) Ammines : Titanium tetrafluoride forms tetra-ammines TiF_44NH_3. Which probably occurs in two isomeric forms. Similar ammines are also formed with the tetrachloride and tetrabromide. These ammines leave compounds such as N ≡ Ti - Cl and N ≡ Ti - Br when the ammonium salt is removed.

(ii) Oxygen Complexes : Just like stannic halides, titanic chloride and bromide readily combine with one or two molecules of oxygen compounds like ketone and esters and also with phosphorous oxychloride.

(iii) Chelate Complexes : All the elements from titanium to thorium have a strong tendency to form derivatives of b-ketones, the structure of which would depends on the co-valency maximum titanium forms salts of the type TiA_3X very much similar to siliconium salts which the titanium has the maximum covalence of 6; X is usually a complex anion as $FeCl_4^-$.

The formation of these Chelate rings may be followed thus: - The ordinary Chelate ketone ring (which we shall write as MA) would use one valency of the metal M and would give it covalency 2.

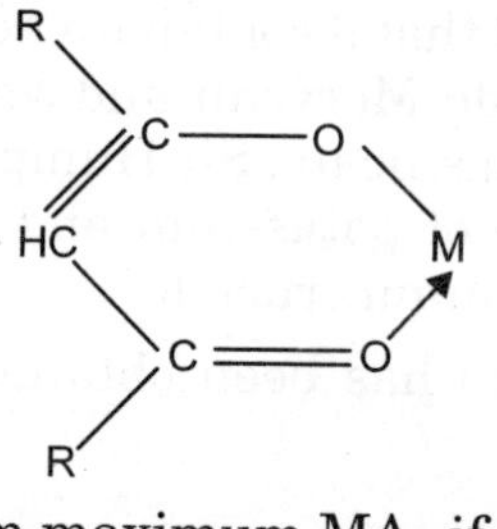

A trivalent element can form maximum MA_4 if it could have valency maximum of 8. If the covalency were restricted to maximum of 6, the chelate formed would be $[MA_3]'$. Where one valency is still unsatisfactory and it may form MA_3X. It might as well choose to take chelate rings, forming the divalent radical MA_2'' in the derivative.

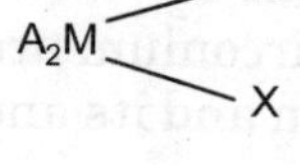

The group IV elements form chelate compounds of all these types :

$[SiA_3]X$; $[GeA_3]x$; GeA_2X_2; SnA_2X_2 $[TiA_3]X$; $[ZrA_3]X$; ZrA_4; HtA_4; ThA_4

(iv) *Oxalate Complex :* Titanium forms no oxalate of the formula $Ti(COO)_2$, it does not even form titanyl oxalate, but it forms complex titanyl oxalate.

$$K_2[Ti(C_2O_4)_2]\ 12H_2O$$

(v) *Catechol Compounds :* When titanic acid is dissolved in alkaline solution of Catechol $C_6H_4(OH)_2$ a complex type $M_2[Ti(C_6H_4)_3]H_2O$ is formed where M stands for Rb and Cs, depending upon the alkali used.

(vi) *Halide Complex $M_2[TiX_6]$:* All the four halogen produce these complexes of these fluorides are most unstable. From the conductivity of hydrogen fluoride and titanium tetrafluoride mixtures in water, it has been shown only in solution and they could not be isolated. Raman lines have indicated the presence of H_2TiCl_6 also in solution. A yellow solution of titanium tetrabromide in water turns blood red on the addition of concentrated hydrogen bromide, which has produced dark red crystals of $(NH_4)_2[TiBr_6]2H_2O$, the Diammonium titanium hexabromide.

ZIRCONIUM

Melting Point : 1,530°C
Boiling Point : 2,900°C

Atomic Number : 40
Atomic Weight : 91.224

History

Zirconium metals are widely distributed in nature and have been used for centuries; ancients have used zirconium for centuries. Zircon has been used by ancients those are called as *hyacinth (jacinth)* and *jargon* have been used as precious stones but the presence in them an unknown metal was not suspected until near the end of the 18th century. It was previously confused with aluminium because of its similarity. In 1789, Kalporth analyzed a specimen from ceylon and realized that it contains a new earth (about 70%) called *zerconite* or *zerconia* (or jargonia). Guyton de Morveau and Vaquelin confirmed his results. This mineral is now known as zirconium silicate. Sir Humphrey Davy failed to get the metal in 1829 by heating the clay mixture of potassium and potassium zirconium fluoride in a closed iron tube place inside a platinum crucible.

Later on the highly pure metal has been obtained by Lidwig and Wess (1910) and before that Moissan (1893).

Occurrence

Though it does not occur nature, it is found widely distributed a silicate or oxide. It is almost an absorbent a carbon, consisting nearly 0.017% of the earth's lithosphere. It's large workable deposit ar rare. It occurs in nature as an oxide in the mineral, baddaleyite (80% ZrO_2) complex zirconite as *Zerconite* (Ca, Fe)O. 2(ZrTi,Th)O_2 which has 50% ZrO_2; *Uhiligite* $Ca(Zr.Ti)_2O_5.Al_2TiO_5$ (33%,ZrO_2) and Wöhlerite $12R(Si.Zr)O_3R(Cb_2O_5)$(15-23%ZrO_3). The important silicates of zirconium are zircon, $ZrSiO_4$ (61 -67% ZnO_2) and its varieties as azorite, hyacinth and jargon and its and its altered forms as adelpholite, alvite orvilite etc. It also occurs in various tantalum and columbium minerals as challoamprite (5%) and columbite (1.11 %) forgusonite (2%), samorskite (0.4%).

Zirconium occurs in the alluvial soil sands and Ceylon and Trivancore.

Extraction

The process employed for extraction of zirconium from the native and impure oxides or silicates, depends on the degree of purity desired. The mineral is first pulverized to a fine state of sub division by heating and suddenly cooling the mineral (i) 90- 95% silicon can be removed by heating a mixture of the ore and carbon in an electric furnace at 2,220°C .Secondly sulphuric acid or hydrochloric acid or aqua regia will remove most of the iron. Silica from zirconium silicate cannot be hydrofluoric acid or aqua regia. If silica is to be removed fully, it is necessary to open up the material by sintering or fusion with suitable fluxes.

1. ***Fusion with Alkali Hydroxides or Carbonates :*** Kalporth fused the mineral with 5 times its weight of potassium hydroxide. During such a fusion, silica zirconia and titania are perhaps changes respectively the alkali silicate. Zirconite, titanate and iron are converted to ferric oxide. Zircon may also form a double alkali zirconium silicate. The fused mass is cooled and leached with water. The insoluble residue consists of hydroxides of iron, zirconium and titanium (and alkali zirconium silicate). When treated with hydrochloric acid, most of it dissolves, leaving a little silica. The solution is repeatedly evaporated with hydrochloric acid (silica and hydrofluoric acid). The zirconium oxychloride crystallizes as :

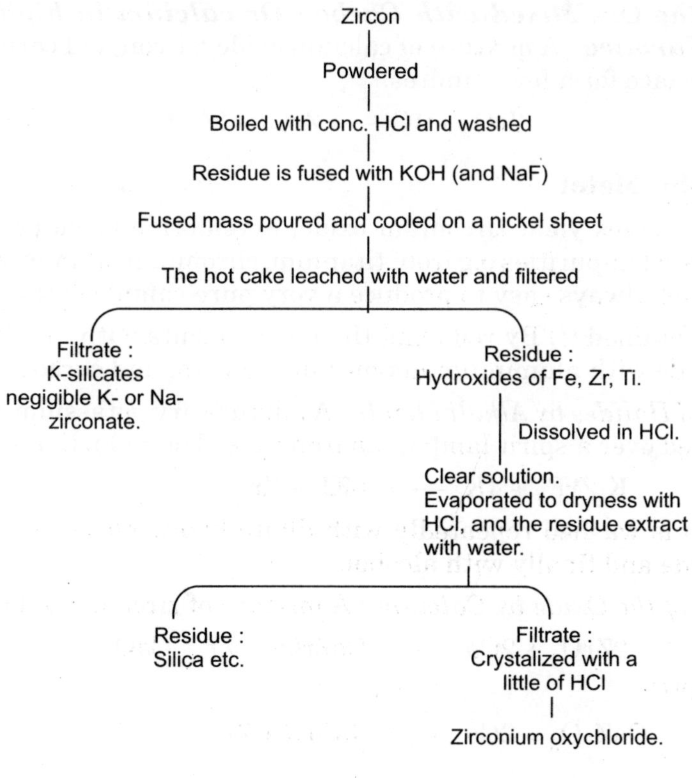

2. ***Fusion Alkaline Earths Or Lead Oxide :*** (BAYER'S METHOD) The powdered mineral is heated with barium carbonate in a graphite crucible. The barium zirconite is extracted from the cold mass with hydrochloric acid and the acid liquid is evaporated to dryness to make the silica insoluble. The residue is leached with hydrochloric acid, from which barium chloride is crystallized out and finally, the crystals of zirconium oxychloride, $ZrOCl_2\ 8H_2O$ also separates out.
3. ***Fusion With Alkali Pyrosulphate Or Hydrosulphate :*** The mineral is fused with potassium hydrogen sulphate (or sodium pyrosulphate) in a graphite, fused silica, or cast iron crucible. The cold cake is then treated with hot sulphuric acid, till all is dissolved and when cold, poured into cold water. The clear solution is decanted or filtered from the silica, which separates and is then ready for the recovery of zirconium. If may be precipitated as the phosphate and may be redissolved in hydrofluoric acid and reprecipitated as hydroxide.
4. ***Fusion with Alkali Hydrofluoride :*** The mineral fused with potassium hydrofluoride.

$$ZrO_2 + 4KHF_2 \longrightarrow K_2ZrF_6 + 2H_2O + 2KF$$

The cold cake is boiled with water containing hydrofluoric acid and the insoluble alkali fluozirconite are filtered off. On cooling the hot solution, the crystals of alkline flurozirconate are deposited. These are treated with sulphuric acid and evaporated to dryness and potassium sulphate removed by lixivation with water.

5. ***Heating the Ore Mixed with Carbon Or calcium in higher Temperature Electric Furnace :*** A mixture of calcium oxide, zircon and carbon is heated in an electric furnace for a few minutes.

$$ZrO_2 + 4C \longrightarrow ZrC_2 + 2CO$$

Preparation of the Metal

The extraction processes yield zirconium oxide, oxychloride or sulphate, together with various contaminated impurities e.g iron, titanium, chromium, aluminium, vanadium and rare earths. It is not always easy to produce a very pure sample of the metal.

The metal is obtained (*i*) By reducing the halogen salts with alkali metals as (*ii*) By reducing the dioxide with aluminium magnesium calcium and carbon.

(i) Reduction of Halides by Alkali Metals : A mixture dry potassium flurozirconate and potassium is heated over a spirit lamp in an iron tube. The reduction proceeds quietly.

$$K_2ZrF_6 + 4K \longrightarrow 6KF + Zr$$

The cold mass is washed repeatedly with dilute hydrochloric acid and a solution of ammonium chloride and finally with alcohol.

(ii) Reduction of the Oxide by Calcium : A mixture of zirconia and calcium is heated.

$$2ZrO_2 + 2Ca \longrightarrow CaZrO_3 + Zr + CaO$$

and again at 1050°C.

$$CaZrO_3 + 2Ca \longrightarrow 3CaO + Zr$$

The cold product is powdered and washed with water acetic acid, dilute hydrochloric acid, water and finally with acetone and then dried at an ordinary temperature, then in vacuum at 250°C and finally in an evacuated porcelain tube at 1000°C

(iii) Reduction of Halides by Aluminium : (*a*) The vapour of zirconium tetrachloride is passed over heated aluminium.

$$3ZrCl_4 + 4Al \longrightarrow 3Zr + 4AlCl_3$$

(*b*) A mixture of potassium fluozirconate, K_2ZrF_6 is heated with aluminium in an crucible to the melting point of iron, the cold mass is treated with hydrochloric acid to extract the metal (*c*) By the thermite process

$$ZrO_2 + 2Al + BaO_2 \longrightarrow Al_2O_3 + BaO + Zr$$

(iv) Reduction of Zirconia by Hydrogen Calcium Carbide Boron or Silicon : The following reaction may be used.

$$ZrSiO_4 + 2CaC_2 \longrightarrow ZrC + SiC + CaO + 2CO$$

The reduction with carbon been begins at 1400°C. Moissan stated that when zircon was heated in a carbon crucible in the electric furnace, he obtained a metal containing 5% of carbon and no nitrogen. The silica passes off as white fumes, with boron in an electric furnace, the metal thus obtained contained a small percentage of the boride and carbide.

(v) Electrolytic Reduction : A solution of zirconium tetrachloride mixed with a small proportion of ferric chloride (in the cathode compartment) is electrolyzed. Zirconium collects or the platinum wire cathode.

Physical Properties

Melting Point	1530°C
Boiling Point	2900°C
Density	
Specific Gravity	4.15
Specific Heat	0.0660
Entropy	9.5 at 25°C
Magnetic Susceptibility	
Hardness	4.7

Chemical Properties

Zirconium absorbs gases very firmly but zirconium does not form hydrides, though zirconium hydride ZrH_2 has been reported as grey deposit, when a current and crystalline zirconium are stable in air at an ordinary temperature. The metal, however, burns to an oxide if heated below redness. It has been report that the crystalline zirconium resists the action of air at red heat.

Amorphous zirconium and crystalline zirconium, contaminated with calcium oxide, are rapidly dissolved by hydrofluoric acid. Chlorine or bromine vapour attacks the heated metal forming the tetrahalides; liquid bromine acts slowly, where as iodine vapour is active at a high temperature only. Cold hydrochloric acid or sulphuric acid does not react on zirconium. But strong sulphuric acid evolves sulphur dioxide with the metal. Hydrochloric acid vapour forms the tetrachloride.

Zirconium combines with nitrogen to form the nitride at a temperature not exceeding 800°C; the metal is not attacked by hot or cold nitric acid also zirconium is dissolved in few minutes when heated in phosphoric acid. Zircon forms with carbon a carbide, with silicon a silicide and with boron a boride.

Alkali or its solution does not attract Very pure zirconium, but the ordinary zirconium evolves hydrogen. The amorphous zirconium is heated with an alkali carbonate it combines with the oxygen of the carbon dioxide with a slight explosion.

Zirconium does not appear to alloy with copper, silver, gold, magnesium, mercury and aluminium but in the latter case the compounds like $ZrAl_2$ and Zr_3Al_4 have been identified.

The alloys with lead, tungsten, nickel and cobalt are also reported.

Valency

The hydride ZrH_2 shows that zirconium may be bivalent, though the existence of ZrO is doubtful. The vapour density of the tetrachloride, $ZrCl_4$ shows the tetravalency; ZrO_2 also shows the tetravalency. Molecular weight has also been determined by the boiling point elevation method of zirconium acetonate.

Atomic Weight

The atomic weight is 91.224, as determined on the basis of such analytical ratios: $ZrBr_4$: 4AgBr; $ZrBr_4$: 4Ag. The atomic number is 40. The isotopes ar 90, 92 and 94. The electronic structure is

$$1s^2.2s^2.2p^6.3s^2.3p^6.3d^{10}.4s^2.4p^6.4d^2.5s^2.$$

Analytical Techniques

1. Zirconium salts give a white precipitate of zirconium hydroxide with reagents like ammonium sulphide and alkali hydroxide (the precipitate does not dissolve in an excess of the alkali, thus differing from aluminium and beryllium).
2. Ammonium carbonate gives a white smooth precipitate of boric carbonate, soluble in excess of the reagent, but reappearing on boiling the solution.
3. Ammonium oxalate and oxalic acid gives a crystalline precipitate of zirconium chloride (not given by Al or Ba).
4. A cold concentrated solution potassium sulphate slowly precipitates all zirconium as a complex, which unlike aluminium or beryllium salts, is insoluble in an excess of precipitant. The precipitates produced with the cold solution is readily soluble in dilute hydrochloric acid, but if produced in a boiling solution, a basic sulphate is precipitated, which unlike the thorium and cerium salts is not soluble in dilute hydrochloric acid.

5. Unlike thorium and yttrium salt solution, zirconium salt solutions give out no precipitate with hydrofluoric acid.
6. A concentrated solution of hydrogen peroxide precipitate zirconium as peroxide, $Zr(OH)_3HO_2$ from slightly acid solutions.
7. Sodium Azide completely precipitates zirconium as a hydroxide from cold solution.
8. A white precipitate of zirconium phosphate is obtained on the addition of zirconium sodium phosphate to a strongly acidified solution (acidified with hydrochloric acid or sulphuric acid) and warming.
9. Zirconium and aluminium can be quantitatively separated Cupferron (ammonium nitroso-phenyl hydroxyl ammine, $C_6H_5N\text{-}(NO)ONH_4$) Nitroso-β-napthol separates iron and zirconium).

USES

Zirconia is used in the manufacture of highly refractory wire crucibles, which not only withstands high temperature but also withstands many fluxes, slags and also fused alkalis. Zirconium carbide has been proposed as an abrasive since it is as hard a topaz. Zirconium oxide, silicate phosphate and basic sulphate are used in paints as good covering and non-toxicating substance.

Zirconium is used in manufacture of firebricks but has characteristics too, because zirconium forms carbide therefore cannot be used in steel producing furnace. Zirconium oxide has replace bismuth nitrate, for defining the intensities of X-Ray photographs in abdomens. Zircon is used in making knife-edges.

Zirconium oxide has also replaced lime cylinders of the limelight. The early incandescent gas mantles of Von Welsback contained much zirconia and so did the filament of Nernst lamp. The rod in the filament does not conduct at low temperature, but conducts only when heated.

Zirconium if mixed with clay and then molded into bricks they do not easily breaks and cracks.

Compounds of Zirconium

Oxides : The four oxides of zirconium are, ZnO, Zr_2O_2, ZrO_2, Zr_2O_5 and ZrO_3.

Zirconia, ZrO_2: It consists of ZnO_2 80.54, SiO_2 6.21 and TiO_2 6.12 and Al_2O_3 3.07 a Fe_2O_3 9.03%.

It is obtained

(*i*) the metal, hydride, nitride or carbide of the metal is air or oxygen.

(*ii*) By heating boric oxide with zirconium fluoride in carbon crucible.

$$3ZrF_4 + 2Br_2O_3 \longrightarrow 3ZrO_2 + 4BF_3$$

(born fluoride is volatile) and

(*iii*) By heating zirconium nitrate. Zirconium oxide is white; the presence of ferric oxide gives shades of colour to the pure zirconia. It has a specific gravity 5.006 its melting point of 2,585°C in an electric arc furnace.

When heated with carbon and chlorine, it produces the tetrachloride, which is also obtained by heating zirconium with chlorine and sulphur chloride mixture. The dioxide, which has been heated to incandescence, does not dissolve in sulphuric acid. The ignited oxide dissolves slowly in hydrochloric acid. The oxide, when moistened with a cobalt salt solution and calcined, it assumes a dirty violet colour.

Zirconium Hydroxide, $Zn(OH)_4$: It comes down as a white voluminous gelatenous precipitate when zirconium salt solution is treated with alkali solution, or even with ammonia, ammonium sulphate and potassium cyanide.

Zirconium nitrate or oxychloride when dialyzed produce colloidal solution of zirconium oxide. The jellies of zirconium borate, molybdate phosphate and arsenate can be prepared from zirconium hydroxide.

A ruby-red jelly of zirconium sulphosalicylic acid has been prepared by the action of zirconium oxychloride and sulphosalicylic acid.

Zirconium hydroxide contains varying amount of water corresponding with ZrO_2 H_2O or metazirconic acid H_2ZrO_3 or zirconyl hydroxide $ZrO(OH)_2$. It exhibits tendencies to polymerize.

When ammonia is added to a solution of zirconium sulphate containing hydrogen peroxide, a gelatinous hydrated zirconium peroxide $Zr(OH)_3O.OH$ is precipitated.

$$Zr(OH)_4 + H_2O_2 \longrightarrow H_2O + Zr(HO_2)(OH)_2$$

The hydrated peroxide dissolves n potassium hydroxide solution consisting hydrogen peroxide and the addition of alcohol to the solution at 3° precipitate potassium perzirconate, $K_4Zr_2O_{11}9H_2O$. The hydrated peroxide has been regarded as derived from the pentoxide Zr_2O_5 $4H_2O$ or a trioxide ZrO_3nH_2O.

Zirconates : Hydrated zirconium oxide is amphoteric, which, gives (*i*) normal salts with acids as $ZrCl_4$ (*ii*) a series of basic salts with bivalent ZrO, zirconyl radical as $ZrOCl_2$ and (*iii*) a series of zirconites or rather metazirconites in which zirconyl hydroxide $ZrO(OH)_2$ acts as an acid H_2ZrO_3, metazirconic acid.

When zirconia is melted with potassium hydroxide, potassium zirconate, $K_2O(ZrO_2)_3$ is formed. This is a soluble substance may be regarded as potassium perzirconate $K_2Zr_3O_7$. Sodium zirconate obtained under similar conditions, which also may be sodium orthozirconates, are $BaZrO_3$ $MgZrO_3$ $PbZrO_3$ etc.

Zirconium Fluoride, ZrF_4: It is obtained (i) by dissolving zirconia in hydrofluoric acid (ii) by passing hydrogen fluoride over white-hot zircon (iii) by heating a mixture of zirconium dioxide with twice its weight of ammonium fluoride or ammonium hydrofluoride. (iv) By the action of anhydrous hydrogen fluoride or zirconium tetrachloride.

$$ZrCl_4 + 4HF \longrightarrow ZrF_4 + 4HCl$$

It is a snow-white crystalline substance, which can be sublimed; Sp. gravity 4.43 at 16°C. It is sparingly soluble in water. This usual hydrate is ZrF_43H_2O. With liquid ammonia, it forms a white powder of $5ZrF_42NH_3$.

Zirconium fluoride complex is formed with other metallic fluorides, which are known as fluozirconates R_2ZrF_6. Zirconium fluoride absorbs hydrogen fluoride in cold forming fluozirconic acid, H_2ZrF_6.

Ammonium Fluozirconate, $(NH_4)_2ZrF_6$: This compound is obtained when a mixed solution of ammonium fluoride and zirconium fluoride is evaporated with excess of ammonium fluoride another hexafluozirconate $3NH_4F.ZrF_4$ or $(NH_4)_3ZrF_2$ is obtained. Normal sodium fluozirconate is obtained by mixing 1-2 parts of sodium fluoride with 14 parts of zirconium fluoride. Berzelius prepared potassium pentafluozirconate $KZrF_4H_2O$ by mixing solutions of potassium fluoride and zirconium fluoride.

Other fluozirconates of importance are $BaZrF_6$. $MgZrF_6\,6H_2O$. $MnZrF_4\,5H_2O$, $NiZrH_6\,6H_2O$ etc.

Zirconium Chloride : The Zirconium chlorides known are $ZrCl_2$,$ZrCl_3$ and $ZrCl_4$

When zirconium tetrachloride is reduced by aluminium magnesium zinc tin etc, the product of reaction at 250°C to 300°C is zirconium trichloride. Which is a brown solid; above 300°C it is converted to chloride $ZrCl_2$ and the tetrachloride:

$$ZrCl_3 \longrightarrow ZrCl_2 + ZrCl_4$$

When zirconium dichloride is heated above 600°C it produce zirconium.

$$2ZrCl_2 \longrightarrow ZrCl_4 + Zr$$

When zirconium burns in chlorine gas, zirconium tetrachloride is produced. It is also obtained by

(*i*) passing hydrogen chloride over hot zirconium

(*ii*) passing chlorine over hot mixture of zirconia and carbon.

$$ZrO_2 + 2C + 2Cl_2 \longrightarrow ZrCl_4 + 2CO$$

(*iii*) passing silicon tetrachloride over heated zirconia

$$ZrO_2 + SiCl_4 \longrightarrow ZrCl_4 + SiO_2$$

(*iv*) passing chloride over zirconium nitride at dark-redness or zirconium carbide at 500°C.

(*v*) treating zirconia with carbonyl chloride at 400°C or carbon tetrachloride and chlorine mixture at 400°C.

The tetrachloride forms white crystals from a strong hydrochloric acid solution. Its aqueous solution is hydrolyzed. It dissolves in water ir in alcohol with hissing sound. The alcoholic solution on heating gives off ethyl chloride and with hydrochloride bases it produces complexes of the type $M_2H_2ZrCl_6$ where M = pyridine Quinoline etc. The tetrachloride of zirconium is soluble in ether also and in the ethereal solution it forms complex compounds with ammonia and organic bases.

The tetrachloride of zirconium absorbs ammonia forming $ZrCl_44NH_3$; $4ZrCl_42NH_3$; $ZrCl_48NH_3$ etc. With phosphorous pentachloride; it gives $2ZrCl_45PCl_5$; with phosphorous oxychloride $2ZrCl_4.POCl_3$.

Chlorozirconates : When the vapour of zirconium tetrachloride is passed over fused potassium chloride, potassium chlorozirconate K_2ZrCl_6 is obtained; the corresponding sodium salt Na_2ZrCl_6 is also similarly obtained.

Zirconyl Chloride or zirconium Oxychloride, $ZrOCl_2$: When the aqueous solution of zirconium tetrachloride or the solution of zirconium hydroxide in hydrochloric acid is evaporated by concentration, small colourless needle lie crystals of zirconyl chloride $ZrCl_28H_2O$ are obtained. Its other hydrates are not known. When heated 100-110°C, zirconyl chloride becomes less soluble in water. Rosenheim and Frank gave the formula $Zr_2O_3Cl_23H_2O$, for the air dried substance, it is represented as

$$Zr_2O_3Cl_2.3H_2O \text{ or } ZrOCl_2ZrO_2.3H_2O$$

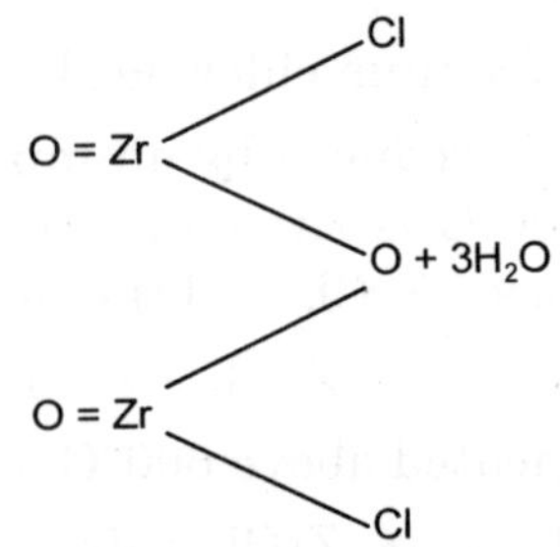

Bromides : Mellis prepared anhydrous zirconium tetrabromide, $ZrBr_4$, by passing carbon dioxide saturated with bromine vapour over an intimate mixture of zirconia and sugar charcoal. Bailey obtained it by heating zirconium in bromine. It is also obtained (i) by passing bromine vapour and carbon dioxide over zirconium carbide on red-hot state (ii) by reacting zirconium nitride with bromine.

The tetrabromide is a mirocrystalline, white, hydroscopic powder. When treated with water, it forms zirconyl bromide.

$$ZrBr_4 + H_2O \longrightarrow ZrOBr_2 + 2HBr$$

It is soluble in alcohol and in ether. With dry ammonia, it forms $ZrBr_4\ 10NH_3$, which is a white hydroscopic substance. It forms addition compounds with bases as ethylammine, aniline, pyridine etc. dissolved in hydrogen bromide (alcoholic solution) or pyridine hydrozirconate.

$$2\ C_5H_5NHBr + ZrBr_4 \longrightarrow (C_5H_6NH_2)_2ZrBr_6$$

The solution of zirconium hydroxide in hydrochloric acid when evaporated to dryness furnishes crystals of zirconyl bromide or zirconium oxybromide, $ZrOBr_2$

Iodide : Stahler and Denk prepared zirconium tetraiodide by heating zirconium or its carbide in a current of dry hydrogen iodide, the sublimate obtained is brown in colour instead of white. It may be $ZrI_4.I_2$. It forms addition products with ammonia as ZrI_48NH_3 and other organic bases. Its basic salt, as zirconyl hydrochloride $ZrO(OH)Cl_2\ 2H_2O$, has been obtained by the double decomposition of barium iodide and zirconyl sulphate.

Sulphides : The main sulphide is ZrS_2. It is obtained

(*i*) When the metal is heated with sulphur in an evacuated vessel, or in an atmosphere of hydrogen

(*ii*) Zirconium tetrachloride in sublimed in an atmosphere of hydrogen sulphide.

(*iii*) Zirconium carbide is treated with sulphur vapour at a dull red-heat and

(*iv*) Zirconia is heated in an atmosphere of carbon disulphide.

(*v*) Zirconium tetrachloride is treated with hydrogen sulphide a cinnamon-brown coloured compound is obtained. The normal sulphate is soluble in hydrochloric acid. When zirconium sulphide is fused with potassium hydroxide, it gives zirconia and potassium sulphide.

When zirconium sulphate is heated at 380°-400°C in a current of hydrogen sulphide and then cooled in the same gas, a light yellow powder is obtained which is zirconyl sulphide ZrOS; it spontaneously takes fire in air.

Sulphates : Berzelius prepared the normal zirconium sulphate $Zr(SO_4)_2$ by dissolving zirconia or zirconium hydroxide in an excess of concentrated sulphuric acid. It crystallizes out with 4 or 5 molecules of water. It produces hexagonal plates of the rhombic system. When heated to 200°C - 300°C it forms a basic sulphate, $2ZrO.SO_3$; at 900°C it gives off must of its sulphur.

Ruer showed that a solution of normal zirconium sulphate does not give the characteristic reaction exhibited by the chloride or nitrate when treated with oxalic acid or ammonium oxalate. Nor do these reactions occur with solutions of zirconyl chloride to which ammonium sulphate has been added. Therefore, he describes the normal sulphate as the structure of the zirconyl hydrosulphate, which is

$$ZrSO_4\ H_2SO_4 \text{ or } ZrO(HSO_4)_2$$

When potassium sulphate in crystals is added to a neutral solution of zirconium salts, the whole of the zirconium in solution is gradually precipitated as a crystalline basic double sulphate. The composition of such basic double sulphate is $3ZrO_3.SO_3xH_2SO_4$ or Zr_2O_3 $(KSO_4)_2$ $8H_2O$.

Nitrates : Berzelius evaporated a solution of zirconium hydroxide in nitric acid and obtained a yellow sticky mass. Weibull evaporated the solution at 65°C and obtained a white powder of $ZrO(NO_3)_2 2H_2O$, which may be zirconyl nitrate. The same substance is obtained, when the solution is evaporated over sulphuric acid. It is doubtful whether the normal nitrate could have ever been obtained. The commercial zirconium nitrate is some form of zirconyl nitrate. As reported earlier the hydrolysis of its aqueous leads to the formation of colloid solution; which may yield gels under proper conditions.

Phosphates : The colloid sols of zirconium phosphate and their gels in fact the basic phosphates. When zirconium is heated with phosphoric acid, about 2% solution is obtained. On cooling two kinds of octahedral crystals separates out, one double refracting and the other singly refracting. The empirical formula corresponds to the zirconyl pyrophosphate, $(ZrO_2)_2P_2O_7$. Zirconia when fused with sodium ammonium phosphate gives crystals of the metaphosphate $ZrO_2P_2O_5$ or $ZrO(PO_3)_2$ (zirconyl metaphosphate). If fusion takes place in the presence of sodium chloride sodium zirconium tetraorthophosphate, $Zr(Na_2PO_4)_4$ may be formed.

Complex Formation

(i) *Ammines :* Stahler and Denk have showed that the ammonium compounds of zirconium tetrahalides contain ammonium salt but the constitution of these products has not been studied.

(ii) *Oxygen Complex :* With the conception of fluoride zirconium halide readily combine of oxygen; compounds such as ketones, esters and ethers, but the products are usually stable at low temperature, above these they react internally with the elemination of hydrogen halide.

(iii) *Cellate Complexes :* Bitz and Clinch prepared zirconium acetylacetonate by the action of alkali or ammonia in a zirconium salt and acetylacetone in water. It takes up 10 moles of water during crystallization. It can be further crystallized from alcohol, when it produces the anhydrous substance. $Zr(CH_3COCHCOCH_3)_4$.

$$Zr(OH)_4 + 4CH_3COCH_2COCH_3 = Zr(CH_3COCHCOCH_3)_4 + 4H_2O$$

In Cellate complexes, the maximum covalency 8 is shown by zirconium.

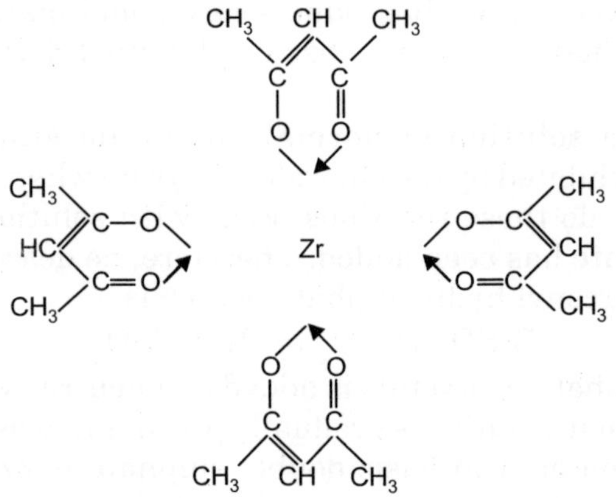

Actually the formation of the chellate takes place in several stages. At first zirconium tetrachloride forms an addition compounds with acetylacetone at -10°C which goes at 0°C to form $ZrCl_2A_2$ and this on boiling with benzene changes into $ZrCl_2A_3$; this further reacts with more acetylacetone in presence of ammonia to furnish ZrA_4

(iv) *Oxalate Complexes :* A number of complexes 8-covalent oxalates of the composition $M_4[ZrOx_2].xH_2O$ have been prepared (Na, $4H_2O$; K $4H_2O$; NH_4 $6H_2O$).

(v) *Catechol Complex :* Rosenheim prepared zirconium compounds with Catechol of crystallization as in $(NH_4)_2$ $[Zr(cat)_2]$ H_2cat $1.5H_2O$.

(vi) *Sulphate Complex :* Zirconium sulphate, $Zr(SO_4)_2$ obtained from concentrated sulphuric acid yields a good amount of heat of solution in water and a compound of the composition, $Zr(SO_4)_2$ $4H_2O$ separates out. This compound must be a complex since it has been found that on its electrolysis, zirconium gives to the anode. The hydrated sulphate has probably the complex structure:

$$H_2[O = Zr(SO_4)_2]\ 3H_2O$$

(v) *Halide Complexes :* Numerous complex fluorides are known which are sufficiently stable (though less than their titanium analogous). They contain 5 to 8 atoms of fluorine to 1 Zr. The pentafluoro salts contain one molecule of water. $M[ZrF_6(OH_2)]$, the hexafluro salts $M_2[ZrF_6]$ are particularly all anhydrous (M may be Li, Na, k and NH_4). The hepta fluorides $M_3[ZrF_7]$ are true 7-covalent compounds. The structure of anion being derived from the octahedron by the consequent distortion.

The salts are important in the separation of zirconium from hafnium. The octahedral $M_4[ZrF_8]$ are formed especially by divalent elements as Ni, Cu Zn, Cd and Mn.

HAFNIUM

Melting Point : 2150°C	Atomic Number : 72
Boiling Point : 5500°C	Atomic Weight : 178.49

History

Moseley stated that within the limits of his researches which covered all the elements between aluminium and gold, there spaces for three missing ones No 43, 61 and 75. It was then believed that the celtium whose arc spectrum was described by Professor Urban in 1911, the name of the element is number 72. However when Moseley and Urban examined the rare earth residue, reported to contain new elements, they found only about 10 lines all which could be attributed to ytterbium. In 1922 Urban resumed his research for the element 72 in the same rare earth residues, which he and Moseley had examined before war; an improved method of X-Ray analysis revealed two faint lines which almost coincided with those predicted for element 72. Where as Mendeleef predicted that this new element would be available in titanium ores, it was however, in zirconium ore those large quantities of element 72 were finally revealed.

Niels Bohr from his atomic theory, could see that the element 72 would not be a rare earth with trivalency and celtium could not be element to be tetravalent like zirconium. Bohr, therefore, advised, Georg von Hevesey to search for it in the zirconium ores. It was in Jan 1923 Drick Coster and Von Hevesey in Copenhagen brought their search for the new element in the zirconium firmly to a successful conclusion. They named it hafnium of the city of Copenhagen, through neither of them was Danish. Hafnium had laid hidden for untold centuries and when Von Hevesey examined some historic museum specimens of Julius Thomson, Morganac, Ramelsberg and others he found that they contained one to five present of these new element. It could not be discovered so far on account of its very dose analogy with zirconium.

Occurrence

Coster and Von Hevesey found upto 30% of hafnium in the different zirconium minerals obtainable from Norway, Greenland, Russia, France Ceylon and Brazil. Typical thorium mineral are free from hafnium content is in such secondary mineral like *alvite* which is *zirconium thorium hafnium orthosilicate, (Zr Hf Th)*O_2SiO_2 and in cyrtolite negeite and malacone. In the earth's crust it is available up to .002%.

Extraction

(i) Von Hevesey's Method : The mineral alvite is fused with ammonium hydrofluoride and extracted repeatedly with boiling water, which dissolves very considerable amounts of ammonium fluosilicate and zirconium hafnium, titanium, niobium, tantalum, vanadium, germanium, iron and manganese. Thorium and also the rare earth elements present remains insoluble, together with the unattacked mineral and insoluble zirconium and hafnium fluoride compounds such as sodium fluozirconate. Zirconium and hafnium fluoride compounds can be separated by methods based on differences in (a) the solubilities or (b) vapour pressure of the corresponding compounds and (c) partly of differences on their basicities.

According to Von Hevesey the most practical method of separation is the crystallization of the double fluoride of the potassium or ammonium. The mineral is melted with potassium hydrofluoride and the hexafluozirconate is extracted with boiling water in which this compound is soluble to the extent of 25%. On cooling the greater part of the salt crystallizes out.

This product is then converted into the sulphate and then into then oxychloride, which is recrystallised or treated with sulphur dioxide, which precipitates hafnium and zirconium from the boiling solution leaving most of the iron in solution. The double fluorides are best obtained by worming the dioxides with ammonium fluoride and hydrochloric acid. The salt resulting from the fusion is unstable. The zirconium double fluoride being less stable (26% at 20°C) than the hafnium double fluoride (nearly 3% at 20°C), the latter accumulates in the mother liquor. Still more convenient is the crystallization of the highly soluble $(NH_4)_2ZrF_6$ soluble at 20°C to the extent of 26%. The ammonium double fluoride of hafnium crystallizes in prismatic crystals. The hafnium content is raised from 2% to 38% by fractionation of the ammonium compounds and the subsequent fractionation of the potassium compounds produces a hafnium salt of 99.9% purity. The separation is controlled by X-Ray spectroscopy and the density determination of the oxides.

[Hafnia content $X = \dfrac{(D - 5.7\%)}{0.394}$

Where D is the absorbed density of the oxide]

(ii) From Malacone (Marquis and Urban) : The mineral is fused with potassium hydrogen sulphate (or decomposed with sulphuric acid). The soluble sulphates are extracted with water, leaving silica. The filtrate is concentrated and then a strong solution of potassium sulphate is added. Zirconium and hafnium can be separated on the basic of solubility differences (*a*) the solubilities of the double sulphates of the two are different (*b*) with sodium carbonate, both of them form double carbonate of which the zirconium double carbonate is extractable first. (*c*) Alkali precipitates zirconium before hafnium from simple salt mixtures and hafnium before zirconium forms a mixture of complex salts. A tedious repetition of some of these processes, specially no (*b*), finally concentrates hafnium at one end of the series, free from zirconium as seen in the spectrum.

(iii) Hafnium and Zirconium can be separated in the form of double oxalates : The mixed salts are precipitated with ammonia and then dissolved in oxalic acid. Thus we

obtain a mixture of ammonium zirconium oxalate and ammonium hafnium oxalate. When the mixture of the two is crystallized, hafnium accumulates in the mother liquor.

(iv) In the form of oxychloride : Hafnium oxychloride is less soluble than zirconium oxychloride. The oxychloride must be crystallized from 12N HCl

(v) Hafnium is more Basic than Zirconium : (*a*) When precipitating a solution of zirconium salt with ammonia, the first precipitate contains less hafnium than the successive ones. (*b*) By boiling a solution of zirconium and hafnium with sodium thiosulphate, zirconium being more strongly hydrolyzed, the first precipitate contains a larger amount of zirconium than the successive ones. (*c*) In the fractional crystallization, it has been found that it is better to work with ammonium fluozirconate and fluohafniate, than with the potassium salts. The ammonium fluohafniate collects in the mother liquor and the ammonium fluosilicate and the ammonium fluozirconate with the crystals.

(vi) It has been found that the compounds of hafnium tetrachloride with phosphorous pentachloride and oxychloride are more volatile than is the case with the corresponding zirconium compounds and hence hafnium concentrates in the earlier fractions.

(vii) Zirconium citrate is relatively insoluble in pure water whilst hafnium citrate is quite soluble.

Metallic Hafnium

It has been obtained (*i*) by passing the vapour of the tetraiodide, HfI_4 over a heated tungsten filament and (*ii*) by reducing the hafnium tetrachloride, $HfCl_4$ by

$$HfCl_4 + 4Na \longrightarrow Hf + 4NaCl$$

The metal has a crystalline structure similar to that of zirconium crystals belong to the hexagonal system. Hafnium is denser than zirconium and also possesses a higher melting point and higher boiling point it is diamagnetic.

Physical Properties

Melting Point	2150°C
Boiling Point	5500°C
Density	13.1
Heat of Capacity	6.15 cal deg^{-1} mol^{-1}
Standard of Formation	13.1 cal deg^{-1}mol^{-1}
Atomic volume	13.6 cc
Atomic Radius	15.8Å
Covalent Radius	14.8Å
First Ionization Energy	127 Kcal/mol
Electronegativity	13
Stable oxidation state	+4
Heat of Fusion	Kcal/gm atom
Heat of Vapourization	155 kcal/gm atom

Compounds of Hafnium

The properties of hafnium lie between those of zirconium and thorium. It is so similar to zirconium, that in fact any reaction characteristic of zirconium is also characteristic of hafnium. Not the slightest difference could precipitation with potassium oxalate, rubidium chloride, ammonium fluoride, picric acid and α nitroso β napthol.

Hafnia, HfO_2: It has a high melting point and in light emissivity it resembles zirconia. This oxide has been prepared by igniting the sulphate of hafnium at 1000°C. The specific gravity, the heat of formation and the molecular volume may be compared as follows: -

	TiO_2	ZrO_2	HfO_2	ThO_2
Specific Gravity	4.26	5.73	9.68	10.22
Molecular Volume	18.8	21.5	21.7	25.8
Heat of Formation	214.4	255	290	228

Hafnia melts at 2774°C its magnetic susceptibility is -0.11×10^{-6}. Hafnium gives peroxide of the type $HfO_3 2H_2O$ or $Hf(OH)_3$ O.OH.

The solubilities of some hafnium compounds are greater than those of thorium and zirconium e.g ammonium hexafluohafniate $(NH_4)_2HfF_6$ and ammonium heptafluohafniate $(NH_4)_3HfF_7$. The solubility of potassium hexafluohafniate K_2HfF_6 in 0.125 N at 20°C is also compared below.

Element	K_2TiF_6	K_2ZrF_6	K_2TiF_6	K_2HfF_6
Solubility	0.0483	0.0055	0.1008	insoluble

Hafnium Tetrachloride, $HfCl_4$: This compound is obtained at 250°C approximately and the first distil to contain rather more zirconium than the original mixture, Hafnium iodide HfI_4 has also been prepared.

Hafnyl Chloride, $HfOCl_4$: It is less soluble than the zirconyl chloride $ZrOCl_2$. It either be added to an alcoholic solution of hafnyl chloride a higher basic hafnium oxychloride $Hf_2O_3Cl_2$ $5H_2O$ is formed less soluble than the corresponding zirconium compounds. The indications are of the formation of H_2HfOCl_4.

Hafnium Sulphide, HfS : Van Archel and De Boer have obtained this compound by passing the vapours of the tetrachloride of hafnium and sulphur over hot tungsten filament.

Hafnyl Hydrophosphate, $HfO(HPO_4)_2$: This compound is prepared by the precipitation of the hafnyl dihydrogen phosphate $HfO(H_2PO_4)$ by 6N HCl. On ignition, it forms hafnyl metaphosphate $HfO(PO_3)_2$.

Atomic Weight

The atomic weight as suggested by Coster and Hevesey is about 178. The electronic configuration is given below:

$$1s^2\ 2s^2\ 2p^6\ 3s^2\ 3p^6\ 3d^{10}\ 4s^2\ 4p^6\ 4d^{10}\ 4d^{16}\ 4d^{14}\ 5s^2\ 5p^6\ 5d^2\ 6s^2$$

MOLYBDENUM

Melting Point : 2,622°C
Boiling Point : 3,560°C

Atomic Number : 42
Atomic Weight : 85.93

History

Until the later part of the eighteenth century, with native molybdenous sulphide and graphite, which look very much alike, were solid under the same name *molybdan* and *molybdenum*. It was in 1778, that Scheele showed that graphite and molybdenum mineral are extremely different substances. While nitric acid has no effect on graphite, it reacts with " molybdenite" or the so called "molybdenum" to give sulphuric acid and a white solid, which Scheele named as molybdic acid. Bergmann suggested this molybdic acid might lie in the oxide a new metal; but since Scheele did not have a suitable furnace, the work of the reduction of this oxide was entrusted to Hjlem. Hjlem attempted to reduce the oxide (molybdic acid) with carbon having pulverized and mixed the two with linseed oil to form a paste. Hjlem was successful getting the metal in 1781. In 1785, Pelletier proved that the ore, which mineral gets used to call *molybdenum* is a sulphide of that metal.

Occurrence

Molybdenum, does not occur in the native state in the elementary form. It is usually found as molybdenite MoS_2 and Wolfenite $PbMoO_4$ Torsile is the colloid form of molybdenite. Other molybdate minerals are Powellite, $CaMoO_4$ Belnonite, $MgMoO_4$, molybdic Ochre or molybdite a mixture of MoO_3 nH_2O and $Fe(MoO_4)_3$ nH_2O. Ochrematite, $3[3Pb(AsO_4)_2 PbCl_2]4Pb, MoO_5$; Puterite, impure $CoMoO_4$ and ilsemannite MoO_4MoO_3 or Mo_3O_8 nH_2O.

In India, molybdenite occurs in the Chotanagpur, Rajasthan, South Trivancore and Mahabagh and Urmi (Bengal). In the earth's crust it is only $n \times 10^{-6}$ %.

Extraction

Since Wolfenite ores have a high specific gravity and are easily wetted, water separation can be employed to remove the gangue. In case of molybdenum which is a flaky material and it is not easily wetted, electrostatic methods are used for concentration of this ore. Oil floatation process is also very helpful.

For the extraction of molybdate or molybdic acid forms the ores the following process have been suggested:

1. *Wöler and Berzelius Method :* The sulphide ore is roasted in a current of air and the sublimate of molybdenum trioxide is collected.
2. *Brunner Method :* An intimate mixture of sulphide ore with sand is roasted with stirring until the mass become yellow; the molybdic acid is then extracted with ammonia; the unoxidized residue is roasted again. The ammonical solution is then treated with a little ammonium hydrosulphide to precipitate copper and the filtrate is evaporated to dryness, again extracted with ammonia and crystallized with ammonium molybdate.

3. *Cristl's Process :* A mixture of powdered Wulfenite and sodium carbonate is melted and the molten mass is powdered out so as to separate as far as possible the alkali molybdate from lead oxide.

Lead Molybdate	+	Sodium Carbonate	$\longrightarrow$	Sodium Molybdate	+	Lead Oxide	+	Carbon Dioxide
$PbMoO_4$	+	Na_2CO_3	$\longrightarrow$	Na_2MoO_4	+	PbO	+	CO_2

The cold cake is extracted with hot water. The filtrate is treated with cold ammonia and ammonium carbonate to precipitate lead carbonate and the filtrate is mixed with nitric acid and evaporated to dryness. Sodium nitrate is removed from the molybdic acid by extraction with water.

Preparation of Elementary Molybdenum

(i) The molybdenum trioxide MoO_3 may be heated with hydrogen to a temperature (since the trioxide is volatile at high temperature) it is first reduced at a low temperature by hydrogen to form non volatile lower oxides; the temperature is then raised high.

$$MoO_3 + H_2 \longrightarrow MoO_2 + H_2O \text{ (300°C)};$$

$$MoO_2 + 2H_2 \longrightarrow Mo + 2H_2O \text{ (500°C – 600°C)}$$

(ii) Most of the molybdenum compounds are reduced by carbon at a high temperature to yield the metal of a high degree of purity containing only traces of molybdenum carbide).

$$MoO_3 + 3C \longrightarrow Mo + 3Co$$

(iii) A mixture of carbon and lime in an electric furnace generates molybdenite to molybdenum metal.

$$MoS_2 + 2CaO + 2C \longrightarrow Mo + 2CaS + 2CO$$

Molybdenite may also be heated with calcium oxide and calcium fluoride (used or flux).

(iv) Instead of carbon, boron, silicon or calcium vapour may also used for the reduction of the trioxide to the metal. Aluminium or even mischmetal has also been suggested as reducing agent.

(v) Ammonium or sodium molybdate or a mixture of molybdenum trioxide in molten potassium cyanide may be subjected to electro reduction and the metal is obtained. The best cathode for this purpose is mercury; 0.6 - 1.0 ampere and acidity 1.2 to 1.4N H_2SO_4 and a solution of sodium molybdate.

Ductile Molybdenum : The pressed rods from the powdered oxide are heated to 1,100°C - 1,150°C in a current of hydrogen, then heated electrically to 2,600°C - 2,700°C in a current of hydrogen. The resulted rods of the metal are ductile.

Colloidal Molybdenum : No suitable method for its preparation is known. T. Svedberg obtained it by electrical dispersion process in isobutyl alcohol. Küzel obtained the colloidal metal by grinding or by cathode disintegration. Attempts were made to have it by the reduction of its tri or dioxide with zinc.

Physical Properties

Melting Point	2,622°C
Boiling Point	3,560°C
Sp. Gravity	8.6
Ionic Radius	1.54Å
Hardness (Moh's Scale)	8.5
Tensile Strength	250 - 270 Kg
Torsion Modulus	13.5×10^{11}
Elastic Modulus	34.5×10^{11} dynes/ sqcm
Atomic Heat	1.35
Chemical Content	4.4
Entropy at 25°C	166.14 cal
Heat of Oxidation	7.5
Specific Heat	0.0141
Electric Resistance	5.6×10^{-6} ohm/cm

Colourless aqueous solution of molybdenum salt gives no characteristic absorption spectrum, a solution of the chloride in absolute alcohol gives a green solution, which if concentrated shows absorption bands in the red, blue and violet but no lines.

Chemical Properties

Like chromium and tungsten, molybdenum as one of these metals which according to its previous treatment can behave either as a *noble* (inactive) metal or as an *ignoble* (active) metal (Beckar and Healberg 1925). Where as the potential of the extreme positive form is +0.66 volt with respect to the normal hydrogen electrode; it is –0.74 volt for the extreme form.

The metal dos not form a hydride nor it absorbs hydrogen. It looses its lusture when exposed in air. Pyrotropic molybdenum if heated to 400°C is not inflammable in air. The metal burns in air oxygen at 500°C- 600°C to form MoO_2. Water has only a slow action on it, imparting a blue film on the metal. Fluorine does not act in the compact metal, but acts on the coarsely powdered metal with incandescence; so does bromine, forming several chlorides and bromides. Iodine has no action on red-hot metal. Where as hydrochloric acid slowly attacks it at 110°C forming some black oxide perhaps Mo_2O_3. Fused potassium chlorate attacks the metal violently, where as sulphur has no action on the metal at 440°C. Hydrogen sulphide attacks it at 1,200 forming a sulphide. Ethereal solution of sulphur monochloride readily attacks molybdenum and so does its vapour, the hot metals forms a complex $Mo_5S_8Cl_4$. When the metal is heated in sulphur dioxide vapour an oxide MoO_2 and sulphur is sublimed. Nitrogen and also ammonia have no action on the metal even when hot. When the metal is heated with nitrous oxide or nitric oxide, its trioxide MoO_3 is formed (and a little of MoO_2 is also produced) and so is the action of nitric acid. Whereas

70% nitric acid only slowly yields trioxide, the 50% acid yields reddish solution as residue. The solution reduces the permanganate solution, so that it may be inferred that the metal is not at once oxidized to the trioxide, but some molybdenum nitrate is also formed. Fused sodium nitrate oxidizes molybdenum rapidly. Molybdenum does not directly combine with phosphorous. The metal slightly dissolves in phosphoric acid. When heated in an electric arc furnace, with boron it forms a boride and with nitrogen in similar condition it yields a nitride. Molybdenum reduces carbon dioxide to carbon monoxide.

$$3CO_2 + Mo \longrightarrow MoO_3 + 3CO$$

Fused caustic potash slowly oxidizes the metal to trioxide.

Molybdenum dissolves in a solution of silver nitrate with the separation of silver. It precipitates gold from gold chloride solution and reduces the normal solution of mercuric chloride to mercurous chloride.

Analytical Reactions

1. A concentrated solution of alkali molybdate gives a white precipitate when heated with dilute mineral acids. The precipitate is soluble in the excess of the reagent.
2. If a trace of the molybdenum salt is mixed with concentrated sulphuric acid and evaporated nearly to dryness an intense blue coloured mass is obtained (if vanadic acid is present, a green coloured is obtained.

Uses

Molybdenum filaments have been prepared for lamps. The metal has been used as an ingredient for the arc light electrodes. Some of the molybdenum preparation have been used in photography is dying silks. It is also used in the preparation of blue colours and for the preparation of glass and glazes. Molybdic acid acts a depolarizing agent in the place of nitric acid. It s various compounds are and as chemical reagents in quantitative tests and qualitative measurements. The trisulphate of molybdenum MoS_3 used for the manufacture of solder for the filament of bulbs. Molybdenite is used in the manufacture of rectifier in the wireless telegraphy. The electro furnaces sometimes use molybdenum wires as resistance. It is used for making “molybdenum steel”. These steels have found use in motor car frames shafts, rods etc.

Valency

In matters of valency, molybdenum resembles chromium. The common valencies are 2, 3, 4, 5 and 6.

Bivalent : As in $MoCl_2$ (or Mo_3Cl_6)

Tetravalent : As in Mo_2O_3, $Mo(OH)_3$, $MoCl_3$ also in $Mo(CNS)_3$, $3KCNS.4H_2O$

Quadravelent : As in MoO_2, $MoCl_4$ in complexes as $Mo(CN)_4$ $4KCN.2H_2O$ and complex oxalates.

Pentavalent : As in $MOCl_5$ in complexes as Mo_2O_5

Hexavalent : As in $MoOF_4$, MoO_2F_2 and in Mo_2O_3 and the molybdates.

Atomic Weight

It has been calculated from each anhydride ratio's as $Pb(NO_3)_2$ $PbMoO_4$; MoS_2 MoO_3 $3(NH_4)_2$ MoO_4Mo; Mo_2 MoO_4; 2NaCl. The best value for the atomic weight is 76. The seven isotopes are 93, 94, 95, 96, 97, 98 and 100. The atomic number is 42 and the following is the electronic configuration.

$$1s^2\ 2s^2\ 2p^6\ 3s^2\ 3p^6\ 3d^{10}\ 4s^2\ 4p^6\ 4d^4\ 5s^2$$

Alloys

Molybdenum forms alloys with copper (at 1600°C in an electric furnace) with aluminium (alumino-thermite process), the alloys are Mo_2Al, Mo_4Al; MoAl etc. and also Al_7Mo and Al_3Mo (molybdenum aluminides).

Compounds of Molybdenum

Oxides : The known oxides of molybdenum are MoO, Mo_2O_3 MoO_2 and MoO_3 with such intermediates as Mo_2O_5 and Mo_3O_8.

Molybdous Oxide, Molybdenum Oxide, MoO : It is formed as a purple or violet powder when zinc and hydrochloric acid are allowed to act for a long time on fused or sublimed molybdenum trioxide. When concentrated potash lye reacts with molybdenum dibromide, a black solid is obtained which may be molybdous hydroxide $Mo(OH)_2$ or $Mo_3(OH)_6$. It gives of hydrogen and yields the trihydroxide $Mo(OH)_3$. It gives off hydrogen and yields the trihydroxide $Mo(OH)_3$.

$$6Mo(OH)_2 + 6H_2O \longrightarrow 6Mo(OH)_3 + 3H_2$$

The same product appears to be obtained by the action of caustic potash on molybdenum dihydroxyltetrabromide

$$2Mo_3(OH)_2Br_4 + KOH + 6H_2O \longrightarrow 6Mo(OH)_2 + 8KBr + 3H_2$$

Molybdenum Sesquioxide or Molybdic Oxide, Mo_2O_3: It is obtained in the form of hydrates $Mo_2O_3\ nH_2O$ for example, when reducing agents act in a solution of a molybdenum salt of molybdic acid. When molybdenum chloride is dissolved in hydrochloric acid, it reacts with liquid potassium amalgam, the solution becomes greenish, then black and finally opaque. This then precipitates the black hydrated sesquioxide. It is a pitch-black substance and is slowly oxidized by exposure to air. It is generally insoluble in acids and alkalis.

Molybdenum Dioxide, MoO_2: It has been obtained by (*i*) gently heating the metal in air for a short time. (*ii*) Heating molybdenum trioxide and (*iii*) heating ammonium molybdate in a covered crucible. Potash lye or hydrofluoric acid may remove the trioxide, which always accompanies during these preparations. Mauro and Panebiano obtained the dioxide by heating a mixture of the trioxide MoO_3, potassium carbonate and basic oxide, B_2O_3 at a high temperature. It is a brownish-black powder. It is a good electrical conductor and is reduced to the metal when heated in a stream of hydrogen. Chlorine converts it to the dioxychloride, MoO_2Cl_2. Carbon tetrachloride converts it to the tetrachloride, $MoCl_4$. It is insoluble in alkali solutions. The salts, derived from it and prepared in an indirect way are known as hypomolybate e.g., magnesium hypomolybate $2MgO.\ 3MoO_2$ is formed by fusing sodium trimolybdate with magnesium oxide and similarly the zinc hypomolybdate, $2ZnO.3MoO_2$.

Molybdenum Trioxide, Mo_3O_8 : Muthmann obtained it by the action of reducing agents on a solution of molybdic acid or by boiling molybdenyl sulphate with molybdenum powder. It is blue coloured oxide, also known as *molybdenum blue*. Molybdenum trioxide also produces a colloidal solution, it is usually mixed with a little tetrahydroxide, $Mo(OH)_4$ also. The various hydrates of this oxide $Mo_3O_8\ nH_2O$ are also known.

Molybdenum Hemipentoxide, $Mo_2O_53H_2O$: This compound is obtained on mixing hydrochloric acid solution of molybdenum dioxide and molybdic acid, the precipitate is washed with an alcoholic solution of ammonium chloride and dried over sulphuric acid. It resembles freshly prepared ferric hydroxide in appearance and is also prepared by the action of ammonia on ammonium molybdenyl chloride.

Molybdenyl Trioxide, MoO_3: It is obtained (*i*) by the final oxidation of molybdenum (*ii*) by roasting molybdenum sulphide in air (*iii*) when steam acts on the metal or on the lower oxide of the metal (*iv*) when potassium hydroxide reacts with the metal (hydrogen is evolved in this reaction) (*v*) when nitric acid acts on the metal (*vi*) by roasting the ammonium molybdate in a current of oxygen. Or (*vii*) by sublimation of ammonium phosphomolybdenum. The trioxide is purified by sublimation in a current of air quartz tube. Thus it is freed from metallic impurities, e.g., tungsten trioxide, silica and phosphates. The trioxide has a platy or talcose structure and colourless in appearance. The fused mass is grayish or yellowish white it has a melting point of 795°C and boiling point 1155°C, when heated it assumes lemon-yellow colour on cooling, the original colour is again restored. When the trioxide is introduced in the oxy-hydrogen flame, the flame gives a bright lustrous metallic mirror on a cold surface. The sublimation of the trioxide begins in the proximity of the melting point.

The crystal structure of the trioxide shows that the oxygen atom forms colloidal octahedra round the molybdenum atoms.

The molten trioxide is a good conductor of electricity and on electrolysis produces crystals of trioxide, Mo_3O_8 out of the cathode. The electrolysis of a hydrochloric acid solution of molybdenum trioxide or of a solution of tetravalent molybdenum chloride, $MoCl_3$ or sulphate, $Mo_2(SO_4)_3$.

Hydrogen reduces the trioxide to MoO_2 between 300°C and 470°C and to the metal above 500°C. The trioxide on reducing with chlorine gives MoO_2Cl_2. Hydrochloric acid at 150°C to 200°C forms the white dioxide, MoO_32HCl. The boiling hydrochloric acid forms a blue solution with the evolution of oxygen and with boiling hydrochloric acid iodine separation and a green and then a blue solution is formed. Stannous chloride reduces the trioxide to a blue oxide. Potassium dichromate reacts with this oxide as follows :

$2K_2CrO_4 + MoO_3 \longrightarrow K_2Cr_2O_7 + K_2MoO_4$ and with molten potassium dichromate, potassium chromomolybdate is produced.

$$K_2Cr_2O_7 + MoO_3 \longrightarrow K_2Cr_2MoO_4 + 3O_2$$

Colloidal Molybdic Acid : When an excess of hydrochloric acid is added to a solution of sodium molybdate, molybdic acid is not at once precipitated. It can be dialyzed to a colloidal solution.

$$Na_2MoO_4 + 2HCl \longrightarrow 2NaCl + H_2MoO_4 \text{ [metamolybdic acid]}$$

Molybdic Acid, or mohydrate of molybdic oxide, MoO_3 : It has been prepared by the action of nitric acid on magnesium molybdate, or on ammonium molybdate, or on lithium molybdate. It is known to exist in two varities namely the α and β. The alpha form is found in the asbestos-white needles and the beta form constitutes small white needles.

When a solution of alkali molybdate in nitric acid is allowed to stand for a time, dihydrate of molybdic acid MoO_3 $2H_2O$ or H_4MoO_5 and is known as *paramolybdic acid*. The *orthomolybdic acid* would be then MoO_3 (this is assumed).

That is H_6MoO_6, which however does not exist in the solid form. Among these solid form of molybdic acid; the yellow acid H_4MoO_5 has the greatest solubility in water. The dihydrate has also been considered as octomolybdic acid. $H_2O.8MoO_3$ or $H_2MoO_8O_{25}$.

Molybdates : The molybdates are classified would be seen from the following a table (where R = Monovalent base).

Name of the class	General formula	Example
Normal molybdate	$R_2MoO_4.nH_2O$ ($R_2O.MoO_3$)	$(NH_4)_2MoO_4$: Na_2MoO_4; $ZnMoO_4$
Dimolybdate	$R_2Mo_2O_7.nH_2O$ ($R_2O.2MoO_3$)	$(NH_4)_2Mo_2O_7$; $Ag_2Mo_2O_7$
Paramolybdate	$3R_2O.7MoO_3.nH_2O$	$3(NH_4)_2O.7MoO_3.4H_2O$; $3K_2O.7MoO_3.4H_2O$; $3Li_2O.7MoO_3.12H_2O$
Trimolybdate	$R_2Mo_3O_{10}.nH_2O$ ($R_2O.3MoO_3$)	$Na_2O.3MoO_3.7H_2O$; $CaO.3MoO_3.6H_2O$
Tetramolybdate	$R_2Mo_4O_{13}.nH_2O$ ($R_2O.3MoO_3$)	$(NH_4)_2O.4MoO_3.2H_2O$; $CaMo_4.O_{13}.9H_2O$
Hexamolybdate	$R_2Mo_6O_{19}.nH_2O$ ($R_2O.6MoO_3$)	$(NH_4)_2O.6MoO_3.5H_2O$
Octomolybdate	$R_2Mo_8O_{25}.nH_2O$ ($R_2O.8MoO_3$)	$Li_2O.8MoO_3.11H_2O$; $Na_2O.8MoO_3.17H_2O$
Decamolybdate	$R_2Mo_{10}O_{31}.nH_2O$ ($R_2O.10MoO_3$)	$(NH_4)_2O.10MoO_3.13H_2O$; $K_2O.10MoO_3.9H_2O$
Dodecamolybdate	$R_2Mo_{12}O_{37}.nH_2O$ ($R_2O.12MoO_3$)	$CaO.12MoO_3.23H_2O$; $3K_2O.Al_2O_3.12MoO_3.20H_2O$
Tetradecamolybdate	$R_2Mo_{14}O_{43}.nH_2O$ ($xR_2O.14MoO_3$)	$5NiO.14MoO_3.70H_2O$

1. *Normal Ammonium Molybdate $(NH_4)_2MoO_4$:* The compound crystallizes from the concentrated solution of molybdenum trioxide in warm concentrated aqueous ammonia during or on adding alcohol.
2. *Anhydrous Normal Sodium Molybdate (Na_2) NoO_4:* This is obtained by melting together equimolecular sodium carbonate and molybdenum trioxide.
3. *Normal Potassium Molybdate K_2MoO_4:* This compound is prepared by evaporating for crystallization of a solution of molybdic acid or an acid molybdate in an excess of potassium hydrpode or carbonate. It has three transformations α and β at 479°C β and γ at 454°C and gold at 327°C.
4. *Normal Silver Molybdate Ag_2MoO_4:* This compoud is produced by the addition of silver nitrate to a solution of molybdate.
5. *Calcium Molybdate, $CaMoO_4$:* Calcium chloride when reacts with sodium molybdate produces this compound.
6. *Ammonium Dimolybdate, $(NH_4)_2Mo_2O_7$:* This is obtained by evaporating the mother liquor for the preparation of the ammonium molybdate to dryness.
7. *Sodium Dimolybdate, Na_2O2MoO_2 :* This has been obtained by fusing the require proportion of sodium carbonate or nitrate and molybdenum trioxide. The potassium salt can also be prepared similarly.
8. *Ammonium Paramolybdate, $3(NH_4)_2O$. $7MoO_34H_2O$:* This compound is obtained by the evaporation of a solution of molybdic acid in aqueous ammonia.
9. *Potassium Permolybdate, $3K_2O7MoO_34H_2O$:* This can be prepared by mixing a solution of molybdic acid with potassium carbonate drop by drop, with hydrochloric acid or nitric acid until a permanent turbidity is obtained, the solution soon deposits crystals of the salt.
10. *Ammonium Trimolybdate $(NH_4)_2O3MoO_3$, nH_2O :* This compound is formed during the spontaneously evaporation of a solution of ammonium paramolybdate.
11. *Sodium Trimolybdate, Na_2O $3MoO_3$ $3H_2O$:* This compound is obtained by treating a concentrated solution of molybdic acid and sodium carbonate, in the presence of nitric acid. The solution then gradually deposits crystals when allowed to evaporate spontaneously.
12. *Potassium Trimolybdate, $K_2O3MoO_33H_2O$:* This compound can be prepared by the action of the paramolybdate with cold water and by treating a solution of molybdic acid and potassium carbonate with nitric acid.
13. *Ammonium Tetramolybdate, $(NH_4)_2OMoO_32H_2O$:* This is supposed to be formed by heating the paramolybdate at 140°C- 150°C.
14. *Sodium Tetramolybdate, Na_2O4MoO_3 $5H_2O$:* This compound is prepared by putting drops of a hot concentrated solution of trimolybdate on a glass plate and allowing it to evaporate spontaneously. The potassium salt K_2O $4MoO_37H_2O$ has been obtained from the liquid produced by adding nitric oxide to a solution of molybdic acid in potassium carbonate.

15. *Ammonium Hexamlybdate, $(NH_4)_2O6MoO_3\ 5H_2O$:* This has been obtained by adding calculated quantity of N-HCl to a cold solution of ammonium paramolybdate and recrystallizing.
16. *Sodium Octamolybdate, $Na_2O\ 8MoO_3\ 17H_2O$:* This has been prepared by mixing a solution of normal sodium molybdate and a calculated amount of hydrochloric acid and spontaneously evaporating the clear liquid.
17. *Potassium Octamolybdate, $K_2O8H_2O\ 13H_2O$:* This has be prepared by the trimoybdate in a solution of molybdic acid. The precipitate dissolves when the liquid is warmed and when the filtered solution is cooled, the salt separates out as crystals.
18. *Ammonium Decamolybdate, $(NH_4)_2O10MoO_313H_2O$:* Klason prepared this compound by making a reaction between N-HCl and with ammonium paramolybdate.
19. *Potassium Aluminium Dodecamolybdate, $3K_2O$, $Al_2O_3\ 12MoO_3$:* When a freshly prepared aluminium hydroxide is bonded with potassium trimolybdate.
20. *Nickel Tetradecamolybdate, $5NiO\ 14MoO_3\ 70H_2O$:* This compound can be prepared from the mother liquor of nickel Hexamlybdate.

Molybdenum Blue

When an acidified solution of molybdate or molybdic acid is treated with such reducing agents s SO_2 H_2S or Hi, hydrazine glucose or even molybdenum metal, an intense blue colouration is obtained due to the formation of a colloidal substance known as molybdenum blue. Its formation is used as an identification of the quantitative molybdenum.

In molybdenum blue (which is of varying and indefinite composition), the valency of Mo is between 5 and 6. The product is Mo_9O_{26} and Mo_8O_{23}. Glemser did obtain some of its hydrates in crystalline form.

(*i*) $Mo_8O_{23}8H_2O$ (*ii*) $Mo_4O\ H_2O$ and (*iii*) $Mo_2O_5\ H_2O$. Since water in these compounds is constitutional. Glemser represented them as follows:

(*i*) $Mo_8O_{15}(OH)_{16}$;
(*ii*) $Mo_4\ O_{10}(OH)_2$ and
(*iii*) $Mo_2O_4(OH)_2$.

The compounds (*ii*) and (*iii*) are also obtained by the action of atomic hydrogen on MoO_3.

The reaction between $LiAlH_4$ and MoO_3 forms a red compound of composition $Mo_5O_7(OH)_8$. Tredwell (1946) prepared a readily soluble molybdenum blue Mo_3O_8OH by the electrolytic reduction of ammonium paramolybdate in presence of sulphuric acid.

Permolybdic acid and Permolybdate : Werther observed that the acid solution of the molybdates, when treated with hydrogen peroxide gives a yellow colouration, which cannot be extracted by ether. Fairley showed that the colour is due to the formation of permolybdic acid. This yellow or orange solution depends on insoluble yellow compound also. Fechard said that permolybdic acid $HMoO_4$ or $Mo_2O_7\ 5H_2O$ is isolated (i) by decomposing the barium salt with sulphuric acid. (*ii*) By heating dihydrate molybdic trioxide with a solution of hydrogen peroxide (*iii*) by the action of hydrogen peroxide on molybdenum or its blue oxide, when the solution of permolybdic acid thus obtained is evaporated in vacuum at ordinary temperature, it yields a yellow crystalline powder soluble in water. When heated with hydrochloric acid, it produces chlorine.

It is very likely that there are two permolybdic acids, *permonomolybdic acid,* H_2MoO_5 *or* $MoO_3\ H_2O_2$ and *perdimolybdic acid,* H_2MoO_6 *or* $MoO_3\ 2H_2O_2$ (i.e. $H_4MoO_7 - H_2O = H_2MoO_6$) with the following constitutions (Pissarjewisty).

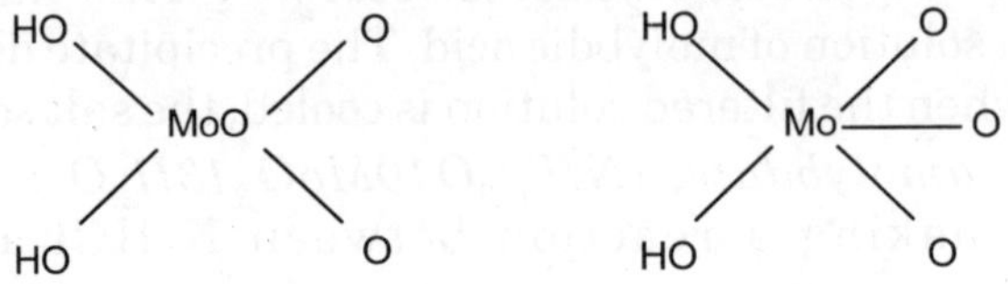

Permono Acid Peridi Acid

21. *Potassium Monopermolybdate,* K_2MoO_5: If potassium dihydroarsen-atomolybdate is dissolved in water and treated with hydrogen peroxide and the liquid allowed to stand for some times, an orange yellow powder of the molybdate is formed. It is not soluble in cold water but dissolves in warm water.
22. *Potassium Dipermoybdate,* $K_2MoO_6\ H_2O$: When the cold solution of potassium monopermolybdate is treated with a solution of potassium hydroxide and hydrogen peroxide and alcohol is added to the liquid it settles down to a reddish precipitate, but becomes yellow, when exposed to air and oxygen is give off. This compound also explodes when triturated and gives off oxygen when acted upon by water.
23. *Potassium Permolybdate,* $KMoO_4$ *or* K_2MoO_8 : The compound is obtained when 20% hydrogen peroxide reacts with powdered fused potassium trimolybdate. It is sparingly soluble in cold water, but more soluble in hot water. The aqueous solution, when boiled, deposits potassium dimolybdate.

Sodium Permolybdate, $Na_2Mo_2O_8$: This compound is produced by dissolving sodium dimolybdate in hydrogen peroxide and evaporating the solution in vacuum. It produces yellow prismatic crystals, which begin to melt in their water of crystallization and give off oxygen at 200°C.

Ammonium Permolybdate, NH_4MoO_4 ***or*** $(NH_4)_2\ MoO_8$: This compound is formed by the action of hydrogen peroxide on ammonium paramolybdate and drying the compound in vacuum.

Ammonium Perparamolybdate, $3(NH_4)_2O\ 7\ MoO_4\ 12H_2O\ (NH_4)_2Mo_2O_8$: This compound can be prepared from the reaction between 20% solution of hydrogen peroxide and ammonium permoybdate and then red crystals which completely decompose at 106°C.

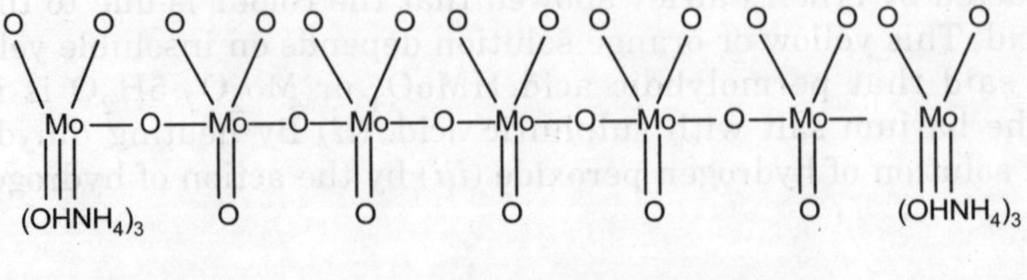

Potassium Pertrimolybdate, $K_2O2MoO_3MoO_43\ H_2O$: It has been formed by the evaporation over P_2O_5 of a solution of potassium trimolybdate in 20% hydrogen peroxide.

Fluorides

Molybdenum Trifluoride, MoF_3: This is obtained by dissolving hydrated molybdic oxide and hydrofluoric acid. It forms a sticky water-soluble purple-red mass, when heated it changes into an insoluble brown mass.

When a solution of molybdenum trioxide in hydrochloric acid is electrolytically reduced, and potassium fluoride is added to the liquor a violet crystalline precipitate of potassium molybdenum tetrafluoride is produced, when ammonium fluoride is added to the liquor then ammonium molybdenum tetrafluoride, $(NH_4)MoF_4$ is produced.

Molybdenum Tetrafluoride, MoF_4: This has been obtained as a red solution by the action of hydrofluoric acid on hydrated molybdenum dioxide. On evaporation the, liquor becomes blue and finely produces a black precipitate.

Molybdenum Hexafluoride, MoF_6: This has been obtained by the action of (*i*) hydrofluoric acid on molybdenum pentachloride, or of (*ii*) antimony pentachloride on molybdenum pentachloride. (*iii*) Direct action of fluorine on molybdenum at 60-70°C and collecting the product in a glass receiver at 70°C. It forms a white crystalline mass, which has a melting point of 17.5 and a boiling point of 35°C. It is immediately hydrolyzed by moisture otherwise it is quite stable not affected by dry air, chloride or sulphur dioxide.

> Molybdenum Pentafly-fuoride has not been obtained.

Oxyfluoride

The oxyfluorides are known as $MoOF.4H_2O$ $Mo_2O_3F_4$, $MoOF_4$ and MoO_2F_2. *Molybdenyl fluoride, $MoOF.4H_2O$* has been obtained by the double decomposition of the oxychloride, MoOCl and ammonium fluoride. It produces a buff coloured precipitate. Molybdenum trioxy-tetrachlroide $Mo_2O_3F_4$ has been obtained in red-needle like crystals by the action of hydrogen fluoride on molybdenum trioxide at 300-400°C. It decomposes to yield finally the blue oxide. Molybdenum oxytetrafluoride, $MoOF_4$ is formed (*i*) in the action of fluorine on a molybdenum oxide; or (*ii*) by distilling hydrogen fluoride into a well-cooled vessel containing molybdenum oxytetrachloride. The product is wormed to 30°C and when the reaction ends it is distilled from a sulphuric acid bath at 230°C. it is a white hydroscopic solid which becomes blue when exposed to air.

Molybdenum Dioxydifluoride, MoO_2F_2: This compound of molybdenum is obtained by the action of hydrogen fluoride on dioxydichloride of molybdenum. It forms a white crystalline hydroscopic sublimate with a faint blue tinge. It sublimes between 265°C and 271°C

Oxyfluomolybdate or Fluoromolybdate : If the normal molybdate is $(RO)_2\ MoO_2$, the five types of the oxyfluomolybdate would be :

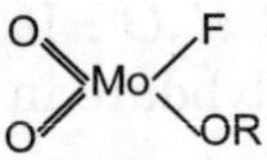

$RF.MoO_3$
I

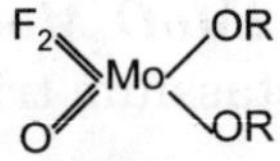

$2RF.MoO_2$
II

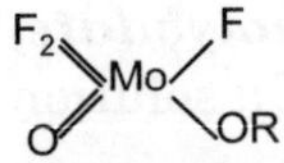

$R_2F.\ MoO_2F_2$
III

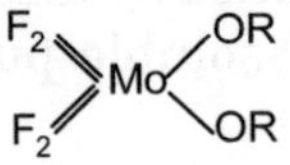

$2RF.\ MoO_2F_2$
IV

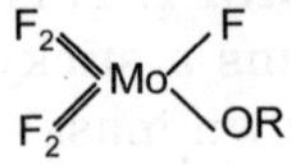

$RF.\ MoOF_4$
V

Werner has reported the five types as follows :

$$M_2\left[M_0{}^{vi}\ {}^{O_3}_{F_2}\right];\quad M_2\left[M_0{}^{v}\ {}^{O_2}_{F_3}\right];\quad M_2\left[M_0{}^{vi}\ {}^{O_2}_{F_4}\right]$$
I II III

$$M_2\left[M_0{}^{iv}\ {}^{O_3}_{F}\right];\quad M_2\left[M_0{}^{vi}\ {}^{O_3}_{F_3}\right]$$
IV V

In type I M = NH_4, in type II M = NH_4, K, Tl' in type III M = NH_4 Na, K, Tl' Cd, Zn, Ni, Ca″, Co″, in type IV, M = NH_4; and in type V M = NH_4

The well-known oxyfluomolybdates are :

(*a*) Ammonium Trioxydifluomolybdate, $2NH_4FMoO_3$ - type II

(*b*) Thallous Dioxytrifluomolybdate, TiF, MoO_2F_2 - type III

(*c*) Ammonium Dioxytetrafluomolybdate, $2NH_4F.Mo_2O_2F_2$ - type IV

Representatives of type I have not been obtained, where as there are various other fluomolybdates of the type nRF, MoO_2F_2 or nRFn' MoO_2F_2

(*a*) Ammonium Dioxypentafluomolybdate, $3NH_4FMoO_2F_2$: - type iv

(*e*) Ammonium hexaoxyhenafluomolybdate, $5Nh_4F.3MoO_2F_2$

Chlorides

The following chlorides and oxychlorides of molybdenum are known :

Chlorides		Oxychlorides	
Molybdous	$MoCl_2$	Molybdenyl Chloride	MoOCl
Chloride Molybdenum Trichloride	$MoCl_3$		
Molybdenum Tetrachloride	$MoCl_4$		
Molybdenum Pentachloride	$MoCl_5$		
Molybdenum Hexachloride	$MoCl_6$		

Molybdous Chloride, $MoCl_2$: It has been prepared (*i*) by heating the trichloride in a current of dry carbon dioxide or, (*ii*) by heating a mixture of molybdenum and mercurous chloride or (*iii*) by passing chlorine, highly mixed with carbon dioxide over moderately heated metal. Or (iv) by passing vapours of carbonyl chloride on powdered metal at 600°C-620°C. It is a by-product in the preparation of other chlorides also. It is an amorphous substance becomes yellow when treated with nitric acid. It does not easily volatilize and is stable in air. It is also represented by the triple formula $(MoCl_2)_3$ or Mo_3Cl_6.

Molybdic Chloride, or Molybdenum Chloride, $MoCl_3$: It is prepared (*i*) by passing the vapour of molybdenum pentachloride over heated molybdenum (*ii*) by passing vapour of the pentachloride, mixed with little hydrogen through a hot tube at 250°C. (*iii*) By the electrolytic reduction of a hydrochloric acid solution of molybdenum trioxide, using a mercury cathode. The trichloride forms a dark red powder, resembling red phosphorous. When heated to dull redness in an atmosphere of carbon dioxide it is decomposed into the di and tetrachloride.

Molybdenum tetrachloride $MoCl_4$: When molybdenum trichloride is headed to redness in a slow current of carbon dioxide it is resolved into di and tetrachloride, the trichloride latter appears as a yellow vapour, which condenses into a brown crystalline power. It is also obtained by the action of chlorine as a heated mixture of carbon and molybdenum dioxide and carbon tetrachloride in a sealed tube at 250°C. it is a brown semi-crystalline powder.

Molybdenum Pentachloride, $MoCl_5$: It is obtained (*i*) by the action of hot chlorine (free from oxygen or air) on the molybdenum surface or (*ii*) by passing chlorine over heated mixture of molybdenum and air-free chlorine over a mixture of molybdenum or sugar charcoal. Or (*iii*) by passing dried air free chlorine over a mixture of molybdenum dioxide, or disulphide and charcoal or (*iv*) by passing vapour of carbon tetrachloride over molybdenum trioxide at 570°C. The pentachloride produces iodine-like black or grayish-black crystals, melting at 194°C and boiling at 268°C. In contact with air, it forms an oxychloride. It dissolves in water so vigorously that the liquid starts to boil.

Molybdenum Hexachloride, $MoCl_6$: Smith and Burr found that metallic molybdenum rapidly acts on ferric salt and dissolves as a hexad and they tried to prepare the hexachloride by driving off the oxygen from an oxychloride of hexavalent molybdenum. Thus molybdenyl oxychloride MoO_2Cl_2 was mixed with an equivalent quantity of phosphorous pentachloride and heated in a sealed tube with an excess of chlorine at 170°C. This produces a green black crystalline compound, the hexachloride of the molybdenum.

Oxychlorides

Molybdenum Oxychloride, $MoOCl_4H_2O$: It has been obtained from the electrolysis of a solution of molybdenum trioxide in hydrochloric acid in a diaphragm cell with platinum electrodes. It is obtained in the cathode solution, which in its formation gives a buff precipitate when acetone is added. The first stage is the formation of potassium molybdenum hexachloride, K_3MoCl_6, which undergoes successive hydrolytic reactions :

$$K_3[MoCl_6] + H_2O \longrightarrow K_2[MoCl_5(H_2O)] + KCl$$

$$K_2[MoCl_5(H_2O)] + H_2O \longrightarrow MoCl_33H_2O + KCl$$

$$MoCl_3nH_2O \longrightarrow MoCl_2(OH) + HCl + (n\text{-}1)H_2O$$
$$MoCl_2(OH) \longrightarrow MoOCl + HCl$$

This oxychloride is insoluble in acetone or ether. It reduces cupric and mercuric chloride to their *-ous* salts. When its solution is boiled with sulphur, it yields hydrogen sulphide. Silver nitrate is reduced to the metallic state. In air free it undergoes double decomposition with sodium benzoate, tartrate and phosphate:

$$MoOCl + C_6H_5COONa \longrightarrow NaCl + C_6H_5COO.MoO$$

Molybdenyl Trichloride, $MoOCl_3$: It is known only in the complex salts as $RCl.MoOCl_3$ for e.g. $2NH_4Cl\ MoOCl_3$ which has been crystallized out from a solution of potassium molybdenyl oxalate and ammonium chloride in hydrochloric acid.

Molybdenum Dioxychloride, MoO_2Cl_2 : This has been prepared (*i*) as a sublimate by the action of dry chlorine on heated molybdenum dioxide and also (*ii*) by heating a mixture of molybdenum trioxide, potassium hydrogen sulphate and sodium chloride or (*iii*) by the action of oxygen on molybdenum chloride or oxychloride. It produces scaly crystals.

$$MoO_2Cl_2 + H_2O \longrightarrow MoO_3 + 2HCl$$

Molybdenum Oxytetrachloride, $MoOCl_4$: Püttbach has obtained this by the action of molybdenum pentachloride or molybdenum dichloride.

Bromide and Oxybromide

Molybdous Bromide, $MoBr_2$: This has been prepared by heating molybdenum tribromide in redness or by heating molybdenum in the vapour of bromine (it is left as non-volatile residue). It produces an infusible red product. It is completely decomposed by alkali lye.

Molybdic Bromide or Molybdenum Tribromide $MoBr_3$: It has been obtained (*i*) by passing the vapour of bromine over (*a*) molybdenum; (*b*) a mixture of molybdenum dioxide and carbon heated by a spirit lamp. Or (*ii*) by passing dry carbon dioxide with dry bromine over warm molybdenum. It is obtained as dark green or black pointed crystals. On heating it decomposes to bromine and the dibromide $MoBr_2$. It forms double salts s $(NH_4)_2MoBr_5$ K_2MoBr_5 etc.

Blomstrand obtained a small yield of *molybdenum tetrabromide $MoBr_4$*, as a product of the action of bromine on heating molybdenum, the dibromide sublimes first, the green tribromide appears near the hot zone and most distant from the heated metal occur black needles of the tetrabromide. This salt is fusible and volatile and produces brownish-red vapour.

A solution of molybdous bromide in an alkali-lye is decomposed by air forming golden-yellow crystals of *molybdenum dioxychloride bromide, $Mo_3Br_4(OH)_2\ 8H_2O$*. It is alkaline solution yields a dark reddish-yellow precipitate of molybdenyl bromide, $MoOBr\ H_2O$, when treated with ammonium molybdate and acetic acid. This oxybromide is also obtained by precipitating a concentrated solution of a tetravalent bromide in acetone, or by the electrolysis of a solution of molybdenum tribromide in hydrobromic acid in diaphragm cell with platinum electrodes. It is found in the cathode solution, which gives a buff precipitate when added to acetone. It has the composition $MoOBr\ 4H_2O$.

By the action of hydrogen bromide on molybdenum trioxide is also obtained *molybdenum trixoytetrabromide $Mo_2O_3Br_4$* resembling potassium permanganate in physical appearance.

Molybdenum Dioxydibromide, $MoBr_2$: This has been prepared firstly by heating molybdenum dioxide in bromine vapour and secondly by melting molybdenum trioxide with boric acid and potassium bromide.

$$MoO_3 + B_2O_3 + 2KBr \longrightarrow 2KBO_2 + MoO_2Br_2$$

It produces yellowish red tubular crystals, forming a colourless solution in water.

Iodides

Guichard observed that when dry molybdenum pentachloride is treated in a current of dry hydrogen iodide, iodine is liberated and brown amorphous molybdenum diiodide or molybdous iodide MoI_2, is formed. It is insoluble in water and alcohol and decomposes to a trioxide when heated. Guichard also obtained molybdenum tetraiodide MoI_4, by the action of liquid hydrogen iodide under presence of molybdenum tetraiodide.

The mixed halides of molybdenum $Mo_3F_{10}Cl_5$, $Mo_3Br_4F_2\ 3H_2O$, $Mo_3Br_4Cl_4\ 3H_2O$ and $Mo_3Cl_4Br_26H_2O$ have also been prepared.

Sulphides

The sulphides known are Mo_2S_2, MoS_2 Mo_2S_5, MoS_3 and MoS_4.

Molybdic Sulphide, Mo_2S_3: Guichard prepared this compound by heating molybdenum disulphide in an electric arc furnace for 2 - 9 minutes. It produces steel grey needles, which dissociate at 1100°C and volatilise at 1200°C.

Molybdenum Disulphide, MoS_2: It occurs in nature as molybdenite. It can be prepared (*i*) by heating one of the higher sulphides in a closed vessel; (*ii*) by heating the trioxide mixed with sulphur in a current of hydrogen sulphide. It can also be prepared by heating a mixture of potassium carbonate, sulphur and molybdenum dioxide at high temperature, then potassium polysulphide is washed out with water. (iv) By heating a mixture of ammonium molybdate, sulphur and lampblack; (v) by the action of potassium thiomolybdate on molybdenum oxide at a high temperature. The disulphide is obtained as a black powder.

Molybdenum Pentasulphide, $Mo_2S_5\ 3H_2O$: This compound is obtained by reducing with zinc a solution of ammonium molybdate (with 20% sulphuric acid) until the colour is dark red, then diluting, filtering and saturating with hydrogen sulphide. When heated in an atmosphere of carbon dioxide, it produces the anhydrous mass, which is black.

Molybdenum Trisulphide, MoS_3: Berzileus prepared it (*i*) by saturating a solution of an alkali molybdate with hydrogen sulphide, which converts the salt to the thiomolybdate, $Na_2[MoS_4]$; this when treated with hydrochloric acid produces a dark brown precipitate of the trisulphide. Or (*ii*) by dissolving molybdenum trioxide in a solution of an alkaline sulphide and decomposing the sulphomolybdate so formed by dilute mineral acid. The reddish brown precipitate becomes dark brown when dried. The trisulphide is soluble in alkalis, ammonium and other alkaline sulphide forming the thiomolybdate.

***Molybdenum Tetrasulphide, MoS_4* :** Berzelius also prepared molybdenum pentasulphide by saturating potassium paramolybdate with hydrogen sulphide. Krüs also prepared the tetrasulphide by melting molybdic acid with potassium carbonate, extracting the melt with water heated to the boiling point. It is obtained as black or brown powder.

Sulphomolybdates

1. *Ammonium Sulphomolybdate $(NH_4)_2MoS_4$*: This can be prepared by (*i*) evaporating a solution of ammonium molybdate saturated with hydrogen sulphide; (*ii*) by evaporating a solution of molybdic acid in ammonium sulphide (*iii*) by dissolving ammonium paramolybdate in aqueous ammonia and passing hydrogen sulphide, till a deep red solution id obtained which would deposit red coloured crystals.
2. *Potassium Sulphomolybdate, K_2MoS_4*: It is obtained (*i*) by heating to bright redness a number of molybdenum disulphide, potassium carbonate sulphur and charcoal, extracting the cold mass with water, evaporating the aqueous solution and finally purifying by adding alcohol; or (*ii*) by digesting molybdenite with a solution of potassium pentasulphide.
3. *Potassium Sulphodimolybdate, $K_2Mo_2S_7$*: This is obtained as a dark grey substance on adding a mineral acid or better acetic acid, to an aqueous solution of the normal sulphomolybdate.

$$2K_2MoS_4 + HCl \longrightarrow 2KCl + K_2Mo_2S_7 + H_2S$$

4. *Permonosulphomolybdic Acid, H_2MoS_5* : This can be obtained by adding acetic acid to the acid salt $KHMoS_5$ (which is obtained as blood red solution , crystals also obtained when molybdenum tetrasulphide is prepared). Its salts are ammonium permonosulphomolybdate $(NH_4)_2MoS_5$ and potassium permonosulphomolybdate, K_2MoS_5 has been also produced.

Sulphates

***Molybdic Sulphate, $Mo_2(SO_4)_3\ nH_2O$* :** This is obtained as a black mass, when hydrated molybdic oxide, $Mo(OH)_3$ is rubbed in a mortar with an equivalent amount of sulphuric acid. It is decomposed by water, into a soluble and insoluble basic salt.

Wardlaw prepared molybdic oxydisulphate, $Mo_2O(SO_4)_2\ 5H_2O$ from a solution of 40g of molybdenum trioxide in 81cc of boiling sulphuric acid, Sp. 1.84 mixed with a little nitric acid and electrolysed in a diaphragm cell with smooth platinum electrodes. It is a hydroscopic substance readily hydrolyzed. This fresh solution does not give precipitate with barium chloride but does give on warming.

***Sulphomolybdic Acid, $H_2[MoO(SO_4)_2(MoO_4)]$* :** This has been by cooling a solution of molybdenum trioxide in hot concentrated sulphuric acid. The prolonged reduction of molybdic acid by alcohol in the presence of sulphuric acid at the ordinary temperature produces molybdosic sulphate, $7MoO_3$ $2MoO_3.7SO_3nH_2O$ in black needles. From the trioxide and sulphuric acid, crystals of molybdenum dioxysulpahte $MoO_2(SO_4)$ are also obtained.

Nitrates

Hydrated molybdic oxide, when dissolved in dilute nitric acid, yields a purple solution of molybdenum nitrate.

Phosphate

When molybdenyl monochloride is heated with an alkaline phosphate, molybdenyl phosphate $(MoO_3)PO_412H_2O$, is formed. When molybdenum dioxide is heated to a dull redness with metaphosporic acid, molybdic metaphosphate, $Mo_2O_33P_2O_5$, is obtained.

Phospatomolybdic Acid

Berzelius discovered them. He found that when freshly precipitated a moist hydrated molybdenum trioxide is digested with phosphoric acid, a lemon yellow insoluble complex is formed. This complex dissolve in an excess hot phosphoric acid; the solution thus obtained yields an evaporation of the amorphous transparent mass soluble in water an alcohol.

The phosphatomolybdic acid with a high precipitation of molybdate with phosphorous, the ratio is P: Mo = 1 : 9 to 1 : 12 are yellow in colour and those with lower ratios are either white or transparent in colour. Rosenheim, regards the yellow ones as the derivative of the heptavalent Po_6-anion as for e.g. H_7PO_6 acid with bivalent Mo_2O_7 radical in the presence of oxygen; while the colourless series is derived from the tetravalent PO_4 anion or H_2PO_4 acid with the bivalent MoO_4 radical.

Yellow	Phosphate-Dodecamolybdic acid	
	$H_7[P(Mo_2O_7)_6]$	P_2O_5 $24MoO_3$ nH_2O
	Phosphate-Henamolybdic Acid	
	$H_{12}[P_2(Mo_2O_7)_{11}]$	P_2O_5 $22MoO_3$ nH_2O
	Phospahtodecamolybdic Acid	
	$H_6[P(OH)(Mo_2O_7)_5]$	P_2O_5 $20MoO_3$ nH_2O
Colourless	Phospahte-dimolybdic Acid	
	$H_3[PO_2(MoO_4)_2]$	$P_2O_52MoO_3nH_2O$

Posphatododeca Molybdic Acid, $P_2O_524MoO_3$: This compound has been obtained by adding yellow molybdenum trioxide to a boiling aqueous solution of phosphoric acid; evaporating and cooling the solution. Its solution in large excess of nitric acid produces doubly refracting crystals of the composition, P_2O_5-$24MoO_332H_2O$. The basicity of the acid is the difference between the negative valency of the acid group and the positive valency of the central atom. This makes the phosphododecamolybdic acid as hexavalent. Its silver salt supports the below mentioned formula.

$$7Ag_2O\ P_2O_524MoO_3$$

When ammonia is added to this acid ammonium phosphatododecamolybdate is produced, $(NH_4)_2OP_2O_5$ $24MoO_3$ nH_2O is the formula.

Debray obtained phosphatomolybdic acid by the action of aqua regia on ammonium phosphate molybdate. Its composition is P_2O_5 $20MoO_3nH_2O$ (n = 24, 41, 51, and 52). From a solution of molybdic acid, ammonia, orthophosphoric acid and a little nitric acid is obtained ammonium phosphate decamolybdate $8(NH_4)_3H_2O(P_2O_520MoO_3)_3$, $12H_2O$ as a yellow precipitate.

Complexes : The pentavalent molybdenum forms complex as (*i*) the oxyhalids like $M_2[MoOCl_5]$, where M = K, NH_4 etc. and $M[MoOBr_4]$ ad $M[MoOBr_5]$. (*ii*) The cyanides as M_3 [Mo(CN)} where M =K or H (*iii*)The thionates belonging to the series : (*a*) reddish brown $M_2[MoO_2(SCN)_3]$ (*b*) Dark green $M_2[MoO(SCN)_5]$. (*iv*) The chelate organic derivatives as those of catechol, $M_3[MoO_{cat3}]$.x and $M_2[Mo(OH)_3cat_{\ 2}$.x

The tetravalent molybdenum complexes are incarcerated to the (i) cyanide as $M_4[MoCN_8]$ yellow and $M_4[MoO_2CN_4]$ (reddish-violet and (ii) a few oxalate-compounds.

The trivalent molybdenum produces complex halides as $K_4[Mo(CN)_7H_2O$. The divalent molybdenum also produces the dihalide complex of the compositions $M_2[Mo_6Cl_{14}(H_2O)_2]$ $4H_2O$ etc.

Molybdenum Compounds

	Mo^{ii}	Mo^{iii}	Mo^{iv}	Mo^{v}	Mo^{vi}
Fluoride and oxyfluoride		MoF_3? $KMoF_4$ MoOF	MoF_4	... $K_2[MoOF_5]$	MoF_6 $MoOF_4$, MoO_2F_2
Chloride and oxychloride	Mo_6Cl_{12} $MoCl_2$	$MoCl_3$, M_3MoCl_6 MoOCl. $4H_2O$	$MoCl_4$	$MoCl_5$ $M_3[MoOCl_5]$ $M_2[MoOCl_5$	$MoOCl_2$
Oxide	MoO (?)	$Mo(OH)_3$, Mo_2O_3	MoO_2	$MoO(OH)_3$ Mo_2O_5	MoO_3
Sulphide		Mo_2S_3	MoS_2	Mo_2S_5	MoS_3
Thiomolybdate					M_2MoS_4
Molybdic acid and molybdates					H_2MoO_4 M_2MoO_4
Sulphate		$Mo_2O(SO_4)_2$		Mo_2O_3 $(SO_4)_2$?	MoO_2SO_4
Cyanides		K_1MoCy_7.-H_2O	M_4[Mo Cys]. yellow, $M_4[MoO_2$– $Cy_4]$ red-violet	K_3[MoCys]	...
Carbonyl	...	...	...	...	$Mo(CO)_6$

Diaphragm Cell

This explanation has just now became an absolute necessary because the extraction and synthesis of some molybdenum compounds is undergoing and electrolysis, which is conducted in this type of cells. The cell consists of two separate compartments the anode and cathode which are separated by a membrane to avoid the contamination of the electrolyte. This membrane is called diaphragm; it is made of asbestos, now days a new material called *naflon*, is used which is persulphonic acid. The typical example of these of cells is Vorce cell.

Vorce Cell (see Fig. 5)

(*a*) Cast iron cylinder tank 26 inch is diameter and 2.5ft in height. The cells are generally joined in series.

(*b*) 24 graphite electrodes work as anode.

(*c*) The cathode is a perforated steel cylinder fixed around the anodes at a little distance. Both anodes and cathodes are insulated from the tank. The diaphragm rests on the thin cylindrical screen separator, which lies against the cathode, alternatively. The cylindrical diaphragm made either from asbestos or from naflon against the cathode (between the anode and cathode) the anode compartment is within the diaphragm.

(*d*) Both the compartments are provided with exit pipes for gases.

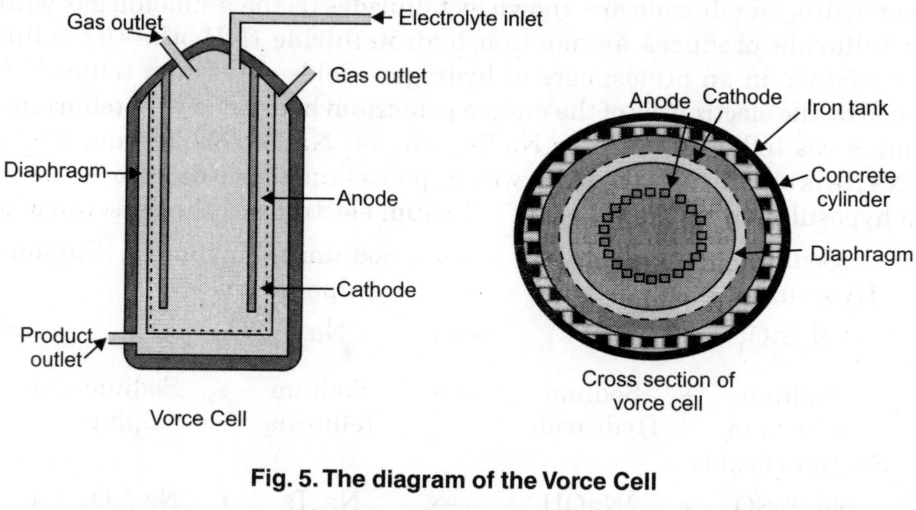

Fig. 5. The diagram of the Vorce Cell

This cell is generally used for the preparation of the molybdenum oxychlorides and other chlorides. The electrolyte for this purpose generally used is hydrochloric acid.

More Tellurium

This sub-head must be studied after reading the chapter of heavy metals in the tome 2 of the architecture of chemistry, the subhead tellurium.

Hydrogen Telluride, H_2Te : The apparatus use here for the preparation of hydrogen telluride is E. Earnyei's apparatus (see Fig. 6). The apparatus is actually a diaphragm cell, consisting a tellurium cathode, the electrolyte is sulphuric acid with a concentration of 50%, the temprature is kept at – 15°C to –20°C. The cell consists of a diaphragm of parchment paper to retard the movements of the electrolyte. Two glass tubes rise from the cell the one over the positive pole is fitted with a two-way cork; funnel and a bent tube (as seen in the Fig.).

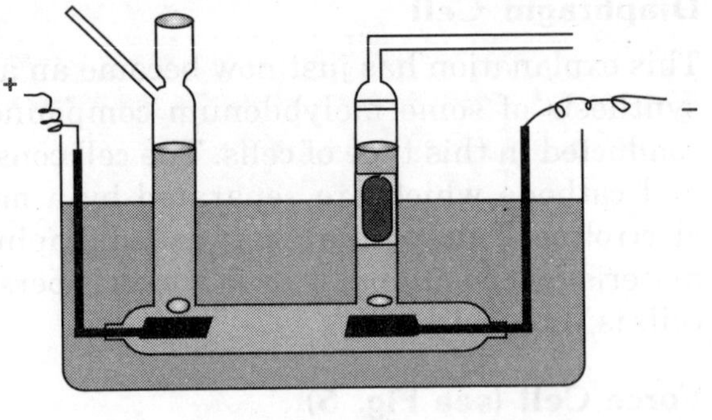

Fig. 6. The Ernyei's Apparatus

The funnel is to introduce liquid into the cell and the bent tube is to allow air to be washed from the apparatus by a current of hydrogen before the electrolyte is introduced and afterwards for the exit of the anode gases. The tube over the negative pole has a partition above which calcium chloride and phosphorous pentoxide can be placed for drying the gas. The reaction vessel is cooled with solid carbon dioxide and alcohol. The fittings are all made up of glass and not for cork or rubber.

Other Tellurides

The salts of hydrogen telluride are known as tellurides (*i*) the ammonia gas with excess of hydrogen telluride produces ammonium hydrotelluride (NH_4)the (*ii*) Tellurium and potassium mixture in an atmosphere of hydrogen yields potassium telluride K_2Te. It is also obtained by the electrolysis of the molten potassium hydroxide with tellurium electrode. (*iii*) Sodium gives tellurides such as Na_3Te_2; Na_3Te_7 Na_2Te_2 Na_2Te_4 and Na_2Te. Sodium telluride Na_2Te is obtained in the same way as potassium telluride, also by the electrolysis of sodium hyposulphite with the help of tellurium electrode in the presence of air.

Sodium Hyposulphite + Tellurium ⟶ Sodium Tellurium Sulpho Dioxide + Sulphur Dioxide

$$N_2S_2O_4 + Te \longrightarrow Na_2TeSO_2 + SO_2$$

Sodium Tellurium Sulpho Dioxide + Sodium Hydroxide ⟶ Sodium Telluride + Sodium Sulphite + Water

$$Na_2TeSO_2 + 2NaOH \longrightarrow Na_2Te + Na_2SO_3 + H_2O$$

The reaction between sodium and tellurium in liquid ammonia gives sodium ditelluride, N_2Te_2. (*iv*) Silver telluride, occur in mineral *Stulzite*, *Ag_4Te Hessite Ag_2Te* and *Empressite AgTe*. (*v*) When gold sulphotelluride, $Au_2S_3TeS_2$, is heated so as to drive off sulphur, one gets gold telluride, Au_2Te. Other tellurides of gold include Au_2Te_3 (Au, Ag)Te etc (*vi*) Zinc when heated with tellurium gives zinc telluride ZnTe (*vii*) F. Wöhler obtained aluminium telluride, Al_2Te_3, by heating the mixture of the powdered elements (viii) Platinum ditelluride, $PtTe_2$, is obtained by the action of sodium telluride on platinum tetrachloride.

Oxides

Tellurium Monoxide, TeO : Tellurium monoxide this is obtained by heating tellurium sulpho trioxide $TeSO_3$ at 180 - 230°C, in vacuum and the product is washed with a dilute solution of sodium carbonate to separate the unchanged tellurium sulpho trioxide.

$$TeSO_3 \longrightarrow TeO + SO_2$$

It is also produced by pouring a red solution of tellurium in sulphuric acid into a large excess of water. This oxide is produced in the form of a brown mass or a black amorphous powder. It is stable in cold dry air but decomposes when heated in vacuum.

$$TeO_2 + Te \longrightarrow TeO$$

In moist air, it is oxidised to the dioxide. Sulphur dioxide reduces it slowly to tellurium. With concentrated sulphuric acid it forms a red solution from which tellurium sulphate crystallises.

$$2TeO + 3H_2SO_4 \longrightarrow Te(SO_4)_2 + TeSO_3 + 3H_2O$$

It reduces potassium permanganate to potassium manganate.

Tellurites

The first dissociation constant of tellurous acid is 0.6×10^{-5}. Its salts are known as tellurites. Sodium tellurite Na_2TeO_3 it is obtained by fusing tellurium dioxide and sodium carbonate.

$$Na_2CO_3 + TeO_2 \longrightarrow Na_2TeO_3 + CO_2$$

It forms many hydrates, e.g. Na_2TeO_3 $5H_2O$. On further fusion it produces sodium pyrotellurite, $Na_2Te_2O_5$; Potassium tellurite, when treated with acids or alkalis tellurium separates out. Sodium tellurite and silver nitrate yields silver tellurite Ag_2TeO_3, by double decomposition appearing as a yellowish white precipitate. Calcium oxide when heated with tellurium oxide yields calcium tellurite, $CaTeO_3$. By double decomposition, tellurites of manganese, $MnTeO_3$, Nickel $NiTeO_3H_2O$ may also be obtained.

Tellurates

They are the salts of metatelluric acid, H_2TeO_4. Though other tellurates are also known, e.g orthotellurates Cu_3TeO_6 $AgTeO_6$ and $ZnTeO_6$, pyrotelurates as $LiHTe_2O_7$ $NH_4Te_2O_7$ $KHTe_2O_7$ and tetratellurates as $Li_2Te_4O_{13}$ and $K_2Te_4O_{12}$.

Ammonium Tellurate, $(NH_4)_2TeO_4 2H_2O$: This is obtained by the action of ammonia on powdered telluric acid. Ammonium hydrotellurate, NH_4HTeO_4 has been obtained by the action of ammonium chloride on sodium hydrotellurate. The action of ammonium chloride on sodium hydropyrotellurate produces ammonium hydropyrotellurate, $NH_4.H.Te_2O_7$.

Sodium Tellurate, Na_2TeO_4: This is obtained when tellurium dioxide is fused with sodium nitrate, Concentrated solution of telluric acid when supersaturated with sodium hydroxide yields *sodium dihydro-orthotellurate $Na_4H_2TeO_6 7H_2O$*. Sodium hydrotellurate is obtained by dissolving telluric acid in a boiling solution of sodium carbonate and adding acetic acid after cooling. A solution of 4 mols of orthotelluric acid and 1 mol of sodium carbonate produces a sticky mass of sodium hydropyrotellurate $NaHTe_2O_7$ when spontaneously evaporated. When heated, it produces insoluble sodium tetratellurate, $Na_2Te_4O_{13}$.

Potassium Tellurate, K_2TeO_4: This is similar to potassium sulphate. Berzelius prepared the pentahydrate $K_2TeO_4\ 5H_2O$ from a solution of potassium hydroxide and telluric acid. A glutenous mass is obtained in cold which, dissolves on warning and when cooled to 0°C. It produces crystals of the pentahydrate. Potassium carbonate solution and ortho-telluric acid produces potassium hydrotellurate, $KHTeO_4\ 11H_2O$. Potassium hydropyrotellurate $KHTe_2O_7\ 11H_2O$ has been obtained by slowly cooling a boiling solution of 4 mols of orthotelluric acid and a mol of potassium carbonate. This on calcination gives potassium tetratellurates, $K_2Te_4O_{13}$. The tetratellurates is also obtained by gently heating tellurium dioxide with potassium chlorate or by the action of chlorine on a hot solution of potassium tellurite.

When barium chloride solution is added to sodium tellurate, a voluminous white heavy precipitate, $BasTeO_4\ 3H_2O$ appears, which is sparingly soluble in cold water, but little more in hot water. It is soluble in hydrochloric acid and also in nitric acid. Precipitates of barium hydro tellurate, $Ba(HTeO_4)_2\ 2H_2O$ and barium hydropyrotellurate $Ba(Hte_2O_7)_2\ nH_2O$, have been obtained by double decomposition from the corresponding sodium salts.

Fluorides : Tellurium combines with fluorine with incandescence, yielding white fumes of tellurium tetrafluoride, TeF_4; which is also prepared when tellurium dioxide dissolves in hydrofluoric acid.

When fluorine acts on tellurium at –78°C, white crystalline tellurium hexafluoride, TeF_6, is formed, which has a melting point –36°C boiling point –35.5°C.

Various fluotellurites as $CsTeF_5$ or $CsF.TeF_4$; $KTeF_5$ and $Ba(TeF_5)_2\ H_2O$ are known. Meitner observed that tellurium dioxide dissolves in 45% hydrofluoric acid and on cooling to -20°C, produces long transparent crystals of an oxyfluoride, $2TeF_4.3TeO_2\ 6H_2O$. Other oxyfluorides from tellurous acids and hydrofluoric acid, as $TeO_2 2HF$ ie $2TeO_2\ 4HF$ or $TeOF_2\ H_2O$ are also known.

Chlorides : Tellurium froms two chlorides; $TeCl_2$ and $TeCl_4$ tellurium dichloride, $TeCl_2$ is obtained (*i*) when a slow current of chlorine is passed over a strongly heated tellurium (*ii*) by passing a vapour of sulphur monochloride over strongly heated tellurium,

$$Te + S_2Cl_2 \longrightarrow TeCl_2 + 2S$$

and (*iii*) by heating tellurium in vapours of carbonyl chloride. It is a black amorphous hydroscopic solid; with a liquid is black.

Its ethereal solution absorbs bromine forming a tellurium chloro-bromide and also iodine forming the chloro-iodide.

The chloride is fairly reactive; its attracts water from the air; with water it reacts as follows:

$$2TeCl_2 + 3H_2O \longrightarrow H_2TeO_3 + 4HCl + Te$$

With oxygen it reacts as follows,

$$2TeCl_2 + O_2 \longrightarrow TeO_2 + TeCl_4$$

Tellurium Tetrachloride, $TeCl_4$: This is formed (*i*) When an excess of chlorine acts on heated tellurium (the dichloride is also produced.) and (*ii*) by passing the vapour of sulphur monochloride on heated tellurium. It is a snow-white crystalline mass, melting to an amber yellow liquid at 229°C, which becomes dark red at the boiling point 414°C. At 500°C dissociates as

$$TeCl_4 \longrightarrow TeCl_2$$

It is stable in dry air, in moist air it leaves the residue of an oxychloride. It hydrolyses in hot solution to the oxide and oxychloride. With concentrated hydrogen chloride at –30°C, yields yellow needles, $TeCl_4HCl$; in ammonia gas at 0°C, it produces a white compound $TeCl_4$ $6NH_3$. With tellurium hydride it reacts as:

$$TeCl_4 + 2H_2Te \longrightarrow 3Te + 4HCl$$

When ammonia is over tellurium tetrachloride at 220°C, nitrogen evolves.

$$3TeCl_4 + 16NH_3 \longrightarrow 3Te + 12NH_4Cl + 2N_2$$

The molten chloride dissolves phosphorous pentachloride at 220.5°C forming $2TeCl_4PCl_5$ (which is lemon yellow in colour).

A solution of ammonium chloride and tellurium in hydrochloric acid furnishes orange yellow crystals of an ammonium hexachlorotellurite $(NH_4)_2TeCl_6$. With potassium chloride K_2TeCl_6 is produced.

Oxychlorides : By the action of cold water or boiling water on the tetrachloride, a number of oxychloride are obtained. They are also obtained by the action of aqua regia on tellurium; or by dissolving tellurium dioxide in the molten tetra chloride. The oxychloride are $TeOCl_2$, TeO_2 3HCl, TeO_2. 2HCl etc.

Bromides

Tellurium Bromide $TeBr_2$: This compound is obtained (*i*) by the action of bromine on tellurium at an ordinary temperature; or by the sublimation of form a mixture of tellurium and its tetrabromide. It is a greenish-black crystalline mass. It is decomposed to barium tetrabromide.

$$2TeBr_2 \longrightarrow TeBr_4 + Te$$

In presence of tartaric acid it produces a brown solution.

Tellurium Tetrabromide, $TeBr_4$: This is obtained (*i*) by introducing bromine into a glass tube, sealed at one end and cooled in ice, adding bromine to a mixture of tellurium and dilute hydrobromic acid; potassium bromide and phosphoric acid also added. Ruby red crystal are obtained with melting 363°C and boiling point of 418°C. Its vapour is yellow, with hydrobromic acid, it exist with an evolution of heat and forms ammonium bromide, it unites with an evolution of heat and forms $TeBr_4$ HBr $5H_2O$. It reacts with ammonia to give $TeBr_4$ $6NH_3$. With ammonium bromide it gives ammonium hexabromotellurite $(NH_4)_2TeBr_6$. The similar potassium salt is K_2TeBr_6.

Oxybromides of indefinite composition are also available, when the moist air or water acts on the tetrabromide of tellurium, tellurium dioxide and hydrobromic acid gives such oxybromides as $TeO_2 3HBr$; $TeO_2 2HBr$, that is the tri-oxybromide and di-oxybromide.

Iodides : Tellurium can be melted with iodine in all proportions, with a little excess of iodine, one gets *tellurium diiodide, TeI_2*, which sublimes as a black substance. *Tellurium tetraiodide, TeI_4*, is generally obtained by (*i*) digesting tellurium powder for a considerable time with iodine and water. (*ii*) By digesting tellurium dioxide with hydroiodic acid. (iii) By the action of hydroiodic acid on telluric acid. It forms a grey powder, with a melting point 280°C. When reacts with water it produces oxychloride and then the dioxide.

$$TeI_4 + 2H_2O \longrightarrow TeO_2 + 4HI$$

With concentrated hydrochloric acid, it produces crystals of $TeI_4.HI.8H_2O$. Its solution in hydrochloric acid when neutralised with ammonium hydroxide gives ammonium hexaiodotellurite $(NH_4)_2TeI_6$ and similarly with potassium iodide it gives $K_2TeI_6 2H_2O$.

Complex Halo-Tellurites

The two types of complex tellurites are known as $MTe(hal)_5$ and $M_2Te(hal)_6$. The following free acid, $HTeCl_5$ $5H_2O$, which produces orange crystals with a melting point of 20°C and $HTeI_5$. It is a black substance with a melting point of 55°C. When exposed in air all of them loose halogen acids only the chloride seems to be known to give salts.

The chlorides, bromides and iodides of the type $M_2Te(Hal)_6$ are known thought the acid, $H_2Te(hal)_6$ corresponding to this salt is not known. The hexachloride ar yellow, the hexabromide are orange in colour and hexaiodides are black. They are stable in strong solutions in water, gets hydrolyzed in dilute solutions. The ultimate product of hydrolysis is tellurium dioxide.

The alkali hexachloro and hexabromotellurates like the selenites, are isomorphous with the other salts of the type M_2XBr_6 (where X = Se, Pb, Pt and Sn) and also with M_2SiF_6, showing that they have an octahedral arrangement with a valency group of 14(12, 2).

Sulphides

When hydrogen sulphide is passed through a solution of tellurium in hydrochloric acid, or through an acidified solution of an alkali tellurite, an unstable tellurium disulphide TeS_2 is produced (it is stable below 20°C). On dialysis it produces a colloidal solution. When tellurous acid reacts with hydrogen sulphide, tellurium monosulphide TeS is produced. It is unstable.

$$2H_2s + H_2TeO_3 \longrightarrow TeS + S + 3H_2O$$

If a dilute aqueous solution of telluric acid is saturated with hydrogen sulphide it turns brown and then the colour disappears and is deposited as greyish-black flakes of tellurium trisulphide, TeS_3.

Sulphotellurates : The salts of hypothetical H_6TeS_5 are called sulphotellurates. They may be obtained by saturating the solution of corresponding tellurite with hydrogen sulphide.

$$3Na_2TeO_3 + 9H_2S \longrightarrow Na_6TeS_5 + 2Te + 4s + 9H_2O$$

They are also produced (*i*) by the action of a boiling solution of an alkali hydrosulphide on tellurium dioxide or (*ii*) by dissolving tellurium disulphide in alkali lye.

$$5TeS_2 + 18KOH \longrightarrow 3K_2TeO_3 + 2K_6TeS_5 + 9H_2O$$

The representative sulphotellurates Na_6TeS_5, K_6TeS_5 Cu_3TeS_5 etc.

Tellurium Sulphotrioxide, $TeSO_3$: This is found when sulphur trioxide reacts with tellurium. It is a truly deep red solid. It is also obtained by (*i*) the action of hydrogen tellurate on tellurium dioxide.

$$2H_2Te + H_2SO_4 + TeO_2 \longrightarrow 3TeSO_3 + 5H_2O$$

Or, (*ii*) by the action of sulphuric acid or tellurium monoxide.

$$2TeO + 3H_2SO_4 \longrightarrow TeSO_3 + Te(SO_4)_2 + 3H_2O$$

It is an amorphous solid of beautiful red colour, when heated sufficiently, its colour changes to a bright greyish-brown. The two substances may have the structures as shown below :

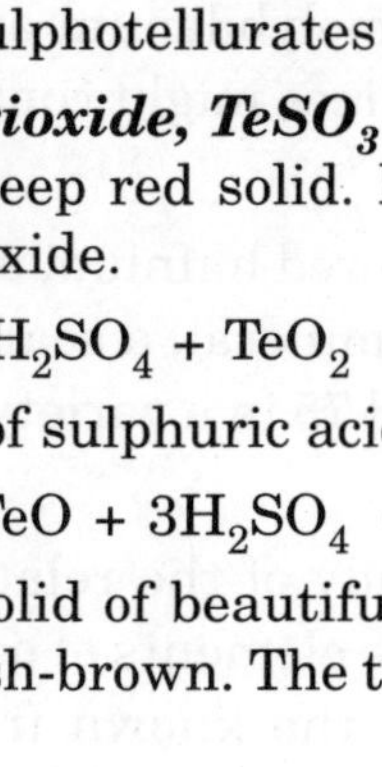

Red

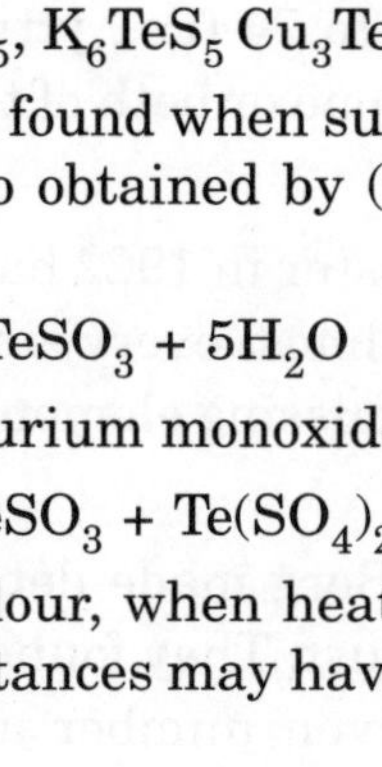

Brown

Telluric acid with potassium nitrate gives the crystals of potassium dinitrate tellurate $2KNO_3H_6TeO_62H_2O$

Parandtl suggested the structure below:

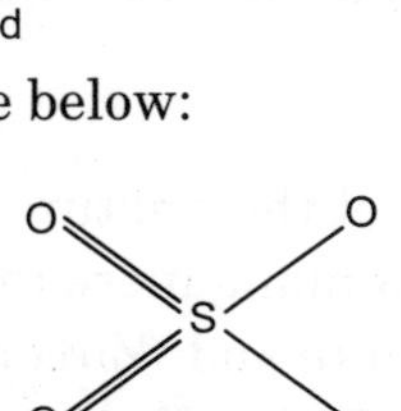

Nitrate : Tellurium readily dissolves in nitric acid forming a colourless solution. On evaporation, it yields rhombic crystals, stable in dry air. At 190°C, they begin to have nitrogen oxide, at 330 they furnish the dioxide. Their structure has been suggested $Te_2O_3(OH)(NO_3)$.

RHENIUM

Melting Point : 3180°C
Boiling Point : 5630°C

Atomic Number : 75
Atomic Weight : 186.207

History or Introduction

The German chemists Dr Walter Noddack and Dr Ida Tache of the Physico-Technical Testing Office in the Berlin and Dr Otto Berg of Werner Simens Laboratory announced in June 1925, the discovery of two new elements of the of the manganese group. The discovery was not accidental, but the result of a long search in platinum ores and in the mineral columbite. Platinum ores contain the elements 24 to 29; 44 to 47 and 76 to 79 (i.e., chromium to copper, ruthenium to silver and osmium to gold.). Where as columbite contain number 39 to 42 and 72 to 74 (i.e., yttrium to molybdenum and hafnium to tungsten). Hence, it was hoped that one or both of these sources might contain the missing elements 43 and 75.

After Hevesey and Coster in 1922 had discovered hafnium and had shown that X-Ray lines can be detected in almost every zirconium mineral, a search was made by means of the X-ray spectra for the missing elements 43 and 75 in a variety of manganese ores but in vain.

Noddack Tache and Berg made detailed study of the relative frequencies of known elements in the earth's crust. They found that the elements of odd atomic number are less common than those of even number and from the known frequency of occurrence of platinum ores and of columbite, they obtained approximate idea of the process of carrying out the procedure of extraction. The elements 43 and 75 belong to the manganese group and have been named as *Eka-manganese* and *Dvi-manganese*; and it was easy to predict their characteristics on the basic of periodic law. When they could be separated from columbite, they were named as Masurium and rhenium in the name of Mersurenland and the river Rhine. Dr Noddack and Tache carried out the tedious and cumbersome chemical process of separation. Whilst Dr Berg assisted in the X-Ray spectroscopic examination.

Occurrence

Columbite contains 10^{-6} to 10^{-7} part of the mixture of Eka and dvi-manganese elements in the proportions 1:10. Traces of eka manganese (maximum) have to also been detected in *sperrylite, *Gladolinite *Fergusonite and *Zircon. Traces of dvi-manganese (rhenium) have indication in tantalite and tungstenite. No known mineral contain more than 0.001% of rhenium.

---

*Gladolinite = $BeF_3(YO)_2(SiO_4)_2$ Sperrylite = $PtAs_2$ Fergusonite = Cb_2O_5

Mineral	Rhenium Content
Gladolinite	1.1×10^{-6}
Columbite	0.2×10^{-6}
Tantalite	0.14×10^{-6}
Alvite	0.5×10^{-6}
Lead Selenide	0.8×10^{-6}

Extraction of Rhenium

(i) Isolation from Columbite : Columbite is heated with sodium hydroxide (twice the weight) and 5% sodium nitrate for many hours. The cooled product was exposed to air for many days and then shaken up with water to yield a 5% alkali solution. After standing, the liquid was filtered and the residue washed with 5% caustic soda solution. This dissolved a minimum of columbium and tantalum. The green filtrate contained Manganite and was treated with hydrogen peroxide to precipitate $MnO_2.5H_2O$ and with 55 ammonium chloride with precipitated the mineral acids and silica titania columbia and tantalica. The resulting filtrate was obviously acidified with sulphuric acid and the nitrous acid removed. A small precipitate of tantalica and hafnia but no rhenium came down. Hydrogen sulphide was passed until the solution was saturated with the gas. The sulphide precipitate is filtered off and the filtrate was treated with ammonium hydroxide and ammonium sulphide and reacidified to 5% hydrochloric acid. These sulphides were united to the first precipitate and were heated in a stream of hydrogen to remove sulphur. The residue contains 40% Cb, Ta; 5% Si; 5% Tr, Zr, Hf; 10% Fe, Co, Ni, Mn; 30% W, Mo, Sn, V; and 10% Cu, Pb, As, Sb, Bi, Pt, Ge, Zn, Cd and traces of Tr, Ru, Os. These sulphides are warmed with dilute aqua regia so that the tungstic acid and the sulphur are filtered off. The filtrate was evaporated to drive off the nitric acid, On reduction of the residue in hydrogen. There remained about one gram of substance containing 15% V; 7% Mo; 3%W; 10% Pt; 5% Pb; 15% As; Cu 15%; 20% Sn, Ge, Sb, Bi; 5% Co, Ni; 5% Cb, Ti, Ta; and 0.2% Re.

The Rhenium was identified by the distinct Lα - line, This residue was treated like the osmoiridium and wolfram ores and the alkali fusion products were treated with ammonium chloride. Concentrated sulphuric acid was added to the filtrate and this was again filtered. This left Co, Ni, Cu, Pt, Bi, Cb, Ta, Sn, Ge, Sb, Pb, W and much V remained behind. A further hydrogen sulphide precipitation gave a preparation (0.1) g containing V 20%, N0 70%, As 5%, Cb, Sn and Sb 3% and Re 2%.

When this preparation was dissolved in aqua regia, filtered, freed from nitric acid and made 5% acid and rapidly saturated with hydrogen sulphide, most of the molybdenum precipitates a MoS_3. This precipitate contains only a trace of rhenium. The filtrate is made alkaline with ammonium hydroxide and ammonium sulphide was added to precipitate V, Mo, As and all the rhenium, whilst Sb and Sn remained in solution. This sulphide was composed of 805 V 10% Re 3% Mo and 2% As.

The final treatment included the process of heating the reduced substance in oxygen. Most of the rhenium was driven off as a white oxide, which sublimed, or the cool parts of the tube and was extracted by cutting the tube. Gladolinite and other rare earth minerals such as monazite, cerite orthite alvite or zircon may be similarly treated to get rhenium.

(ii) From Crude Manganese Sulphate : 100 gms of crude manganese sulphate were dissolved in 600 cc of water and treated with 50 gm of ammonium chloride, 25 cc of aqueous ammonia of sp. gr. 0.88 and then with hydrogen sulphide. The mixture was allowed to stand and then filtered. The precipitate consisted of iron sulphide and aluminium and chromium hydroxide. The filtrate is retreated with hydrogen sulphide until there was no further precipitation. This required several filtration processes before a solution was obtained, which no longer gave any precipitate when saturated with the gas and allowed to stand.

The filtrate and the washings are allowed to stand for several days, whilst still alkaline with ammonia and saturated with hydrogen sulphide to make quite sure that al manganese had been removed. The solution was afterwards boiled for sometimes to remove excess of hydrogen sulphide and some ammonia acidified with hydrochloric acid and evaporated to dryness. The residue was strongly heated to drive off the ammonium compounds and burn away the small quantity of sulphur formed.

The final residue was dissolved in acetic acid and ammonium oxalate, in slight excess was added to precipitate the calcium present. The filtrate was evaporated to dryness and ignited strongly to drive off all the ammonium compounds. The residue was dissolved in hot dilute nitric acid. The solution was precipitated with ammonium hydroxide and washed precipitate was dissolved in hot dilute nitric acid and evaporated to dryness again, leaving a white nitrate which was ignited to a very light brown oxide.

(iii) From Molybdenite : The molybdenite is dissolved in nitric acid and this solution is rendered alkaline with ammonia solution. It is then treated with sodium phosphate and nitric acid till the whole of molybdenum has been precipitated as phosphomolybdate. It is filtered out. The filtrate will now contain the whole rhenium and a small part of molybdenum. It made alkaline and evaporated to dryness.

The dry mass is dissolved in water and hydrochloric acid is added to it (so much that the total solution contains 30% by volume of strong hydrochloric acid). The solution is then treated for a number of hours with hydrogen sulphide as to precipitate rhenium and molybdenum sulphides. The total precipitate is dissolved in mixture of hydrogen peroxide and ammonium hydroxide solution. The process of phosphomolybdate and sulphide precipitation is repeated a number of times till the product become rich in rhenium.

The mixed sulphide of molybdenum and rhenium are then reduced to metal in hydrogen at 1000°C. The mixed metals thus obtained ar then oxidised in a stream of air are oxygen to give rhenium heptoxide and molybdenum trioxide an hence the sublimate gets further enriched in rhenium.

The sublimate is dissolved in ammonia solution and evaporated to dryness; thus ammonium pertitanate is obtained which, is then reduced with hydrogen. The metal rhenium thus obtained still contains molybdenum. The sublimation process is further repeated till one gets rhenium of desired purity.

Preparation of the Metal

1. *It has been obtained electrolytically from neutral or acidic solution of its salts,* e.g., potassium perrhenate ($KReO_4$) solution (1%) is acidified with dilute sulphuric acid (1:5) current 0.0025 Amps per sq cm. This in 43 hours yield 0.0232 gm of rhenium.
2. *It can be obtained heating any rhenium compound in hydrogen at 1000°C* e.g
   $2\ KReO_4 + 4H_2 \longrightarrow 2KOH + 3H_2O + 2\ Re$
   The product is washed with water to remove potassium hydroxide.
3. *The Metal can also be obtained by the thermite process* that is heating the metal salt with aluminium and magnesium.
4. *When tungsten filament is heated in the vapour of rhenium halide,* the metal is deposited on the wire.

Physical Properties

Melting Point	3180°C
Boiling Point	5630°C
Density	21.02 g/cc
Specific Heat	0.0346
Specific Resistance	20 × 10^{-6} ohms/cm
Ionic Diameter	0.61Å
First Ionic Potential	7.9 eV
Atomic Volume	8.82
Tensile Strength	50.6 kg/sq mm
Hardness (Brinell)	250
Standard Entropy	10 cal deg^{-1} mol^{-1}
Heat Capacity	6.1 cal deg^{-1} mol^{-1}
Heat of Vapourization	220 Kcal/mol
Co-efficient of Expansion	a axis = 4.67 × 10^{-6} ±8%
Electrical Resistivity	c axis = 12.45 × 10^{-6} ±8%
Magnetic Susceptibility	1.9 × 10^{-6} ohm/cm
Crystal Structure	0.046 × 10^{-6}
	Hexagonal (closely packed)

Chemical Properties

Rhenium inflames oxygen at 300°C forming a white tetraoxide and yellow rhenium hemiheptaoxide, R_2O_7. When heated in air, the metal takes a film at 1000°C. The vapour of water in nitrogen attacks the metal at 1,900°C. If heated in moist hydrogen to about 2,000°C the metal becomes positive towards water vapour even at 2,300°C. The attack then begins at 2,500°C.

When heated in chlorine rhenium produces chloride, which sublimes and similar sublimate of rhenium bromide is also obtained, when heated in bromine. The metal is not attacked by hydrofluoric acid, it slowly dissolves in sulphuric acid. Its nitride has not yet been prepared. It readily dissolves in nitric acid. Rhenium-Copper couple reduces carbon monoxide to methane. When heated in methane it produces carbide. It does not amalgamate with mercury. It forms alloy with tungsten and the indications are to the formation of W_2Re_3.

The metal is not attacked by air-free water or dilute air free solutions of acids, As HF, HCl, HBr and H_2SO_4 and alkalis as KOH or NH_3. But the metal reacts with oxygen and concentrated acids containing an oxidizing agent as H_2O_2, HNO_3 or chlorine water.

Atomic Weight

It has been determined on the basis of such analytical as $AgReO_3$: AgBr or from the analysis of rhenium oxide, hexachloride or the disulphide. The atomic weight of the rhenium is 186.207. The atomic number is 75. The electronic configuration is

$$1s^2\ 2s^2\ 2p^6\ 3s^2\ 3p^6\ 3d^{10}\ 4s^2\ 4p^6\ 4d^{10}\ 4f^{14}\ 5s^2\ 5p^6\ 5d^5\ 6s^2\ (\text{or } 5d^6\ 6s^1)$$

Analytical Techniques

- When a mercury or thallium salt is added to a perrhenate a white precipitate of mercury or thallium perrhenate is produced.
- Rhenium salts give precipitate with brucine or veratrin.
- Methylene blue produces blur needle like crystalline precipitate.

Compounds of Rhenium

Oxides of Rhenium : The oxides of rhenium are heptoxide Re_2O_7; Rhenium trioxide, ReO_3; Rhenium dioxide, ReO_2 and rhenium sesquioxide, Re_2O_3.

Rhenium Heptoxide, R_2O_7: This compound is obtained (*i*) When rhenium compounds are ignited in a good supply of oxygen (*ii*) When an aqueous solution of perrhenic acid, $HReO_4$, is evaporated to dryness. Or (*iii*) When the metal is heated in oxygen to a temperature above 150°C. Rhenium heptoxide appears to occur in two forms the yellow and white. According to Noddack, the white form is obtained by the direct oxidation of rhenium or its dioxide more preferably when a stream of cold oxygen comes in contact with hot rhenium. At 150°C, the white heptoxide is transferred into the yellow variety without any loss of oxygen and therefore, the white form has been assumed to be an isomeric of the yellow one.

The ordinary heptoxide is yellow in colour, though it become almost colourless at a temperature below -80°C. On heating, the colour darkens. In the heptavalent oxides and other compounds rhenium shows dissimilarity with the manganese compounds; Mn_2O_7 is a dark green and permanganates are purple but heptavalent rhenium produces colourless compounds. Perrhenic acid and other compounds derived from the heptavalent rhenium show however, a strong adsorption in the ultraviolet region.

Rhenium heptoxide melts at 304°C to a liquid, which boils at 350°C. It produces colourless vapours on boiling. It is readily soluble in water, alcohol and acetone and to some extent in carbon tetrachloride and ether. It is reduced even in cold, though slowly, by sulphur dioxide or carbon monoxide and more so. When heated it produce coloured lower oxides, if reduced in a current of hydrogen, it produces a bluish-black oxide, ReO_2 at 300°C. Finally, it is reduced to the metallic state at 500°C. When exposed to a current of hydrogen sulphide; it is covered with a layer of the heptasulphide Re_2S_7, which prevents further action.

Rhenium Trioxide, ReO_3 : This oxide is obtained (i) by heating rhenium hepaoxide with metallic rhenium at a temperature of 250C or (ii) by burning the metal in a regulated quantity of oxygen. According to its states of division, it produces many shades of colour from dark blue to copper red.

The trioxide in its crystalline form resembles the trioxide of chromium and tungsten, CrO_3 and WO_3. The trioxide is however not attacked by water, hydrochloric acid (even hot) or dilute alkali solutions. It is oxidised by nitric acid to perrhenic acid, $HReO_4$ and it liberates iodine from and acidified solution of potassium iodide. When treated with hot concentrated solution of an alkali or when heated alone in vacuum to 400°C, it produces a mixture of the heptoxide and dioxide.

$$3\,ReO_3 \longrightarrow Re_2O_7 + ReO_2$$

It must be noted here that the treatment with alkalis does not yield the salt known as rhenates, derived from rhenic acid, H_2ReO_4. The rhenates are only indirectly prepared, say from perrhenates.

Rhenium Dioxide, ReO_2 : Noddack refers to the preparation of this compound by burning of the metal in a sufficient supply of oxygen. It is also prepared when the higher oxides are reduced with hydrogen. The electrolytic reduction of acidic or alkaline perrhenates produce the hydrated form of the oxide. The anhydrous sample of their oxide is a black powder, which shows high capacity towards the absorption of gases and solutes. In absence of oxygen, it is sufficiently stable, but if heated in vacuum to a high temperature, it yields a mixture of the heptoxides and metallic rhenium.

$$7\,ReO_2 \longrightarrow 2Re_2O_7 + 3Re$$

In this concept it differs from MnO_2 which produces oxygen on heating and also produces MnO). When gently warmed in an atmosphere of oxygen, it is oxidised to the heptoxide (white form at 150°C and yellow at higher temperatures).

It forms complexes with hydrogen halides and their salts of the composition M_2ReX_6. When treated with alkalis, it forms rhenites M_2ReO_3. The oxidising agents like nitric acid, chlorine water or hydrogen peroxide convert it to perrhenate.

Rhenium Sesquioxide, Re_2O_3 : It can be prepared in the form of hydrate, which comes down as a black precipitate, when a solution of trivalent rhenium (say rhenium chloride) is with alkali hydrate.

$$2\,ReCl_3 + 3NaOH \longrightarrow Re_2O_3 + 3NaCl + 3HCl$$

Its susceptibility towards oxidation is remarkable, only ten minutes shaking with oxygen in cold transforms it into perrhenate, it is more easily converted than dioxide, ReO_2. It may be remarked that the original precipitate of the sesquioxide contains always more oxygen than in the formula Re_2O_3 (in spite of the attempts to exclude oxygen in the experiment). Geilmenn and Wrigge (1931) have shown that rhenium trichloride will decompose water at 100°C with evolution of hydrogen even in presence of 1% sulphuric acid and much more readily in alkaline solution (salts of trivalent tungsten and ferrous hydroxide).

The sesquioxide, when treated with hot alkali solution undergoes decomposition as follows:

$$3Re^{III} \longrightarrow 2Re^{II} + Re^{V}$$

$$3Re^{V} \longrightarrow 2Re^{IV} + Re^{VII}$$

Adding together the total reaction is obtained.

$$8Re^{III} \longrightarrow 6Re^{II} + 2Re^{IV} + 2Re^{II}$$

There is no evidence that the trivalent rhenium oxide has any acidic properties. No alkali salts of the type Me_3ReO_3 or $MeReO_2$ are known.

Other Oxides : The pentavalent oxide of rhenium, Re_2O_5 has not been prepared. The earlier literature reported the preparation of the rhenium tetraoxide ReO_4, the existence of which is very doubtful. It was supposed to have been formed as a white cloud when rhenium or some of its lower oxides was heated in a current of cold oxygen under 150°C, but it is now identified to be mearly an isomeric form of the yellow heptoxide.

Acids and Salts derived from the Oxides of Rhenium

Oxides	Acids	Salts
Heptoxide, Re_2O_7	Perrhenic Acid, $HReO_4$	Perrhenates
Trioxide, ReO_3	Rhenic Acid, H_2ReO_4	Rhenates, M_2ReO_4
Pentoxide Re_2O_5 (not upheld)	Hyporhenic Acid, $HReO_3$	Hyporhenates, $MReO_3$
Dioxide, ReO_2	Rhenous Acid, H_2ReO_3 (not known)	Rhenites, M_2ReO_3
Sesquioxide, Re_2O_3	—	$ReCl_3$

Perrhenates : The salts of perrhenic acid (aqueous rhenium heptoxide) $HReO_4$, are known as perrhenates. The perrhenic acid solution dissolves metallic hydroxides to produce perrhenates. Sodium perrhenates, $NaReO_4$ is obtained by neutralising perrhenic acid with 0.1N sodium hydroxide with neutral red as an indicator. Potassium perrhenate $KReO_4$, is also similarly obtained. Copper carbonate dissolved in perrhenic gives copper perrhenate, $Ca(ReO_4)_2\ 4H_2O$. Baryta water and perrhenic acid react to form barium perrhenate, $Ba(ReO_4)_2$.

The salts of perrhenates when acidified give reactions of perrhenic acid:

(i) With *nitron it gives a precipitate of perrhenate, $C_{20}H_{16}N_4ReO_4$ (quantitative)

(ii) Sulphurous acid, hydrazine hydrochloride and hypophosphorous acid give a yellow colouration in concentrated solutions of perrhenic acid.

(iii) With potassium thiocyanate solution, a strong solution of perrhenic acid when added gives a yellow red colouration, which can be extracted with ether.

Resemblance between Perrhenates and other Similarly Constituted Salts

The anions like MnO_4' (permanganate) ClO_4' (perchlorate) IO_4' (periodate) BF_4' (borofluoride) and SO_3F (fluorosulphonate) have close similarity and the perrhenate ion also fall in the same category. The close similarity is observable in their crystal structures and the solubility.

(i) Crystal Structure : The structure of the ReO_4' ion is similar to that of MnO_4' or ClO_4' ion with the four oxygen, arranged round the four corners of the rhenium tetrahedron. However, it may be noted that whereas alkali permanganate and borofluoride are rhombic, the perrhenate and periodate become to the tetragonal system but as has been shown by V.M. Goldschmidt, this difference is a direct result of the difference of their ionic size.

(ii) Solubility : In solubilities, perrhenates very closely resemble permanganate, perchlorates and borofluorides. The salts of these anions with univalent cations such as K, Rb, Cs, Tl and Ag (other than Na and Li) exhibit low solubility. The salts of strychnine and nitron bases with perrhenates, perchlorates permanganates etc show also the sparing solubility. The similarity is also seen in the much greater solubility of the salts of sodium and of the simple divalent cations such as Ca, Sr Cu″, Mg, An, Cd, Co″, and Ni″. The salts of univalent metals with perrhenates, perchlorates do not exhibit hydration, where as the salts of polyvalent metals are hydrated.

The solubilities of these salts in water at 20°C are give below for comparison (moles per litre)

Cation	ReO_4'	MnO_4	ClO_4'	IO_4'	BF_4'
Na	Ca 3.5	Deleq	17	0.48	—
K	0.034	0.904	0.121	0.0022	—
Rb	0.031	0.052	0.055	0.03	0.027
NH_4	0.0227	0.060	2.0	0.135	
$Cr(NH_3)_6$	0.00073	0.0035 (at 17°C)	0.0199 (at 17°C)	—	0.0412 (at 17°C)

Nitron : A reagent for the detection of nitrate, perchlorate, boron, and rhodium.

The similarity in the solubilities or in the crystalline structure does not successively involve any similarity in solubility. The perrhenates are remarkably stable. Potassium permanganate begins to loose origin at 200°C, potassium perchlorate at 400°C and potassium periodate at red-hot state (nearly at 600°C) and potassium borofluoride decomposes at 500°C.

None of these salts can be distilled without decomposition even in vacuum; whereas potassium perchlorate melts at 518°C and distils unchanged at 1,370°C under one atmosphere pressure.

Mesoperrhenates : The salts derived from the acid H_3ReO_5 are known as mesoperrhenates. The mesoperrhenates are obtained (*i*) by fusing metallic rhenium or the dioxide ReO_2, with an alkaline hydroxide in excess of air.

$$3\ Na_2O + 2\ ReO_2 + 3O \longrightarrow 2\ Na_3ReO_5$$

(*ii*) By fusing metallic rhenium or its dioxide in presence of an oxidizing agent as peroxide, nitrite or chlorate; or (*iii*) by adding excess of the base to the solution of a normal perrhenate.

Similarly the barium salt, $Ba_3(ReO_5)_2$ separates out as a bright yellow precipitate. When barium hydroxide is added to a solution of barium perrhenate, $Ba(ReO_4)_2$. The mesoperrhenates are yellow when cold and red when hot. When dissolved in water, they are at once hydrolyzed to the normal or metaperrhenate, $MReO_4$ and the solution becomes colourless.

Thioperrhenates : It is the characteristic of the heptavalent rhenium that in its compounds the oxygen can be replaced by sulphur. If hydrogen sulphide is passed into a neutral solution of a perrhenate, the colourless solution becomes greenish-yellow on account of the formation of a thioperrhenate.

$$NaReO_4 + H_2S \longrightarrow NaReO_3S + H_2O$$

The salt produced is contaminated with poly-thio compounds of a dark coloured mixture. But it can be separated from them by the addition of thallus nitrate. The thallus salt $Tl(ReO_3S)$ is fractionally precipitated and can be recrystallised from water and alcohol. These thioperrhanates give a precipitate of silver sulphide with a solution of silver nitrate and also the coloured mixed precipitates of this perrhenates and sulphides with plumbus, cupric and mercuric salts. The thioperrhenates are oxidised by nitric acid bromine water and hydrogen peroxide. When acidified, the thioperrhenates precipitate rhenium heptasulphide

$$7\ HReO_3 \longrightarrow Re_2O_7 + 5\ HReO_4 + H_2O$$

Rhenates : The salts of rhenic acid H_2ReO_4 (similar to permanganic acid) are known as rhenates; M_2ReO_4 similar to mangantes. The aqueous solution of rhenic acid and rhenates ar very unstable; they are immediately decolourized when warmed. Rhenates like manganate are also green. Alkali rhenates ar slightly more stable, their yellow solution become browned on standing in air probably due to the formation of brown colloidal rhenium dioxide. If the yellow alkaline solution is dried over phosphorous pentoxide in vacuum, a yellow crystalline salt potassium rhenate, K_2ReO_4 or sodium rhenate Na_2ReO_4 is also obtained. Barium Perrhenate, $BaREO_4$ is sparingly soluble in water. If Barium perrhenate is fused with rhenium dioxide ReO_2 and caustic soda and the product extracted with alcohol an impure sample of barium rhenate is obtained as a green sparingly soluble mass stable only in presence of excess of alkali. In contact with water it decomposes into the perrhenate, $Ba(ReO_4)_2$ and the rhenate $BaReO_3$.

Alkali rhenates reacts with silver nitrate to give white precipitate of silver rhenate, $AgReO_4$. Silver and barium rhenates are soluble in nitric acid.

Hyporhenates, $MreO_3$: The only hyporhenate up to this time isolated is the sodium salt, $NaReO_3$, which produces pale yellow crystals. Noddack prepared it by forming sodium perrhenate with rhenium dioxide, ReO_2 and caustic soda in absence of air. In the dry state if it is free from alkali, it readily goes over into a mixture of sodium rhenite and sodium perrhenate.

$$3\,NaRe_5O_3 + Na_2O \longrightarrow 2Na_2Re_4O_3 + NaRe_7O_3$$

In water and in acids, sodium hyporhenate is at once converted to rhenium dioxide and perrhenate. Hyporhenates, on oxidation by air produce perrhenates, but they can be retained undecomposed under aqueous or alcoholic soda. A litre of 1.8N caustic soda solution at 0°C will decompose only 0.04 gm of the hyporhenate. The solution is yellow and soon begins to deposit the hydrated dioxide. The black oxide ReO_2 is precipitated on the addition of ammonium chloride. When treated with hydrogen sulphide sodium thioperrhenate forms mixed crystals. With sodium metaniobate, $NaNbO_3$ and metratantalate, $NaTaO_3$ it produces also mixed crystals.

Fluorides : Two fluorides of rhenium are known as ReF_6 and ReF_4. The penta and trifluorides could not be prepared. Ruff (1932) obtained the hexafluoride ReF_6, by reacting fluorine and rhenium powder at 125 in a fluorite (CaF_2) tube (quartz or silica vessels cannot be used). For this preparation it is very necessary that fluorine is fee from oxygen or chlorine. Rhenium hexafluoride is a pale yellow crystalline solid with melting point 18.8°C and boiling point of 47.6°C. Critical temperature 209°C. It is very unstable and at once reacts with fats, glass quartz etc. With quartz it reacts as

$$2ReF_6 + SiO_2 \longrightarrow 2ReOF_4 + SiF_4$$

Or,
$$3ReF_6 + 3SiO_2 \longrightarrow ReF_4 + 2ReO_2F \text{ (Rhenium Oxyfluoride)} + SiO_4$$

With water, it gives perrhenic acid, $HReO_4$, and the hydrated dioxide. The hexafluoride is reduced by hydrogen at 200°C by carbon monoxide at 300°C and sulphur dioxide at 400C. In all such reductions, the tetrafluoride is formed. At temperatures above 550°C, it is reduced to metal. Oxygen and oxidising agents convert the hexafluoride to the oxyfluoride, $ReOF_4$ and ReO_3F.

Rhenium tetrafluoride, ReF_4 is best obtained by passing the vapour of the hexafluoride, ReF_6, along with hydrogen or sulphur dioxide through a hot platinum tube. The tetrafluoride being non-volatile, stays behind. It is dark green colour melting at 124.5°C. It attacks quartz at 83°C, but it can be sublimed in a platinum tube without change in a current of sulphur dioxide at 500°C.

Chlorides : Rhenium forms a number of chlorides as $ReCl_7$, $ReCl_6$, $ReCl_5$, $ReCl_4$ and $ReCl_3$, when heated in chlorine, it forms two volatile chlorides (i) *Rhenium Hexachloride, $ReCl_6$* in brown needles, sublimating at 150°C and (ii) *Rhenium Heptachloride, $ReCl_7$*, which is green and more volatile than the hexachloride. Rhenium tetrachloride, $ReCl_4$ is the primary product of heating rhenium in chlorine. Rhenium when heated with chlorine also forms a pentachloride, $ReCl_5$, which is a green solid and which sublimes to give a brownish red vapour, when treated with water, it hydrolyses to a mixture of chlorine, hydrochloric acid, rhenium dioxide, Chlororhenous acid, H_2ReCl_5 and perrhenic acid. With alkalis it produces the dioxide and perrhenates. When heated in dry air, it is transformed to $ReOCl_4$ and ReO_3Cl. The pentachloride is thermally unstable.

$$ReCl_5 \longleftarrow ReCl_3 + Cl_2$$

Pure rhenium trichloride, $ReCl_3$ is obtained by heating the pentachloride in a stream of dry nitrogen. It is a dark red solid subliming to a heavy green vapour, $(ReCl_3)_n$. It is soluble in water, producing a red solution, which does not conduct electricity nor it gives precipitate with silver nitrate. It is quite stable in absence of moisture. When heated in oxygen, it produces the oxychloride. On standing, the solution is hydrolyzed to the sesquioxide. In hydrochloric acid solution, it exists probably as hypochlorhenous $HReCl_4$, which produces rubidium and caesium salts. The trichloride produces ammoniates with ammonia $ReCl_3$ nNH_3 (n = 6, 7 and 14).

Potassium chlororhenate K_2ReCl_6 and similarly caesium and silver chlororhenate, Cs_2ReCl_6 and Ag_2ReCl_6, have also been obtained. Potassium chlororhenate gives precipitates with silver, mercurous and thallus salts. Chlororhenates are stable towards potassium permanganate (in acids); but are oxidised by hydrogen peroxide in alkaline solution. They are reduced by zinc and sulphuric acid or in sulphuric acid at mercury cathode, producing the evidence of the tetravalent rhenium and when oxidised by air or oxygen indications are to the presence of quiniquivalent rhenium. They undergo autoxidation in alkaline solution.

When mixture of potassium perrhenate and potassium iodide is treated with hydrochloric acid, sp. gr. 1.18 and the mixture boiled until evolution of iodine here created potassium rhenium chloride $K_2[ReCl_6]$ is formed. A large excess of water decomposes it. Thallus rhenium chloride, $Tl_2[ReCl_6]$, is similarly produced through it is unstable.

Bromides and Iodides : When rhenium is heated in the bromine vapour, dark coloured volatile *Rhenium Bromide, $ReBr_6$* is formed and with iodine vapour dark volatile *rhenium iodide ReI_6* is formed. If potassium rhenite, potassium bromide and hydroiodic acid react together, *Potassium Rhenium Bromide is $K_2[ReBr_6]$*is produced. Thallous rhenate and hydrobromic acid gives *Thallous Rhenium Bromide, $Tl_2[ReBr_6]$*. Briscoe and his co-workers prepared *Potassium Rhenium Iodide, $K_2[ReI_6]$* by heating potassium perrhenate with an excess of potassium iodide and aqueous hydrogen iodide, at the boiling point of the latter.

This produces shining dark chocolate brown crystals. It is stable up to 200°C. At 300°C it decomposes at

$$K_2ReI_6 \longrightarrow 2KI + Re + 2I_2$$

Oxyhalides of Rhenium : *(i) From Heptavalent Rhenium :* It is interesting to see that heptavalent rhenium does not yield any binary halides but few of its oxyhalides are known. When the heptoxide Re_2O_7 is heated with an excess of the tetrachloride $ReCl_4$ (or even a mixture of the penta and trichloride, $ReCl_5$, $ReCl_3$) and the product fractionated, the oxychloride, ReO_3Cl is formed. It is a colourless liquid freezing at 4.5°C and boiling at 131°C. in presence of moisture, it hydrolyses to $HReO_4$ and HCl. When exposed to sunlight, it turns reddish-violet in few minutes the colour again fades in the dark.

The oxybromide $ReBrO_3$ has been prepared by heating metallic rhenium in mixture of oxygen and bromine vapour and finally distilling the product over the heptoxide Re_2O_7. It is a pure white solid with a melting point of 39.5 and boiling point of 163°C.

(ii) From hexavalent Rhenium : It is obtained (*a*) by passing a mixture of fluorine and oxygen over metallic rhenium at 125°C – 300°C or (*b*) by the action of rhenium hexafluoride on quartz or glass. The preparation is always contaminated with the hexafluoride, which is removed by fractional distillation, in a high vacuum. Rhenium dioxydifluoride ReO_2F_2 is also formed in the above reaction and it is left behind during the fractional distillation. The oxytetrafluoride is a pure white substance, melting at 39.7°C and boils at 62.7°C. It is reduced by most organic substances but it is stable towards oxidising agents.

The Rhenium dioxydifluoride ReO_2F_2 is also obtained in small amounts when fluorine and oxygen are passed over heated rhenium. It melts with decomposition at 156°C and is hydrolyzed in moist air and turns violet.

Rhenium Oxytetrachloride, $ReOCl_4$: This is obtained by heating the tetrachloride (or a mixture of tri and pentachloride) with oxygen at 150°C. It is a dark brown solid melting point at 28°C, which gives off brown fumes on heating. It dissolves in concentrated hydrochloric acid to form a brown solution of probably the composition, H_2ReOCl_6, which further breaks up

$$3H_2ReOCl_6 + 5H_2O \longrightarrow H_2ReCl_6 + 2HReO_4 + 12HCl$$

The oxychloride reacts violently with gaseous or liquid ammonia.

When heated at 300°C in a stream of ammonia, the ammonium chloride away, leaving a solid, of the $ReO(NH_3)_2Cl_2$.

Complex Rhenium Halides : *(i) From Pentavalent Rhenium :* Hexavalent rhenium gives indication to the formation of only unstable complex compounds of the type K_2ReOCl_6. But more stable complexes are obtained from the pentavalent rhenium, many of them have been isolated also. "It is curious that with pentavalent or with hexavalent rhenium the complex salts are not derived from the binary halides, which are known, but from the binary oxyhalides which in the pentavalent series do not exit".

It has been shown by Jacob and Jezowska from their e.m.f experiments that a solution of potassium perrhenate in hydrochloric acid is rapidly reduced electrolytically to Re^{V} stage; the reduction then proceeds slowly to the Re^{IV} stage.

The pentavalent rhenium solution is greenish. This green solution when mixed with potassium chloride gives a yellowish-green precipitate of the composition $K_2[ReOCl_5]H_2O$ or $K_2(OH)_2Cl_5$. The corresponding ammonium salt $(NH_4)_2[ReOCl_5]$ is yellow. The potassium and ammonium salts are readily oxidised by potassium permanganate, hydrogen peroxide, and nitric acid etc and thus in this respect they differ from the corresponding tetravalent coomplexes like K_2ReCl_5. The solution of the salts in concentrated hydrochloric acid is not oxidised by air, but in boiling, the Re^{V} is converted into a mixture of Re^{VII} and Re^{IV}

$$3Re^{V} \longrightarrow 2Re^{IV} + Re^{VII}$$

(ii) From Tetravalent Rhenium : The complex halides of the tetravalent manganese are very unstable and they do not include bromides or iodides. But very stable complex halides, even including bromides and iodides have been prepared from the tetravalent rhenium.

(a) Complex Fluoride or Hexafluoride, K_2ReF_6 : This has been obtained by reducing potassium perrhenate in hydrofluoric acid solution with potassium iodide.

$$2KReO_4 + 6KI + 6HF \longrightarrow 2K_2ReF_6 + 4KF + 3I_2 + 8H_2O$$

It forms green crystals, giving green solution. The salt is stable and is hydrolyzed by water only on warming.

(b) Complex Chlorides or Hexachlororhenate from the Acid H_2ReCl_6: This acid is formed (*i*) When rhenium tetrachloride is dissolved in hydrochloric acid (*ii*) Rhenium Chloride is boiled with hydrochloric acid, or (*iii*) potassium perrhenate is dissolved in concentrated hydrochloric acid, reduced with potassium iodide and the iodine is removed. The alkali salts of the acid (of K, Rb,Cs and NH_3) are all green when pure and give yellow-green solutions. The silver salt ar orange and the thallous and the manganous salts are yellow.

This hexachloride decompose in the presence of cold alkalis yielding finally trivalent and heptavalent rhenium.

$$4Re^{IV} \longrightarrow 3Re^{III} + Re^{VII}$$

On boiling with alkali solution we get

$$5Re^{IV} \longrightarrow 3Re^{II} = 2Re^{VII}$$

(c) Complex Bromides or Hexabromorhenates, M_2ReBr_6: When rhenium dioxide dissolves in hydrobromic acid a deep yellow solution is obtained which contains likely the acid *H_2ReBr_6, hexabromorhenic acid*. This acid can also be made by reducing perrhenic acid with hydrobromic acid even at 20°C. From this solution, potassium, rubidium and caesium salts have been obtained which produce thick red crystals.

(d) Complex Iodides or Hexaiodorhenate, M_2ReI_6: Briscoe Robinson and Rudge (1931) obtained the potassium salt K_2ReI_6 by reducing potassium perrhenate with potassium iodide in hydroiodic acid solution. It is a black crystalline solid giving a deep violet solution

in water the rubidium and caesium salts are only slightly soluble in water in presence of a large excess of their halides. Thus rhenium chloride can be detected by using rubidium chloride as a precipitant; $RbReCl_4$ is a bright red powder soluble in water and dilute hydrochloric acid. On heating to 550°C it decomposes as below :

$$6RbReCl_4 \longrightarrow 3Rb_3ReCl_6 + 2ReCl_3 + Re$$

The complex bromides of the type $M[ReBr_4]$ have also been prepared.

Sulphides

The sulphides of rhenium are known as Re_2S_7 and ReS_2

Noddack, Geilmann, Müller, Briscoe and others have studied the heptoxide Re_2S_7 in details. It is obtained by the action of hydrogen sulphide on hypo solution on an acidified solution of a perrhenate. When it produces the hydrated type $Re_2S_7H_2O$. If the dry heptoxide is brought in contact with sulphide, the anhydrous form of the heptasulphide may also be obtained. The precipitation of rhenium as the heptasulphide is quantitative if hydrogen sulphide gas is passed through an ammonical solution of perrhenate. The anhydrous heptasulphide is a fine black powder. It is vigorous oxidised by air, sometimes with incandescence. It undergoes decomposition if heated to the stage of melting and then it produces the disulphide and sulphur. The heptoxide is not attacked by hydrochloric acid, but is readily oxidised by nitric acid, hydrogen peroxide or bromine water to perrhenic acid.

Rhenium Disulphide, ReS_2 : Noddack has prepared this by fusing rhenium with sulphur and also by heating the heptasulphide. A tetravalent rhenium compound is solution when treated with hydrogen sulphide produces the anhydrous sulphide, which is a soft black substance, with a layer lattice like that of CdI_2 or MoS_2. It volatilises at about 1000°C with slight decomposition the vapour phase, however does not contain any lower sulphide. The sulphide prepared in the wet way is finely divided and is insoluble in alkali solution and even in sulphide solution and also in acids. The disulphide is oxidised by the usual regents to perrhenic acid. It has been used as a catalyst (like metallic rhenium) in the oxidation of alcohols to aldehydes and ketones.

Selenides, Re_2S_7 and Re_2Se_2 : The hepta selenide is made in the same way as the heptasulphide. It is a fine black powder, which is a strong absorbent. When heated to 325°C in vacuum, it produces the diselenide, $ReSe_2$ and free selenium. The diselenide is stable in air and is only attacked by strong oxidising agents.

Carbonyl Compounds from Rhenium : Hiecher in 1941 has prepared a number of rhenium carbonyl compounds by the action of carbon monoxide under 200 atm pressure on rhenium oxide at 240°C

$$Re_2O_7 + 17CO \longrightarrow [Re(CO)_5]_2 + 7CO_2$$

Potassium perrhenate also reacts in this way though less easily. Rhenium hepta sulphide reacts with difficulty. The pentacarbonyl thus formed is insoluble. The freezing point in depression in camphene and cyclopentanone shows its dimeric formula. The carbonyl

$[Re(CO)_5]_2$ forms small colourless and odourless crystals, which is not decomposed by dilute acids or alkalis. The crystals are however, decomposed by the hot concentrated sulphuric acid or nitric acid. In a tube they melt at 170°C. The carbonyl readily combines with bromine and iodine at 120°C and with chlorine it combines with difficulty, forming carbonyl halides, $Re(CO)_5Cl$. These carbonyl halides or rhenium are more easily formed than the carbonyl halides of nickel or cobalt, but not than those of platinum metals. Rhenium carbonyl iodide is pale yellow, but the bromides and chlorides are colourless. They are stable in air and sublime undecomposed in an atmosphere of carbon monoxide; the iodide decomposes at 90°C and the chlorine at 140°C.

EPITOME–4

Halogen

This appendix must be studied after studying the chapter halogen.

Corson Mackenzie and Segré in 1940 have first obtained the element-85 belonging to the iodine group. Three isotopes of this element ar described in the natural radioactive series.

(*i*) In uranium Series

Radium A → (99.96%) α-particle ⟶ Radium B(Pb^{214})

Radium A → (0.04%) β-particle ⟶ Astatine (At^{218})

Or $_{84}Pb^{218} = {}_{85}At^{218} + {}_{-1}e^{0}$

(*ii*) In the Thorium Series

Thorium A → (99.8%) α-particle ⟶ Astatine (At^{216})

Thorium A → (0.014%) β-particle ⟶ Astatine (At^{216})

Or $_{84}Pb^{216} = {}_{85}At^{216} + {}_{-1}e^{0}$

(*iii*) In the actinium Series

Actinium A → (almost 100%) α-particle ⟶ Radium B(Pb^{214})

Actinium A → (Traces) β-particle ⟶ Astatine (At^{215})

Or $_{84}Pb^{215} \longrightarrow {}_{85}At^{215} + {}_{-1}e^{0}$

Corson Mackenzie and Segré found that when bismuth is irradiated with 32MeV alpha particles from a cyclotron a 7.5 hours activity develops, which can be separated from all the adjacent elements 85. This element is formed in the following nuclear reaction.

$$_{83}Bi^{209} + {}_{2}He^{4} \longrightarrow {}_{85}Ac^{211} + 2{}_{0}n^{1}$$

This new nuclide (At) captures an orbital electron (K-captive and forms polonium known as Actinium C′). Actinium has a very short half-life (5×10^{-3} sec) and it expels a large range of α-particles.

Chemistry of Astatine

Tracer techniques have been used to isolate and study this element. It shows marked differences from that of halogens and yet is in keeping with the regular transition of properties with increasing of atomic number in the group VII-B. The electronegative or acidic and non-metallic properties of halogen decrease steadily in passing from fluorine to iodine, so much so that the element next t iodine in the same family would have even some metallic or electropositive characteristics.

After reduction by sulphur dioxide or zinc, the astatine activity is carried either by silver or thallium iodide, so that it evidently forms sparingly soluble thallus or silver salts and like iodine it is univalently electronegative.

However, astatine is very readily oxidised by bromine and by ferric ions, there being indication of at least two oxidation statuses.

Although there is no indication from migration experiments of the presence of positive astatine ion in solution; astatine deposits on the cathode as well as anode in the electrolysis of oxidised solutions. Element astatine can be volatilised, though not so readily as iodine and it has a specific activity for metallic silver. Though similarly to iodine is also shown by the observation that astatine concentrates in the thyroid glands of animals.

Astatine from Neptunium Series : This artificial radioactive series has been worked out in details and is an achievement in analogy to the three naturally occurring radioactive series (uranium, thorium and actinium). Francium and astatine both occur in this series. (See also transuranium elements, page 183-184). These occur as Astatine 214, [uranium series] Astatine 216 [thorium series] and Astatine 215 [actinium series].

(*i*) *From protactinium 226* : (which is found among the products of bombardment of thorium with 150 MeV deuterons) - (uranium series):

$$_{91}Pa^{238} \xrightarrow{\alpha} {}_{89}Ac^{222} \xrightarrow{\alpha} {}_{87}Fr^{218} \xrightarrow{\alpha} {}_{85}At^{214} \xrightarrow{\alpha} {}_{83}Bi^{210}$$

(*ii*) *Four protactinium 218* (resulting from the as irradiation of thorium with 80 MeV deuterions by the reaction Th^{232} by the reaction Th^{232} (d, 6n) Pa^{238} (Th series):

$$_{91}Pa^{238} \xrightarrow{\alpha} {}_{89}As^{224} \xrightarrow{\alpha} {}_{87}Fr^{220} \xrightarrow{\alpha} {}_{85}At^{216} \xrightarrow{\alpha} {}_{83}Bi^{212}$$

(*iii*) *From protactinium 227*, obtained by the reaction Th^{232} (d, 7n) Pa^{237} - (Ac series)

$$_{91}Pa^{227} \xrightarrow{\alpha} {}_{89}Ac^{223} \xrightarrow{\alpha} {}_{87}Fr^{219} \xrightarrow{\alpha} {}_{85}At^{215} \xrightarrow{\alpha} {}_{83}Bi^{211}$$

CHEMICAL ANALYSIS OF DIFFERENT ELEMENTS

Determination	Reagents
Aluminium	Alizarin Red S
	Aluminon
	Aluminon + Cetyltrimethylammonium bromide Chrome Azurol S
	Chrome Azurol S + Cetyltrimethylammonium bromide
	Chromazol KS + Cetylpyridinium bromide Eriochrome Cyanine R
	Eriochrome Cyanine R + Cetyltrimethylammonium bromide 8-Hydroxyquinoline
Ammonia	Phenol + Sodium hypochlorite
Antimony	Brilliant Green
	Bromopyrogallol Red
	Rhodamine B
Arsenic	Silver diethyldithiocarbamate
Barium	Sulfonazo III
Beryllium	Beryllon II
	Chrome Azurol S
	Chrome Azurol S + Cetyltrimethylammonium bromide
	Eriochrome Cyanine R
	Eriochrome Cyanine R + Cetyltrimethylammonium bromide
Bismuth	Dithizone
	Pyrocatechol Violet
	Pyrocatechol Violet + Cetyltrimethylammonium bromide
	Thiourea
	Xylenol Orange
Boron	Azomethine H
	Carminic acid
	Curcumin
	Methylene Blue
Bromide	Fluorescein
	Phenol Red
Cadmium	2-(5-Bromo-2-pyridylazo)-5-diethylaminophenol Cadion
	Dithizone
Calcium	4-(2-Pyridylazo) resorcinol
	Chlorophosphonazo III
	Glyoxal-bis (2-hydroxyanil)
	Murexide
Cerium	Phthalein Purple
	N-benzoyl-N-phenylhydroxylamine
	8-Hydroxyquinoline

Determination	Reagents
Chlorine	N,N-Diethyl-1,4-Phenylenediamine
Chromium	1,5-Diphenylcarbazide
	4-(2-Pyridylazo) resorcinol
	4-(2-Pyridylazo) resorcinol + Tetradecyldimethyl-benzylammonium chloride
	4-(2-Pyridylazo) resorcinol + Hydrogen peroxide
Cobalt	Nitroso-R salt
	1-Nitroso-2-naphthol
	2-Nitroso-1-naphthol
	4-(2-Pyridylazo) resorcinol
	4-(2-Pyridylazo) resorcinol + Diphenylguanidine
Copper	Bathocuproine
	Bathocuproine disulfonic acid
	Dithizone
	Neocuproine
	Cuprizone
	4-(2-pyridylazo) resorcinol+ Tetradecyldimethyl-benzylammonium chloride
Cyanide	Barbituric Acid + Pyridine
	Barbituric Acid + Pyridine-4-carboxylic acid
Fluoride	Alizarin Fluorine blue + Lanthanum (III) ion
	Eriochrome Cyanine R + Zirconium (IV) ion
Gallium	Pyrocatechol violet + Diphenylguanidine
	8-Hydroxyquinoline
	1-(2-Pyridylazo)-2-naphthol
	4-(2-Pyridylazo) resorcinol
	Rhodamine B
	Xylenol Orange
	Xylenol Orange + 8-Hydroxyquinoline
Germanium	Brilliant Green + Molybdate
	Phenylfluorone
Gold	5-(4-Diethylaminobenzylidene) rhodanine
	Rhodamine B
Hafnium	Arsenazo III
Indium	Bromopyrogallol Red
	Chrome Azurol S
	Chrome Azurol S + Cetyltrimethylammonium bromide
	Dithizone
	8-Hydroxyquinoline
	1-(2-Pyridylazo)-2-naphthol
	4-(2-Pyridylazo) resorcinol

Determination	Reagents
Iodine	Neocuproine + Copper (II)
Iodine	Starch
Iridium	Rhodamine 6G + Tin(II)
	N,N-Dimethyl-4-nitrosoaniline
Iron	Bathophenanthroline
	Bathophenanthroline disulfonic acid
	2,2′-Bipyridyl
	Chrome Azurol S + Cetyltrimethylammonium bromide
	1,10-Phenanthroline
	1,10-Phenanthroline + Bromothymol Blue Ferrozine
Lanthanum	Arsenazo III
Lead	Dithizone
	Sodium diethyldithiocarbamate
	4-(2-Pyridylazo) resorcinol
Lithium	Thoron
Magnesium	Eriochrome Black T
	8-Hydroxyquinoline
	8-Hydroxyquinoline + Butylamine
	Titan Yellow
	Xylidyl Blue
Manganese	Formaldoxime
Mercury	Dithizone
	Michler's thioketone
	Xylenol Orange
Molybdenum	Bromopyrogallol Red + Cetylpyridium chloride Phenylfluorone
	Toluene-3,4-dithiol
Nickel	2-(5-Bromo-2-pyridylazi)-5-diethylaminophenol
	Dimethylglyoxime
	Dimethylglyoxime + Oxidant
	2,2′-Furildioxime
	2-(2-Pyridylazo)-2-naphthol
	4-(2-Pyridylazo) resorcinol
Niobium	N-Benzoyl-N-phenylhydroxylamine
	Pyrocatechol + EDTA or 2,2′ Bipyridyl or
	1-(2-thenoyl)-3,3,3-trifluoroacetone
	Bromopyrogallol red
	Bromopyrogallol red + Cetylpyridinium chloride
	4-(2-Pyridylazo) resorcinol
	Sulfochlorophenol S
	Xylenol Orange

Determination	Reagents
Nitrate	Brucine
	Chromotropic acid
	Sulfanilamide + N-(1-Naphthyl)ethylenediamine dihydrochloride
Nitrite	Sulfanilamide + N-(1-Naphthyl)ethylenediamine dihydrochlorine
	Sulfanilic acid + 1-Naphthylamine
Osmium	1,5-Diphenylcarbazide
Palladium	2-(5 Bromo-2-pyridylazo)-5-diethylaminophenol Dithizone
	2-Nitroso-1-naphthol
	4-(2-Pyridylazo) resorcinol
Phosphate	Rhodamine B + Molybdate
	Malachite Green + Molybdate
Platinum	Sulfochlorophenolazorhodamine
	Dithizone
	2-Mercaptobenzothiazole
Rare Earths	Arsenazo I
	Arsenazo III
	Xylenol Orance
Rhenium	2,2′-Furildioxime
Rhodium	1-(2-Pyridylazo)-2-naphthol
Ruthenium	1,10-Phenanthroline
	Thiourea
	1,4-Diphenylthiosemicarbazide
Scandium	Alizarin red S
	Arsenazo III
	Chrome Azurol S
	Xylenol Orange
Selenium	3,3′-Diaminobenzidine
	2,3-Diaminonaphthaline
Silver	Dithizone
	Eosin + 1,10-Phenanthroline
Sulfate	Methylthymol blue + Barium (II)
Sulfide	*N*,*N*-Dimethyl-1,4-Phenylenediamine
Sulfite	Pararosaniline + Formaldehyde
Tantalum	Methyl Violet
	4-(2-Pyridylazo) resorcinol
	Phenylfluorone
Tellurium	Diethyldithiocarbamate
	Bismuthiol II

Determination	Reagents
Thallium	Brilliant green
	Dithizone
	Rhodamine B
Thorium	Arsenazo III
	Thoron
	Xylenol Orange
	Xylenol Orange + Cetyltrimethylammonium bromide
Tin	Pyrocatechol violet (and + Cetyltrimethylammonium bromide)
	Gallein
	Phenylfluorone
	Toluene-3,4-dithiol + Dispersant
Titanium	Chromotropic acid
	Diantipyrinylmethane
	Tiron
Tungsten	Pyrocatechol Violet
	Tetraphenylarsonium chloride + Thiocyanate
	Toluene-3,5-dithiol
Uranium	Arsenazo III
	2-(5-Bromo-2-pyridylazo) diethylaminophenol
	Chlorophosphonazo III
	1-(2-Pyridylazo)-2-naphthol
Vanadium	*N*-Benzoyl-*N*-phenylhydroxylamine
	8-Hydroxyquinoline
	4-(2-pyridylazo) resorcinol
Yttrium	Alizarin Red S
	Arsenazo III
	Xylenol Orange
Zinc	Dithizone
	1-(2-Pyridylazo)-2-naphthol
	Xylenol Orange
	Zircon
Zirconium	Alizarin Red S
	Arsenazo III
	Pyrocatechol Violet
	Morin
	Xylenol Orange

EPITOME–5

Photography

HISTORY

Sixteenth century Alchemists noticed that silver halide darkens when exposed to light. This was later used in the preparation of photographs. The first photograph was produced by Wed Wood is 1800 when as a permanent photograph was made only by Daguerre in 1839.

Photographic Process

The modern photography has six major steps given below :

(i) Preparation of Sensitive plate : Sensitive emulsion is a paste of colloidal silver bromide or silver iodide in gelatine. It is obtained by adding ammonical silver bromide solution containing gelatine.

$$AgNO_3 + KBr \longrightarrow AgBr + KNO_3$$

This is allowed to stand in a warm room (about 39°C) for sometimes when silver bromide particles grow in size and at the saine time become more sensitive to light. The process is called ripening. After sufficient time has been allowed the interval depending upon the derived sensitiveness, the emulsion is applied to glass plates or celluloid film in dust free dark rooms. The glass plates of celluloid films serve mearly to hold the sensitive emulsion.

(ii) Exposure : The sensitive plate or film is mounted in a camera and exposed for a few seconds to the image of a properly focused object. Actual time of exposure depends upon the relative sensitivities of the plate, intensity of light and size of the aperture. Different parts of the body reflected different quantities of light and effect the plate proportionally. Under effects of light the silver bromide decomposes in trace where it is illuminated forming extremely minute particles or nuclei of silver. There is no visible effect but in the affected parts the emulsion becomes more easily reduced to the formation of silver nuclei while the image remains latent.

(iii) Developing : A developer is a weak reducing agent, e.g. ferrous oxalate or an alkaline solution likes quinol, amidol or metol (these are organic developers).

When the plate is dipped in the developers the parts affected by less light remaining unattacked by the developer. The reaction taking place can be written as follows:

$$4AgBr + FeC_2O_4 + 2H_2O \longrightarrow 4Ag + 4HBr + FeC_2O_6$$

The image becomes visible but its shade is in the negative relationship with the shade of the object. The plate is therefore called the negative plate.

(iv) Fixing : The sensitive emulsion still present on the negative plate at white centres is dissolved out in hypo solution ($N_2S_2O_3$) in a fixing bath and the plate is thoroughly washed.

$$AgBr + 2Na_2S_2O_3 \longrightarrow Na_3[Ag(S_2O_3)_2] + NaBr$$

Sesitive Emulsion + Hypo ⟶ Sodium Argento thyosulphate

Now the negative plate which was all along kept in the dark room or faint red light can be taken into light.

(v) Printing : The processes used in the preparation of negative are repeated again in the process of printing. Two types of printing paper are generally used. They are:

(a) *P.O.P (printing out paper) :* Coated with silver chloride and silver citrate mixture. It is placed under the negative and exposed to light in a printing frame. The process being slow can be controlled easily. When the image of required shade is obtained it is fixed and toned.

(b) *Velox brand boride paper :* It is more sensitive, it needs short exposure, followed by developing, fixing and toning

(vi) Toning : The black shade of the photograph can be changed to a purple or fine steel grey by dipping it in the gold chloride or potassium chloroplatinate [K_2PtCl_6] solution respectively. The process is known as toning.

$$AuCl_3 + 3Ag \longrightarrow 3AgCl + Au$$

EPITOME–6

Atomic Molecular and Equivalent Weights

Determination of Atomic Molecular and Equivalent Weights

The concept of equivalence is of fundamental importance in the study of quantitative chemistry. It follows from the law of reciprocal proportions. It is that every element must have a definite weight in which or in integral multiples of which, it combines with other elements and consequently the proportion by weight in which the chemical combination between two elements takes place are known from the same law. In support of these, the results of analysis of some hydrogen-compounds are shown below :

Compounds and formulae		Analytical results
Methane, CH_4	1 part by wt. of hydrogen combines with	3 parts by wt. of carbon
Water, H_2O	1 part by wt. of hydrogen combines with	8 parts by wt. of oxygen
Hydrogen sulphide, H_2S	1 part by part by wt. of hydrogen combines with	16 parts by wt. of sulphur
Hydrogen chloride, HCl	1 part by wt. of hydrogen combines with	35.5 parts by wt. of chlorine
Hydrogen bromide, HBr	1 part by wt. of hydrogen combines with	80 parts by wt. of bromine
Sodium hydride, NaH	1 part by wt. of hydrogen combines with	23 parts by wt. of sodium
Calcium hydride CaH_2	1 part by wt. of hydrogen combines with	20 parts by wt. of calcium
Ammonia. NH_3	1 part by wt. of hydrogen combines with	4.67 parts by wt. of nitrogen

It is seen from the above analytical results that a definite quantity of each of the elements on the right hand side unites chemically with a definite weight of hydrogen (i.e. 1 part by wt. of hydrogen). It is further observed that these elements shall combine (if they combine at all) with each other in proportions of the weights shown or in simple multiples of them.

This will be clearly understood if we note the ratio of the combining weights of the constituent elements in their binary compounds.

Compound	Ratio of wts. of constituent elements
Carbon dioxide (CO_2)	wt. of carbon : wt. of oxygen = 3 : 8
Carbon tetrachloride (CCl_4)	wt. of carbon: wt. of chlorine = 3 : 35.5
Sodium chloride (NaCl)	wt. of sodium: wt. of chlorine = 23 :35.5
Calcium sulphide (CaS)	wt. of calcium : wt. of sulphur = 20 :16

It is evident from these examples that the weights in which the elements carbon oxygen chlorine sodium calcium sulphur separately combine with a fixed weight of hydrogen (1 part) are also the weights in which two of the elements combine with each other. Similarly results of careful analysis of suitable compounds will reveal that all elements are associated with definite numbers expressing the parts by weights in which they combine with definite numbers expressing the parts by weights in which they combine with one another. These numbers representing *weights or combining weights or chemical equivalent* or simple equivalents of the elements (see Fig. 1).

The concept of equivalent weight is not only applicable in chemical combinations but also in the displacement of one element from a compound by another. Thus, it follows from the above discussions that the elements always combine with or displace each other in the ratio of their equivalents.

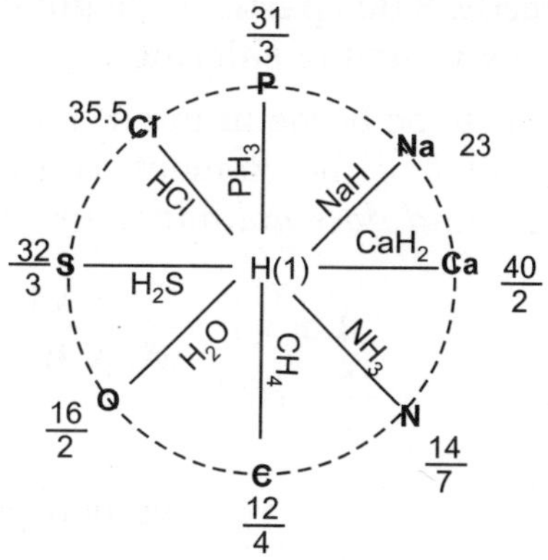

Fig. 1. The Equivalent Weights

The equivalent weight is a quantity to be determined experimentally. So for its numerical expression it is necessary to have a standard of reference. Originally 1 part by weight of hydrogen was chosen as the standard by Dalton (we have so far discussed equivalent weights in terms of hydrogen). But since many elements particularly metals are incapable of forming stable compounds with hydrogen the standard of equivalent weights was changed to oxygen. At present, 8 parts by weight of oxygen has been accepted as the standard of reference. On the basis of the oxygen scale, the equivalent weights of hydrogen is 1.008 and that of chlorine is 35.46. Similarly, minor changes in the values of equivalent weight of other elements will be observed if calculated with reference to oxygen instead of hydrogen. As some elements form well defined chlorides which are easily purified 35.46 parts by weight of chlorine is sometimes taken as the standard. Practically, Nowadays, equivalent weight is expressed in terms of 8 parts by weight of oxygen or its equivalent e.g. 1.008 parts by weight of hydrogen or 35.46 parts by weight of chlorine.

When an element forms more than one compound with another element (say oxygen) the equivalent weights will differ in these: cases but the values will bear a simple integral ratio to one another. For example:

The element carbon (eq. wt. 3.002) and oxygen (eq. wt. 8) will combine to form two different compounds.

In carbon dioxide (CO_2), wt of oxygen: wt of carbon = 8 : 3.002

In carbon monoxide (CO), wt of oxygen: wt of carbon = 8 : 6.004

Again sodium (eq. wt 22.99) will combine chemically with oxygen to produce two oxides.

In sodium monoxide (Na_2O), wt of oxygen: wt of sodium = 8 : 22.99

In sodium peroxide (Na_2O) wt. of oxygen: wt of sodium = 8 : 11.495

Thus, the elements combine with one another in the ratio of either their equivalent weights or simple multiples of them. This is the *law or equivalent proportions* and may be taken as another form of the law of reciprocal proportions, It will be seen later that an element with a variable valency *will* have more than one equivalent weight.

Equivalent Weight or Chemical Equivalent

Definition : The equivalent weight or chemical equivalent of an element is the number of parts of it, which can combine chemically with or can displace from a compound directly or indirectly. 8.000 parts by weight of oxygen (or 1.008 parts by weight of hydrogen or 35.46 parts by weight of chlorine).

It is to be borne in mind that the chemical equivalent of a given element is a ratio of the weight of the element to a definite weight of a standard element. So *it is a pure number and does not have any unit.*

Thus, $$\text{Eq. wt.} \; \frac{\text{wt. of the element} \times 1.008}{\text{wt. of hydrogen combined or displaced}}$$

Or, $$\frac{\text{wt. of the element} \times 8}{\text{wt. of oxygen combined or displaced}}$$

Or $$\frac{\text{wt. of the element} \times 35.46}{\text{wt. of chlorine combined or displaced}}$$

Example: (*a*) On analysis of the compounds sodium hydride (NaH) hydrogen bromide (HBr) and hydrogen sulphide (H_2S) we find that 1.008 parts by weight of hydrogen unite separately with 22.99 parts by weight of sodium. 79.916 parts by weight of bromine and 16 parts by weight of sulphur. So the equivalent weights of sodium bromine and sulphur are 22.99; 79.916 and 16 respectively.

(*b*) In the following equations :

$$\underset{65.38}{Zn} + H_2SO_4 = ZnSO_4 + \underset{2 \times 1.008}{H_2} \; ; \; \underset{24.32}{Mg} + 2HCl = MgCl_2 + \underset{2 \times 1.008}{H_2}$$

We find that 65.38 parts by weight of zinc displaces 2.016 parts by weigh of hydrogen from the acid or 32.69 parts by weight of zinc displace 1.008 parts by weight of hydrogen.

Similarly 12.16 parts by weight of magnesium displace 1.008 parts by weight of hydrogen, Therefore, the equivalent weights of zinc and magnesium are 32.69 and 12.16 respectively.

(*c*) In calcium oxide (CaO), 16 parts by weight of oxygen combine with 40.08 parts by weight of calcium or 8 parts by weight of oxygen are in combination with 20.04 parts by weight of calcium. 20.04 is the equivalent weight of calcium, In carbon dioxide (CO_2) 32 parts by weight of oxygen combine with 12.01 parts by weight of carbon or 8 parts by weight of oxygen combine with 3.002 parts by weight of carbon. Hence equivalent weight of carbon is 3.002.

(*d*) In aluminium chloride ($AlCl_3$) 3×35.46 parts by weight of chlorine combine with 26.96 parts by weight of aluminium. So the equivalent weight of aluminium = 8.99.

Equivalent weight of a radical : We know that a radical behaves like an element. So, the equivalent weight of a radical represents how many parts by weight of it can combine with 1.008 parts by weight of hydrogen or the equivalent of any other element or a radical. It is seen that in nitric acid (HNO_3), (14.008 + 48) or 62.008 parts by weight of nitrate radical (NO_3) are in combination with 1.008 parts by weight of hydrogen.

∴ Equivalent weight of NO_3 = 62.008, In sodium carbonate (Na_2CO_3). 2×22.99 parts by weight of sodium combine with (12.01 + 48) or 60.01 parts by weight of the carbonate radical (CO_3). Or 22.99 parts by weight of sodium (equivalent weight of sodium) are in combination with 30.005 parts by weight of the carbonate radical.

∴ Equivalent weight of CO_3 = 30.005. In ammonium nitrate (NH_4NO_3) 62.008 parts by weight of the nitrate radical (I equivalent of NO_3) are found to combine with 18.04 parts by weight of ammonium radical (NH_4), Hence, the equivalent weight of NH_4 = 18.04.

Equivalent Weight of a Compound

The concept of equivalent weight has also been extended to the compounds.

The equivalent weight of a compound is the number of parts by weight of the compound, which is obtained by adding the equivalent weights of its constituent elements or radicals. Thus :

Compound	Eq. wts of constituents	Eq. wt. of comp.
H_2SO_4	H-1.008 ; SO_4 - 48	49.008
$AgNO_3$	Ag-107.88 ; NO_3 - 62.008,	169.888
$AlCl_3$	Al-8.99; Cl-35.46	44.45

Alternatively, the equivalent weight of a compound represents how many parts by weight of it contain or react with 8.000 parts by weight of active oxygen or 1.008 parts by weight of hydrogen or an equivalent of any other substance. Detailed discussions on eq. wts of acids and bases as well as oxidising agents and reducing agents will be given in proper places.

Gram-equivalent

The equivalent weight expressed in grams is known as the gram-equivalent. The gram-equivalents of oxygen, sodium, silver nitrate sulphuric acid are 8 g of oxygen, 22.92 grams of sodium, 169'888 gms of silver nitrate and 49.008 gms of sulphuric acid respectively, A more elaborate definition of gram-equivalent is also given below :

The gram-equivalent of a substance is that weight of it (in g) which will combine with, displace or contains 1.008 g of hydrogen (i.e. 11.2 litres of hydrogen measured at N. T. P). Or 8 g of oxygen {i.e. 5.6 litres of oxygen at N.T.P) or 35.46 g of chlorine (i.e.11.2 litres of chlorine at N.T.P.) or the known equivalent of any other substance.

Determination of Equivalent Weights of Elements

Different methods employed in determining the equivalent weights of both non-metallic and metallic elements are described below!

(1) Equivalent weights of non-metals : (A) Determination of equivalent weight of oxygen-Duma's method: (by direct union with hydrogen or reduction of a metallic oxide).

Principle 1 A current of pure and dry hydrogen is passed through a weighed quantity of cupric oxide when hydrogen reduces the oxide, to metallic copper and itself combines with oxygen of cupric oxide, to form water.

$$CuO + H_2 = Cu + H_2O$$

The weights of hydrogen and oxygen involved in the combination care determined from the weight of water produced and the loss in weight of cupric oxide. The parts by weight of oxygen combining with 1.008 parts by weight of hydrogen will represent the equivalent weight of oxygen.

Experimental procedure : A small quantity of pure and dry black cupric oxide is taken in a hard glass combustion tube having a bulb in the middle. The bulb tube with its content (CuO) is accurately weighed. One end of the bulb tube is connected with a previously weighed U-tube containing fused calcium chloride. To prevent the entry of water vapour into the U-tube from outside, a guard tube filled with fused calcium chloride is attached with the open end of the U-tube. Then a stream of pure and dry hydrogen is passed through the other end of the combustion tube. When the air from inside the whole apparatus is removed, the tube is strongly heated. At the high temperature hydrogen and cupric oxide interact to produce metallic copper and steam. The steam is driven away by hydrogen and is absorbed in the calcium chloride U-tube. When all the black cupric oxide is completely converted into red copper heating is stopped and the whole apparatus is allowed to cool down to the room temperature in the current of hydrogen. After cooling the bulb tube and the calcium chloride U-tube are detached and separately weighed. The weight of oxygen that has combined with hydrogen to produce water is known from the loss in the weight of the combustion tube while the increase in weight of calcium chloride U-tube gives the weight of water formed.

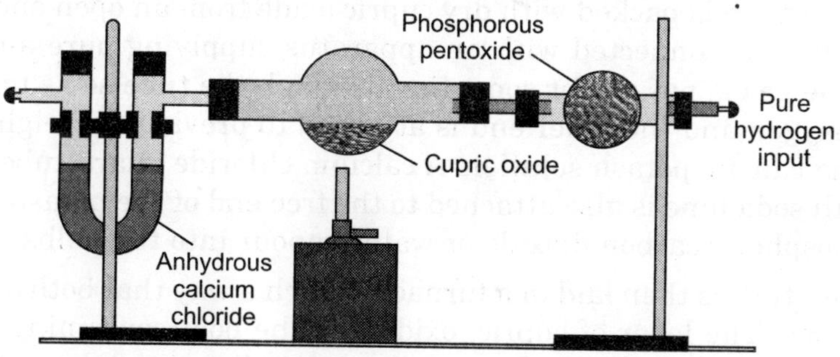

Fig. 2. The lab set up for the Determination of Equivant weight of Oxygen

Calculations : Let

wt. of combustion tube + cupric oxide before exp. = W g

wt. of combustion tube + cupric oxide before exp. = W_1 ..

wt. of oxygen = $(W - W_1)$ g

wt. of $CaCl_2$ U-tube before exp. = W_2 g

wt. of $CaCl_2$ U-tube before exp. + after = W_3 g

∴ wt. of water produced = $(W_3 - W_2)$ g

wt. of hydrogen that takes part in combination = $[(W_3 - W_2) - (W - W_1)]$ g = x gms. (say)

∴ x g of hydrogen combine with $(W - W_1)$ g of oxygen

∴ 1.008 g $= \dfrac{(W - W_1) \times 1.008}{x}$

So, equivalent wt. of oxygen $= \dfrac{(W - W_1) \times 1.008}{(W_3 - W_2) - (W - W_1)}$

(B) Determination of Equivalent Weight of Carbon : (By Direct Combination with Oxygen)

Principle : A weighed quantity of pure carbon is converted into carbon dioxide by heating it in a stream of pure and dry oxygen. Carbon dioxide produced is absorbed in strong caustic potash solution and its weight is taken. The weight of oxygen that has combined with the known weight of carbon can be obtained from the weight of carbon dioxide produced. Thus the equivalent weight is determined from the relation,

$$\text{Equivalent Wt. of carbon} = \frac{\text{Weight of the carbon taken} \times 8}{\text{Wt. of the oxygen combined}}$$

Experimental procedure : The exact weight of a clean and perfectly dry porcelain boat is taken. A small quantity (about 1 g) of pure charcoal (sugar charcoal) is placed into it and weighted again. The boat with charcoal is then introduced into a hard-glass tube nearly one half of which is packed with dry cupric oxide from an open end. The end of the tube nearer the boat is connected with an apparatus supplying pure and dry oxygen. A slow stream of the gas is passed for some time through the tube so as to displace the air from inside by oxygen and the other end is attached to previously weighed potash bulbs containing strong caustic potash solution. A calcium chloride guard-tube together with a U-tube filled with soda lime is also attached to the free end of the potash bulbs to prevent the entry of atmospheric carbon dioxide or water vapour into the bulbs.

The hard glass tube is then laid in a furnace in such a way that both of its ends remain outside the furnace. The layer of cupric, oxide and the boat containing carbon are then strongly heated while a slow stream of oxygen is continually passed through the combustion tube. The charcoal at the high temperature is oxidised by oxygen to carbon dioxide.

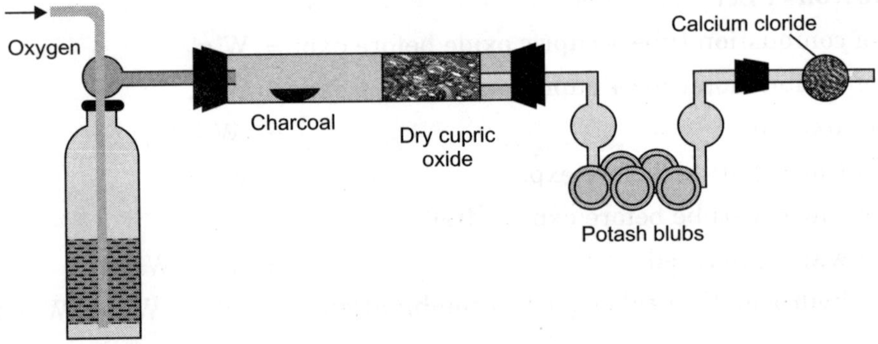

Fig. 3. The determination of equivalent weight of oxygen.

Any carbon monoxide which may simultaneously be formed due to shortage of oxygen is converted into carbon dioxide by the heated cupric oxide.

$$C + O_2 = CO_2\,;\; CO + CuO = CO_2 + Cu$$

The carbon dioxide thus formed is swept away by the oxygen gas and is absorbed in the potash bulbs.

When the charcoal is completely burnt away heating is stopped; but oxygen is passed through the tube for some time to ensure that all the carbon dioxide formed has been driven away and absorbed. When cooled to room temperature, the potash bulbs are re-weighed carefully.

Calculation : Let

wt. of the empty boat	= W g
wt. of boat + charcoal before exp.	= W 1 g
∴ wt of carbon taken	= (W1 – W) g
wt. of potash bulbs before exp.	= W g
wt. of potash bulbs before after exp.	= W 3 g
∴ Wt. of carbon dioxide produced	= (W 3 – W 2) g

and wt. of oxygen = wt. of carbon dioxide - wt. or carbon = {(W3–W2) – (W1–W)} g

$\therefore$ {(W 3 - W 2) - (W 1- W)} gms. of oxygen combine with (W 1 – W) g of carbon.

$$\therefore 8 \text{ g of oxygen combine with } \frac{(W_1 - W) \times 8}{(W_3 - W_2) - (W_1 - W)} \text{ of carbon}$$

This method is generally employed in cases or elements like sulphur, phosphorus, which form volatile gaseous oxides.

(2) Determination of equivalent weights of metals *(a) Hydrogen-displacement method :* This method is applicable to metals like zinc, iron, magnesium etc. which react with dilute mineral acids easily at room temperature with the evolution of hydrogen. A certain known weight of the metal is allowed to dissolve in excess of dilute acid and the weight of liberated hydrogen is found out from its volume. Then the equivalent weight of the given metal can be determined from the relation,

$$\text{Equivalent wt of metal} = \frac{\text{Wt. of the metal} \times 1.008}{\text{Weight if hydrogen displaced}}$$

It is to be remembered that wt, of hydrogen cannot be obtained directly. The wt of hydrogen in gms is calculated as follows :

Firstly, the volume of hydrogen displaced is measured in ml. and reduced to) NTP. Secondly, the vol. at N.T.P. is multiplied by 0.000089 (the wt, or 1 ml. of hydrogen at N.T.P. is 0.000089 g according to Avogadro's hypothesis).

So the wt. of H in g = Vol. in ml. of H at N.T.P. $\times$ 0.000089.

Equivalent weight of zinc :

Procedure : An accurately weighed quantity of zinc (about 0.08 ml.) is taken in a watch glass which is then placed at the bottom of a beaker. Zinc in the watch glass covered with inverted funnel, water is poured till the stem of the funnel remains completely immersed in water. A graduated tube closed completely at one end is filled completely with water and inverted carefully over the stem of the funnel so that the whole stem remains inside the tube. The tube is then clamped with a stand.

A graduated tube at one end is filled completely with water and inverted carefully over the stem of the funnel so that the whole stem remains inside the tube. The tube is then clamped with a stand.

A small amount of conc. sulphuric acid is poured into the beaker. A few drops of copper sulphate solution are also added to water, which is then stirred with a glass rod. Now the acid in dilute condition comes in contact with zinc and the reaction starts at

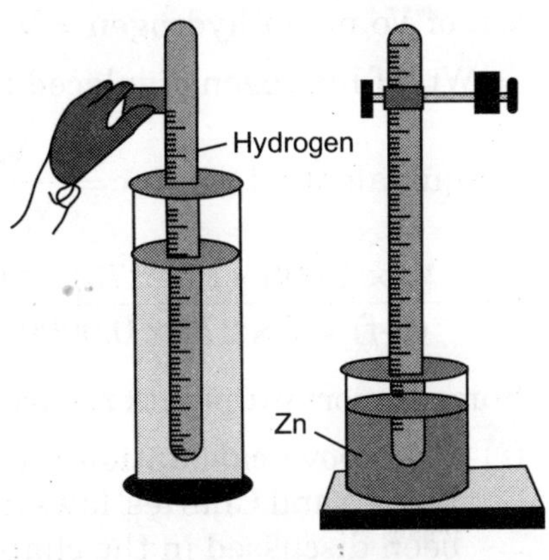

Fig. 4. The lab setup for the determination of equivalent weight of zinc.

once with quick evolution of hydrogen gas, which collects inside the graduated tube by displacement of water.

When the zinc completely dissolves (if necessary more, acid is added) and evolution of hydrogen ceases the tube is carefully taken out by closing the open end with the thumb and placed in a tall jar full of water.

The surface of water inside and outside the tube is then brought to the same level by dipping or raising it in the water of the jar During this operation, the tube is held with a piece of paper and no with the fingers.

Now the volume of the hydrogen gas is noted carefully, so also the temperature of water and the atmospheric pressure.

Calculation : Let

Wt. of zinc used	= W g
Vol. of hydrogen collected	= Vcc
Temp of the room during experiment	= t°C
Atmospheric pressure	= P mm
Tension of aquouus vapour	= (f mm)

∴Actual pressure of dry hydrogen = (P – f mm) (Since hydrogen over water, it is moist and the atmospheric pressure = pressure of hydrogen + tension of water vapour at t°C).

Let the volume of hydrogen (Vml.) at t°C and (P – f) pressure occupies V_0 ml. at N. T. P. Applying the Boyle's and Charles's laws, we get.

$$\frac{(P\text{-}f) \times V}{t + 273} = \frac{V_o \times 760 V}{273} \text{ or } \frac{(P\text{-}f) \times V \times 273 \text{ ml}}{(t + 273) \times 760}$$

Wt. of 1 ml. of hydrogen at N. T. P. (according to Avogadro's hypothesis) = 0.000089 g

wt. of Vo ml. of hydrogen = $V_0 \times$ 0000089 g

= Wt. of hydrogen displaced by w g of zinc.

$$\therefore \text{Equivalent wt. of zinc} = \frac{W \times 1.008}{V_o \times 0.000089}$$

$$\text{Or } \frac{W \times 1.008 \times (t + 273) \times 760}{(P\text{-}f) \times V \times 273 \times 0.000089}$$

Some important points regarding the above method :

(*i*) The above calculation requires the application of the gas equation derived from Boyle's and Charles' laws and also Dalton's law of partial pressure, these laws have been discussed in the chapter 2 of tome 1 the architecture of chemistry.

(*ii*) In the above experiment a few drops of copper sulphate solution are added, as the pure zinc does not react with dil. acid appreciably. Since zinc occupies a higher position than copper in the electro-chemical series. A very small amount of zinc is lost in displacing copper from copper sulphate solution.

$$Zn + Cu^{++} \Rightarrow Zn^{++} + Cu.$$

But this loss is however so small that it can be neglected.

(*iii*) To determine the equivalent wt. of aluminium, dil. hydrochloric acid is used instead of dil. sulphuric acid. It is not necessary to add copper sulphate solution in this case.

(*iv*) In this method, a small amount of the metal must be used so that the hydrogen evolved is conveniently collected in the graduated tube. The wt. of zinc taken in this exp. must not exceed 0.1 g and in case of magnesium, its wt. should be less than 0.05.

(*v*) The gas measuring tube must be held with the help of a strip of paper; and not with the fingers as the warmth of the hand will expand the vol. of the gas.

(C) Oxide Formation Method

A weighed (is either directly or indirectly converted into its oxide. The weight of the combined oxygen can be found out from the weight of the oxide produced. The parts by weight of the metal combining with 8'000 parts by weight of oxygen will give the equivalent wt. of the metal.

Equivalent weight of magnesium (by direct formation of oxide) : The constant weight of a clean and dry porcelain crucible with its lid is taken. A piece of well-cleaned magnesium ribbon is introduced in the crucible, which is weighed again. The crucible with the metal and the lid is placed on a tripod- stand with the help of a clay-pipe triangle [Fig. 5] and heated slowly.

While heating, the lid is put slightly on one side so that air may come in contact with magnesium and magnesium oxide formed does not escape into the atmosphere. When the metal is burnt completely to its oxide, the crucible is cooled in a desiccator and weighed again. The process of heating, cooling and weighing is repeated till the two successive weights are found to be the same.

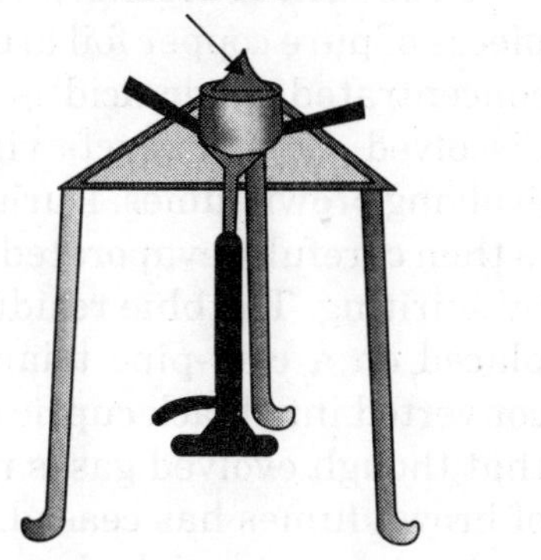

Fig. 5. Determination of Equivalent Weight of Magnesium.

$$2Mg + O_2 = 2MgO.$$

Calculation: Let

wt. of the empty crucible with lid	= W gms.
wt. of the crucible with lid + magnesium	$= W_1$ g
Wt. of magnesium taken	$= (W_1 - W)$ g
Again,	
Wt. of the crucible with lid + magnesium oxide	$= W_2$ g
∴ wt. of oxygen combined	$= (W_2 - W_1)$ g

$\therefore$ $(W_2 - W_1)$ gms. of oxygen combine with $(W_1 - W)$ g of magnesium

$\therefore$ 8 g of oxygen combine with $\dfrac{(W_1 - W)\times 8}{(W_2 - W_1)}$

Equivalent weight of magnesium

In the above process, a portion of magnesium oxide may escape into the air during heating and the heated magnesium may form a little nitride (Mg_3N_2) in addition to its oxide (MgO). Thus the method is not an accurate one and is not used in practice.

Equivalent weight or copper (by indirect formation of oxide) Principle : A weighed quantity of copper is dissolved in concentrated nitric acid and the resulting solution is slowly evaporated to dryness. The solid nitrate thus formed is strongly heated till it i8 completely converted into cupric oxide.

$$Cu + 4HNO_3 = Cu(NO_3)_2 + 2NO_2 + 2H_2O$$

$$2Cu(NO_3)_2 = 2CuO + 4NO_2 + O_2$$

The difference between the weigh is of cupric oxide formed and that of copper taken gives the weight of oxygen, which has combined with the metal. Therefore, according to the definition,

$$\text{Equivalent weight of copper} = \frac{\text{Wt. of the copper taken} \times 8.00}{\text{Wt. of oxygen combined}}$$

Procedure : A clean and dry porcelain crucible with its lid is accurately weighed. A few pieces of pure copper foil are put into the crucible, which is weighed again with its lid. Now concentrated nitric acid is added drop by drop to the copper foil till the metal is fully dissolved. Copper reacts vigorously with nitric acid forming copper nitrate solution and evolving brown fumes. During the reaction, the crucible is nearly kept covered. The solution is then carefully evaporated to dryness on the water-bath so that there is no loss of solution by spiriting. The blue residue left after evaporation is copper nitrate. The crucible is then placed on a clay-pipe triangle and heated strongly when copper nitrato is completely converted into black cupric oxide. While heating the lid is slightly put on ono side in order that though evolved gases might escape. The reaction is complete when though evolution of brown fumes has ceased. Then the crucible with its lid and copper oxide is cooled in a desiccator and weighed again. The process of heating, cooling and weighing is repeated till the weight of the crucible is constant.

Calculation: Let

wt. of the empty crucible with its lid $= W$ g

wt. of the crucible with its lid + copper $= W_1$ g

$\therefore$ Wt. of copper taken $= (W_1 - W)$ g

Wt. of the crucible with lid and cupric oxide $= W_2$ g

$\therefore$ wt. of combined oxygen $= (W_2 - W) - (W_1 - W)$ g

$= (W_2 - W_1)$ g

Equivalent weight of copper

$\therefore$ $(W_2 - W_1)$ g of oxygen combine with $(W_1 - W)$ g of copper

$\therefore$ 8 of oxygen combine with $\frac{(W_1 - W) \times 8}{(W_2 - W_1)}$ Equivalent weight of copper

(d) Method based on reduction of an oxide (Replacing oxygen from an oxide): This method is applicable for determining the equivalent weight of metals like copper, lead iron etc.

A weighed quantity of a metallic oxide is reduced to the metal by heating in a current of hydrogen.

$CuO + H_2 = Cu + H_2O$; $PbO + H_2 = Pb + H_2O$. By calculating the weight of combined oxygen from the difference between the weight of the oxide taken and the weight of the metal obtained after reduction the equivalent weight of the metal can be determined as before.

Equivalent weight of copper : The weight of a clean and dry porcelain boat is taken. A small quantity of dry cupric oxide is put into the boat, which is weighed again. The boat with its content (CuO) is then placed in a combustion tube and heated strongly in a current of hydrogen. At the high temperature, hydrogen reduces the oxide to red metallic copper.

After the completion of the reaction, the tube is allowed to cool while the flow of hydrogen is continued for some time. When cooled to the room temperature the boat containing the metallic copper is taken out and weighed.

Calculations: Let

Wt of empty porcelain boat $= W$ g

Wt of porcelain boat + cupric oxide $= W_1$ g

$\therefore$ Wt of cupric oxide $= (W_1 - W$ g$)$

Wt of porcelain boat + copper(after experiment) $= W_2$ g

$\therefore$ Wt of copper $= (W_2 - W)$ g

Therefore, wt of combined oxygen $= (W_1 - W) - (W_2 - W) = (W_1 - W_2)$

$(W_1 - W_2)$ gms of oxygen combine with $(W_2 - W)$ g of copper

$\therefore$ 8 g of oxygen combine with $\frac{(W_1 - W) \times 8}{(W_2 - W_1)}$

$\therefore$ Equivalent weight of copper

(e) Chloride Formation Method: (combining with chlorine indirectly or replacing chlorine from a chloride):

(a) Equivalent weight of Silver : A weighted piece of clean and pure silver is taken in a beaker and is dissolved completely in dilute nitric acid taking care to see that the solution is slightly acidic. The solution of silver nitrate thus produced is slightly heated and chemically pure dilute hydrochloric acid is added to it drop by drop until all the silver is precipitated as curdy white silver chloride.

$$Ag + 2HNO_3 = AgNO_3 + H_2O + NO_2$$
$$AgNO_3 + HCl = AgCl + HNO_3$$

The precipitate is allowed to settle and is filtered carefully through a tarred filter paper. The precipitated silver chloride is washed first with dilute nitric acid and finally with distilled water. The filter paper with the precipitate is then dried in an air-oven at 130°C. on cooling, it is weighed. The processes of heating, cooling and weighing are repeated till the final weight is constant.

Equivalent weight of Silver

Calculation : Let

wt. of silver = W g and wt. of silver chloride = W_1 g

Wt. of chlorine combined = $(W_1 - W)$ g

$(W_1 - W)$ g of chlorine combine with W g of silver

$\therefore$ 35 46 g of chlorine combine with W g $\dfrac{W \times 35.46}{W_1 - W}$ g of silver

(f) *Equivalent weight of sodium*

A weighed quantity of dry and pure sodium chloride is taken in a clean beaker; and is dissolved completely in distilled water. The solution is made slightly acidic by adding dilute nitric acid and a dilute solution of silver nitrate is then added slowly to it in a little excess with constant stirring with a glass rod. A curdy white precipitate of silver chloride is formed,

$$NaCl + AgNO_3 = AgCl + NaNO_3$$

When further addition of silver nitrate solution forms no more precipitate the content of the beaker is slightly heated and filtered through a previously weighed filter paper. The precipitated silver chloride on the filter paper is repeatedly washed with cold water. It is then dried in an air-oven at 130°C. On cooling the filtre paper with the precipitate is weighed, The processes of heating, cooling and weighing are repeated till a constant weight is obtained.

Calculation : Let the

wt, of sodium chloride = W g

Wt, of silver chloride produced = W_1 g

(107.88+35.46) or 143.34 g of silver chloride contain 35.46 g of chlorine

$\therefore$ W_1 g of silver chloride contain $\dfrac{35.6 \times W_1}{143.34}$ or W_2 g (say) of chlorine

The same amount of chlorine is present in W g of sodium chloride,

$\therefore$ Wt, of sodium in W g of sodium chloride = $(W - W_2)$ g

W_2 g of chlorine combine with $(W - W_2)$ gms, of sodium

$\therefore$ 35.46 g of chlorine combine with $\dfrac{(W - W_2) \times 35.46}{W_2}$ g of sodium

The alternate method of calculation : Let the equivalent weight of sodium be E Then,

$$\frac{\text{Wt. of sodium chloride taken } (W_{NaCl})}{\text{Wt. of silver chloride produced } (W_{AgCl})} = \frac{E + 35.46}{107.88 + 35.46}$$

Or

$$E = \left(143.34 \times \frac{W_{NaCl}}{W_{AgCl}}\right) - 35.46$$

The equivalent weight of potassium can also be calculated by the same method.‡

(g) Method based upon Replacement of a Metal by Another Metal

It is an established fact that a more electropositive metal displaces a less electropositive metal from its aqueous salt solutions.

$$Zn + CuSO_4 = ZnSO_4 + Cu$$
$$Fe + CuSO_4 = FeSO_4 + Cu$$
$$Zn + 2AgNO_3 = Zn(NO_3)_2 + 2Ag$$
$$Cu + 2AgNO_3 = Cu(NO_3)_2 + 2Ag$$

It is further known that this type of displacement occurs in equivalent amounts. Let x stands for the weight of the displacing metal (A) and y for that of the metal displaced (B). Then

$$x/y = E_A/E_B$$

(Where E_A and E_B are the equivalent weights of the metals A and B respectively).

Therefore if the equivalent weight of anyone of the two metals is known the equivalent weight of the other can easily be calculated.

Equivalent weight of zinc (By metal-displacement method) : A weighed quantity of pure zinc foil is added to an excess of aqueous copper sulphate solution taken in a beaker. Zinc gradually goes into the solution and red, heavy copper powder deposits at the bottom of the beaker. The reaction is completed on slight warming. When the zinc foil completely dissolves the solution is filtered carefully through a previously weighed filter paper. The copper powder collected on the filter paper is washed thoroughly with bot water until the washing is no longer blue in colour. Copper powder is next washed twice or thrice with alcohol. The filter paper with its content (i.e. copper) is then dried in an air-oven cooled in a desiccator and weighed. The processes of beating cooling and weighing are repeated till there is no difference between the two subsequent weights.

Calculation: Let

wt. of zinc taken = W g

wt. of copper + filter paper = W_1 g

wt. of filter paper = W_2 g

---

*It is difficult to get sodium or potassium in the pure state therefore the equivalent weight is determined by the above method.

∴ wt. of copper deposited = $(W_1 - W_2)$ g

Now taking the equivalent weight of copper as 31.78,

$(W_1 - W_2)$ g of copper are displaced by W g of zinc

∴ 31.78 g of copper are displaced $\frac{(W \times 31.78}{(W_1 - W_2)}$ g of zinc

Eq. wt. of zinc $\frac{W \times 31.78}{(W_1 - W_2)}$

It is to be noted that if the equivalent weight of zinc is known, eq. wt of copper can be found out by the above method.

The Relation between Equivalent Weight and Atomic Weight

The following simple relationship exists between the equivalent weight and the atomic weight of an element.

Let the atomic wt. of an element = A.

the equivalent wt. .of the *element* = E and

the valency of the element = V

Then. by definition of valency.

V atoms of hydrogen combine with 1 atom of the element.

∴ V × 1.008 parts by wt. of hydrogen combine with A parts by wt. of the element.

∴ 1.008 parts by wt. of hydrogen combine with A/V parts by wt. of the element.

Hence, according to definition,

equivalent weight of the element. E = A/V or A = E × V.

∴ **Atomic weight = equivalent weight × valency**

This relation can also be established on the basis of equivalent weight of oxygen. Let the symbol of the element be X and its atomic weight. equivalent weight and valency are A. E and V respectively.

Now the formula of the oxide of X will be X_2O_v and the molecular weight of the oxide will be 2A + 16v.

So 16v parts by weight of oxygen combine with 2A parts by wt. of X 2A *x* 8 A

∴ 8 parts by weight of oxygen 2A × 8/16V or A/V parts by wt. of *x*.

According to definition the number of parts of an element which can combine with 8 parts by weight of oxygen is the equivalent weight of that element.

∴ E =A/V or A (Atomic weight) = E (Equivalent weight) × V (valency)

This relation indicates that the atomic weight of an element is the same as its equivalent weight when the valency of the given element is 1. Thus the atomic weights of the monovalent elements like sodium, potassium, silver, chlorine and bromine etc. coincide respectively with their equivalent weights. The atomic weight of a bivalent element (like magnesium, calcium etc.) is twice its equivalent weight and atomic weight of a trivalent element like aluminium equals to its equivalent weight multiplied by 3.

Since the valency is always a small whole number the atomic weight of an element is not always constant a simple multiple of its equivalent weight.

Variable Equivalent Weight

It has already been mentioned that equivalent weight of an element is not always constant and an element may have more than one equivalent weight. We know that.

Equivalent weight = $\frac{\text{Atomic Weight}}{\text{Valency}}$ and the atomic weight is fixed according to Dalton

$E \propto 1/V$ or the equivalent weight of an element is inversely proportional to its valency. This led to the conclusion that the element having more than one valency will have different equivalent weights. Some illustrations are given below :

The valency of copper in cuprous oxide (Cu_2O) is 1 and that of copper in cupric oxide (CuO) is 2; the atomic weight of copper is 63.57.

∴ The equivalent weight of copper (ous) = 63.57 /1 = 63.57

and equivalent weight of copper (ic) = 63.57 /2 = 31.78

Similarly, the valencies of iron in ferrous chloride ($FeCl_2$) and in ferric chloride {$FeCl_3$) are 2 and 3 respectively. The atomic weight of iron is 55.85,

So, equivalent weight of iron (ous) = 55.85/2 =27.925 2

and equivalent weight of iron (ic) = 55.85/3 =18616 3

It may be said that "the equivalent weight of an element depends on the reaction it undergoes."

This statement can be explained with reference to the following pair of reactions.

(*i*) $Fe + 2HCl = FeCl_2 + H_2$

(*ii*) $2Fe + 3Cl_2 = 2FeCl_3$

From the quantitative aspects of the above equations, it is seen that in the reaction (i), 55.85 parts by weight of iron liberate.

2.016 parts by weight of hydrogen from hydrochloric acid or combine with 70.92 parts by weight of chlorine to form ferrous chloride, so, according to the definition, the equivalent weight of

iron (ous)= 55.85/2 =27.925

Similarly it can be shown by the reaction (*ii*) that 55.85/3 or, 18.616 parts by weight of iron combine with 35.46 parts by weight of chlorine giving rise to ferric chloride.

∴ Equivalent weight of iron (ic) is 18.616. Determination of atomic weights.

The atomic weights are usually determined by three chief methods given below:

(A) Cannizzaro's method - with the help of Avogadro's hypothesis.

(B) By the application of Dulong and Petit's law,

(C) By the application of Mitscherlich's law of isomorphism,

(A) Cannizzaro's method : This method of determining the atomic weight of an element which can form a large number of stable gaseous or volatile compounds has already been described,

(B) The method based on Dulong and Petit's law : The product of the atomic weight and the specific heat of an element is called the atomic heat. Dulong and Petit after having performed a large number of experiments proved that the atomic heats of the solid elements are nearly the same and approximately equal to 6.4 at ordinary temperature. This is known as Dulong and Petit's law.

In other words, the law states that for any solid element, the product of the atomic weight and the specific heat is nearly constant and numerically equal to 6'4 (approx.) at the ordinary temperature.

Atomic weight × specific heat = Atomic heat = 6.4 (approx.)

$$\therefore \text{Atomic Weight} = \frac{6.4}{\text{Specific heat}}.$$

It is evident that the relation is only approximate and thus an approximate value of the atomic weight of an element can be found out if its specific heat is known.

Limitations : (*i*) This method of estimation of approximate atomic weights applies to solid elements only and is obviously inapplicable to liquid or gaseous elements.

(*ii*) For certain solid elements of low atomic weights and high melting points like carbon, boron, silicon, beryllium, the atomic heat is considerably less than the expected values at the ordinary temperature. The atomic heat in such a case increases gradually with increase in temperature attaining a value of 6.4 calories degree^{-1} gram^{-1} at very high temperatures, Therefore, Dulong and Petit's law does not hold true in the cases of these solid elements.

Utility of the method : In spite of these limitations, this method which is very simple in its application is frequently used as the approximate atomic weights determined can be converted into the exact values with the aid of the following relationship :

$$\text{Atomic wt.} = \text{Valency} \times \text{Equivalent wt.}$$

The steps involved in the determination of actual atomic weight with the help of this method have been given later.

It may be noted here that as specific heats of almost all the solid elements are now known. Dulong and Petit's law may be applied for quick calculation of approximate atomic weights of most solid elements excepting a few like carbon, boron silicon, beryllium etc.

(C) By the Application of Mitscherlich's Law of Isomorphism : Isomorphous Compounds and isomorphism : Chemical compounds having the same crystalline structure and exhibiting similarly in composition (i.e. in the number of atoms which form the compounds) are said to be *isomorphous* with one another and the phenomenon of occurrence of different chemical substances in the same crystalline form is termed as *isomorphism*.

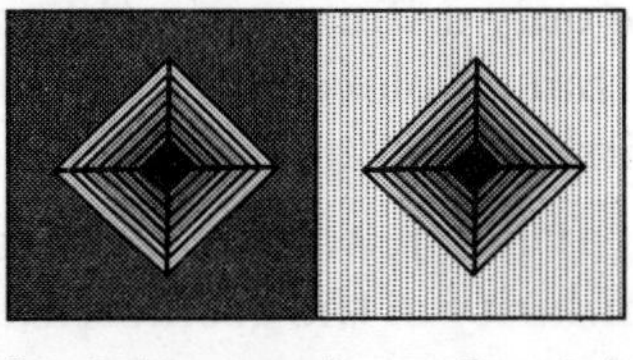

Fig. 6. Isomorphous Crystals

However, the reverse is not necessarily true as the substance possessing similar crystalline structure may not be isomorphous in all cases.

Criteria of Isomorphism

In addition to the identity of crystalline structure the compounds to be isomorphous must satisfy the following conditions :

(i) *Possession of analogous formulae :* Two isomorphous compounds should have similar formulae consisting of the same number of atoms united in the same way.

(ii) *Formation of mixed crystals :* A solution containing two or more isomorphous substances on cooling separates out homogeneous crystals which contain and the substances, in varying proportions depending on the composition of the solution. Such crystals are known as mixed crystals or solid solutions.

Formation of overgrowth : When a crystal of a substance is kept suspended in a saturated solution of its isomorphous the suspended crystal begins to grow in size due to deposition of the latter on its surface. This phenomenon is referred to as formation of overgrowth.

A few common groups of isomorphous compounds are given below. The key elements (the elements which are different) in each group have been underlined:

1. Zinc sulphate ($ZnSO_4$. $7H_2O$), magnesium sulphate ($MgSO_4$, $7H_2O$) and ferrous sulphate $FeSO_4$. $7H_2O$).
2. Copper Sulphide (Cu_2S) and silver sulphide (Ag_2S).
3. Potassium sulphate (K_2SO_4) and potassium chromate (K_2CrO_4).
4. Potassium permanganate ($KMnO_4$) and potassium perchlorate ($KCIO_4$) .
5. Potash alum [$K_2SO_4Al(SO_4)_3$. $24H_2O$ and chrome alum K_2SO_4,$Cr_2(SO_4)_3$.$24H_2O$].

Since zinc sulphate magnesium sulphate and ferrous sulphate are isomorphous substances, they exhibit identity of crystalline forms. When a solution containing zinc sulphate and magnesium sulphate (or ferrous sulphate) is allowed to crystallise, homogeneous crystals known as mixed crystals which contain both the sulphates of magnesium and zinc (or of iron) will separate out. Further, if a crystal of zinc sulphate be placed in a saturated solution of magnesium sulphate (or of ferrous sulphate), magnesium sulphate (or ferrous sulphate) continues to deposit on the crystal of zinc sulphate.

From the formulae of the above substances, it becomes evident that each of the molecules of two isomorphous compounds contains equal number of atoms united in the same manner. More clearly the same number of similarly arranged atoms of the key elements and the remaining identical elements constitute the molecules of the isomorphous substances. Thus, the only difference in the molecules of isomorphous sulphates of zinc magnesium and iron is that in place of zinc in the first one, we have magnesium in the second and iron in the third.

It has now been realised that the phenomenon of isomorphism depends more on the similarity in the geometrical of the crystals rather than the similarity in chemical properties.

Law of isomorphism and determination of atomic weight:

From the examination of a large number of isomorphous compounds, Mitscherlich formulated a law, which states. The equal number of atoms united in the same fashion produce the isomorphous crystals. This is known as *Mitscherlich's law* of isomorphism. This law leads to the concept that crystalline form of a substance depends essentially upon the number of constituent atom and the way in which they are combined and is independent of its chemical nature. In other words, the law of isomorphism may be stated as the isomorphous compounds i.e. the compound having similar crystalline forms and usually similar chemical characteristics can he represented by similar formulae which contain the same number of atoms (chemically similar or dissimilar) united in the same manner. A critical study reveals that in isomorphous compounds two analogous elements (chemically similar) replace each other, atom for atom without disturbing the crystalline structures. More over the weights of two elements replacing each other are in the ratio of their atomic weights. These facts are utilised in the determination, the atomic weight of either of the *two* replacing elements.

Let A and B be two analogous elements present In a pair of isomorphous compounds and let *a* and *b* be the atomic weights a and b respectively. Let us further suppose that W_1, gms of A one compound replace W_2, gms of B from the other compound. Then

the number of atoms of A = $\frac{W_1}{a}$

the number of atoms of B = $\frac{W_2}{b}$

We know that one atom of A will replace one atom of B.

∴ Number of atoms of replacing element (A) = Number of atoms of element replaced (B).

therefore , $w_1/a = W_2/b$ or $W_1/W_2 = a/b$

that is to say,

$$\frac{\text{wt. of the element A}}{\text{wt. of the element B}} = \frac{\text{atomic weight of A}}{\text{atomic weight of B}}$$

It follows directly from the law of isomorphism that in a group of *two* isomorphous compounds, elements substituting each other should have the same valency. So if the valency of one element is known the valency of the other is automatically decided. For example, zinc oxide is isomorphous with beryllium oxide. The formula of zinc oxide is ZnO and the valency of zinc is 2. Therefore according to the law of isomorphism the valency of beryllium must be 2 and beryllium oxide must have the formula BeO. The application of the knowledge of valency in determining the exact atomic weight of an element will be discussed later.

Exceptions of the law of Isomorphism

(*a*) A crystal of sodium chloride and that of a diamond have identical geometrical forms but these substances are not isomorphous because they fail to satisfy the other conditions of isomorphism.

(*b*) Many cases are known where substances with similar formulae and chemical properties are not isomorphous, e.g. sodium nitrate ($NaNO_3$) and potassium nitrate (KNO_3). In certain cases isomorphism is found amongst chemically similar substances having dissimilar formulae differing in the number of atoms in the molecules. For example, Ag_2S is isomorphous with PbS; K_2SO_4 and $(NH_4)_2SO_4$ are isomorphous substances.

(*c*) Compounds having the same number of atoms of entirely different chemical character are sometimes found to be isomorphous. Thus $NaNO_3$ and $CaCO_3$; $KMnO_4$ and $BaSO_4$ are two pairs of isomorphous compounds.

Determination of Exact Atomic Weights

(a) From the knowledge of equivalent weight and valency of the element : We know that the product or the equivalent weight and the valency or an element is numerically equal to its atomic weight. i.e.,

Atomic weight = Equivalent weight × Valency. The exact atomic weight or an element is round out from the above relationship and the steps to be followed in the method are given below :

Step - I The accurate equivalent weight or the element under consideration is determined by a chemical method.

Step - II There is no reliable method by which valency of the element could be derived with accuracy, it is obtained indirectly from the relation-

$$\text{Valency} = \frac{\text{Atomic weight}}{\text{Equivalent weight}}$$

The approximate value of the atomic weight determined by the method based on Dulong and Petit's law or by any other suitable method is then divided by the equivalent weight. The valency thus obtained is changed to the nearest whole number (since valency of an element can never be a fraction).

Step-III Finally, the equivalent weight is multiplied by the valency in order to get the exact atomic weight.

As an illustration, let us determine the exact atomic weight of magnesium.

Equivalent weight of magnesium determined by a chemical method is 12.16, Specific heat of magnesium is 0.262.

According to Dulong and Petit's law, the approximate atomic of magnesium

$= \frac{6.4}{0.262} = 24.42$.

$\therefore$ Valency of magnesium $= \frac{24.42}{12.16} = 2.01 \approx 2$

$\therefore$ Exact atomic weight or magnesium $= 2 \times 12.16 = 24.32$

Graham's Law of Diffusion

Graham established a quantitative relation between the rate of diffusion and the relative density of a gas and expressed it in a rate of diffusion and the relative density of a gas and expressed it in the form of a law known as Graham's Law, the law states that,

At constant temperature and pressure, the rates of diffusion of different gases are inversely proportional to the square roots of their densities.

The rate of diffusion means the volume of the gas diffusing per unit time. If volume V ml of a gas diffuse out in t seconds, then the rate of diffusion of the gas per second is V/t = ml.

It is to be kept in mind that the rate of diffusion increases with increasing temperature and pressure

That is Rate of diffusion $= \frac{\text{Volume of the diffused gas (in ml)}}{\text{Time taken (in seconds)}}$

This while comparing the rates of diffusion, the gases must be under the identical conditions of temprature and pressure. It is also to be noted that Graham's Law holds good whether the gas densities are expressed as normal or relative densities.

Mathematical expression of Graham's Law: If r and d be the rate of diffusion and density of a gas respectively, then under nay specified temprature and pressure,

$r \propto 1/\sqrt{d}$ or $r \propto 1/\sqrt{D}$ [where D is the relative density]

$\therefore r = k/\sqrt{d}$ k is proportionately constant)

If r_1 and r_2 be the rates of diffusion of two gases having densities d_1 and d_2 respectively (temperature and pressure remaining constant), then

$$r_1 = k/\sqrt{d_1},\ r_2 = k/\sqrt{d_2} \text{ or } r_1/r_2 = \sqrt{d_2}/\sqrt{d_1} = \sqrt{D_2}/\sqrt{D_1}$$

It is seen that a lighter gas (*i.e.* of lower density) diffuses faster than the gas, which is relatively heavier.

The graph in Fig. 7 shows the variation of the rates of diffusion with the densities of the gases. This graph also proves the relation.

$$r_1/r_2 = \sqrt{d_2}/\sqrt{d_1}$$

We know that the molecular weight of a gas is equal to twice its vapour or relative density. Therefore, if M_1 and M_2 be the molecular weights of two gases, then Graham's law can be expressed by the following relation :

$$\frac{r_1}{r_2} = \frac{\sqrt{D_2}}{\sqrt{D_1}} = \frac{\sqrt{2\times D_2}}{2\times D_1} = \sqrt{\frac{M_1}{M_2}}$$

If equal volumes of two gases (say) under the same conditions of temperature and pressure diffuse in times t_1 and t_2, then

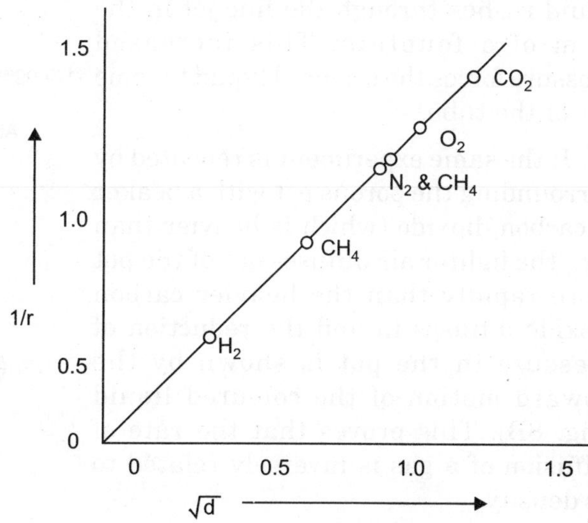

Fig. 7. The Graham's law of Diffusion.

$$r_1 = \frac{v}{t_1},\ r_2 = \frac{v}{t_2} \text{ and } \frac{r_1}{r_2} = \sqrt{\frac{M_2}{M_1}}$$

$$\text{or } \frac{v/t_1}{v/t_2} = \frac{\sqrt{M_2}}{\sqrt{M_1}} \text{ or } \frac{t_2}{t_1} = \frac{\sqrt{M_2}}{\sqrt{M_1}} = \sqrt{\frac{D_2}{D_1}}$$

This is under a given condition of temperature and pressure, the time of diffusion (t) of a gas is directly proportional to the square rot of its relative density (D) or molecular weight (M).

$$t \propto \sqrt{M}$$

Effusion : Experimental determination of the rate of effusion of a gas is comparatively easy than its rate of diffusion. *Effusion is a process in which a gas is forced under pressure to escape through a fine orifice* (pin-hole type) *made in the wall of a closed container.* The difference between diffusion and effusion is that in the former case. The gas spontaneously streams through a porous partition, while in the case of effusion. the gas is forced out through a small hole by applying some external pressure.

However, the two processes are essentially the same and Graham's law is applicable also in the case of effusion.

Experiment to verify Graham's law : The fact that a light gas having lower density diffuses more quickly than a heavy gas having higher density may be demonstrated by the following experiment.

The apparatus consists of a porous clay pot fitted with a rubber stopper through, which passes the longer limb of U-shaped glass tube containing a coloured liquid such as a solution of potassium permanganate. The U-tube has a bulb in the longer limb and its shorter limb ends in a fine jet. The apparatus is clamped in a vertical position (as in Fig. 8] so that the levels of the coloured liquid in both the limbs are same. On inverting a beaker full of hydrogen (which is lighter than air) over the porous pot. It is noticed that the coloured liquid rushes through the fine jet in the form of a fountain. This increased pressure forces the coloured liquid to come out of the tube.

If the same experiment is repeated by surrounding the porous pot with a beaker of carbon dioxide (which is heavier than air); the lighter air diffuses out of the pot more rapidly than the heavier carbon dioxide diffuses in and the reduction of pressure in the pot is shown by the upward motion of the coloured liquid [Fig. 8B]. This proves that the rate of diffusion of a gas is inversely related to its density.

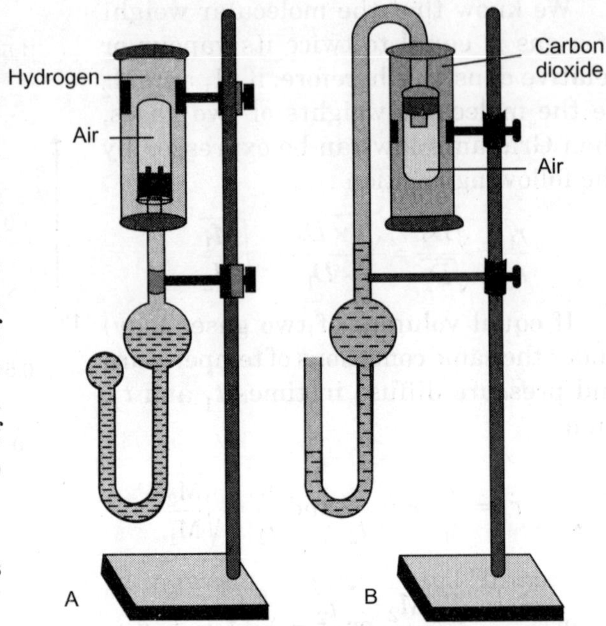

Fig. 8. The Gaseous Diffusion

Practical Applications of Gaseous Diffusion

1. The difference in the rates of diffusion of two gases of different densities can be employed in the separation of the constituents of a gas mixture. This process of separation of gases is called *atomolysis* or pressure analysis. This method has also been applied in the isolation of isotopes.

2. The results of Graham's law of diffusion enable us o determine the molecular weigh of a gas by comparing its time of effusion with that of a gas of known weight.

This can be done by mean of a *Bunsen's effusiometer*. It consists of a long glass tube open at the lower end and closed at the upper end by a stopcock over which a glass tube closed by a piece of thin platinum plate is attached. A very fine hole is pierced at the platinum plate. The stopcock communicates with the free air through the fine aperture made on the metal plate.

The effusiometer is immersed in a cylinder containing water (Fig. 9). The tube is filled with a gas whose rate of effusion is to be compared to a level below A and the tap is opened. The gas streams out through the aperture and the time taken for the level to raise from A to B is noted by a stopwatch. This experiment is repeated with a gas 'of known density *(i.e.* molecular weight) say, oxygen.

Let t_1 and t_2 are the times of effusion of equal volumes (v c.c.) of two gases (*i.e.,* the gas under examination and the gas used for comparison) of densities d_1 and d_2 and molecular weights M_1 and M_2.

If r_1 and r_2 are the rates of effusion, then

$$r_1 = \frac{v}{t_1}\ r_2 = \frac{v}{t_2}$$

$$\frac{r_1}{r_2} = \frac{t_2}{t_1} = \sqrt{\frac{d_1}{d_1}} = \sqrt{\frac{M_2}{M_1}}$$

For gases soluble in water, mercury is used instead of water.

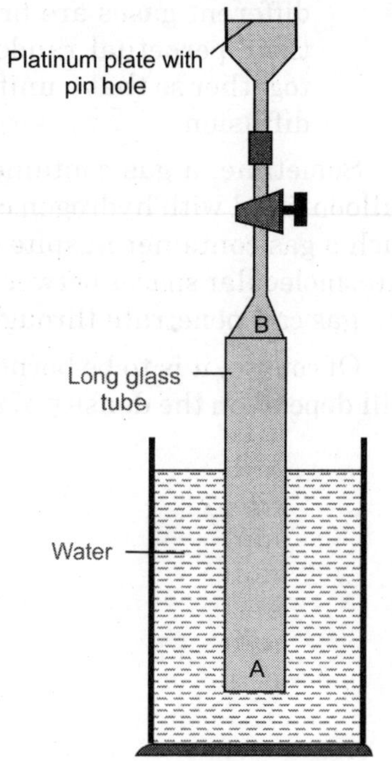

Fig. 9. Bunsen's Effusiometer.

3. We know that marsh gas (methane) cause devastating fire and violent explosion in the coal mines. The electric bell (marsh gas detector) used for giving signal indicating the presence of marsh gas is based on the phenomenon of diffusion of marsh gas.

The apparatus has a porous pot connected by means of a long glass tube and a rubber stopper with one limb of a glass U-tube containing mercury. Two limbs of the tube are provided with two platinum wires one of which is in touch with mercury and the other is just above it. Two platinum wires are connected with the terminals of a battery through an electric bell. A jar is kept inverted over the porous pot. Marsh gas being lighter than air will push the diffuse more quickly into the porous pot and mercury level up to touch the platinum wire. As a result the electric circuit is completed and the bell rings, indicating danger (see Fig. 10).

1. In fact, the phenomenon of diffusion is a spontaneous tendency of any substance gaseous, liquid or even every solid to spread uniformly through out the space available to it. But diffusion of gases is the rapidest process among these three; (here only gaseous diffusion is discussed).
2. The process of diffusion can be satisfactorily explained on the basis of the fact that the molecules of a gas are not in a state of rest but are moving in all directions with all possible velocities. The molecules of a substance in its gaseous state are far apart from each other and consequently, there are large empty spaces between them. When

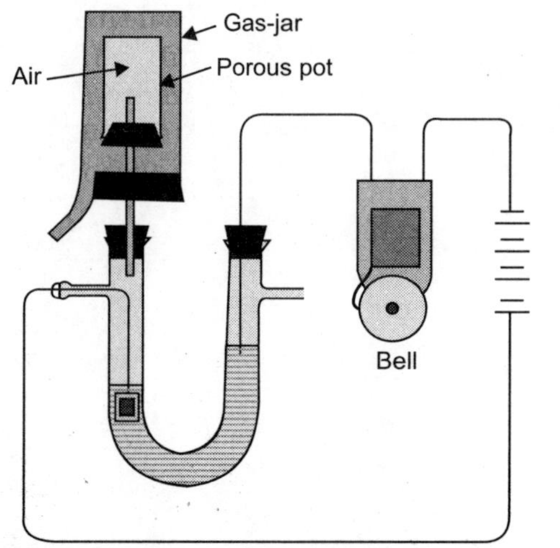

Fig. 10. The Marsh gas Apparatus.

different gases are brought in contact, the molecules of each of the gases due to their perpetual random motion pass through these empty spaces and intermix together so that a uniform mixture results. This is the mechanism of the process of diffusion.

Sometime, a gas contained in a closed vessel is found to come out gradually. A gas balloon filled with hydrogen collapses after a short time. This happens because the wall of such a gas container in spite of being made of the solid substance has porosity. There are intermolecular spaces between the molecule constituting the wall. As a result, molecules of the gas can penetrate through these spaces, and pass out.

Of course, it is to be borne in mind that the rate of passing out of a gas through a wall will depend on the density of the gas as well as on the substance of which the wall is made.

EPITOME–7

More Boron

Introduction

Davy first isolated boron in 1807 by electrolysis of moist boric acid. In 1808, Davy obtained it by heating fused boric acid (boron trioxide, B_2O_3) with potassium.

Occurrence

Boron does not occur free in nature but in combination, it occurs in soils of different species. The most important minerals of boron are : *Borax* ($Na_2B_4O_7$ $10H_2O$) *and Kernite* ($Na_2B_4O_74H_2O$), *Colemanite* ($Ca_2B_6O_{11}.5H,O$), *Ulexite* ($Na_2B_4O_7Ca_2B_6O_{11}.16H_2O$), *Boracite* ($2Mg_3B_8O_{11}.5H_2O$), *Pandermite* $Ca_2B_6O_{11}.3H,O$).

Gaseous substances from volcanic hot springs contain small quantities of boric acid (H_3BO_3). Most of the major deposits of boron are found in Chile (South America), California (USA), Tibet and Turkey. In fact, boron isolated from the minerals of California meets the world's demand of the element.

Preparation of Boron

Boron exists in two allotropic modifications, for example amorphous and crystalline :

Amorphous boron : Amorphous boron is generally prepared by reduction of boron trioxide with metallic magnesium, Boron trioxide is fused with the required amount of magnesium powder in a covered crucible, A violent reaction takes place,

$$B_2O_3 + 3Mg = 3MgO + 2B$$

The fused mass is then cooled and subsequently boiled with dilute hydrochloric acid to dissolve the magnesium oxide, unchanged magnesium and magnesium boride (Mg_3B_2) which may be formed in small amount. Boron is separated from the dissolved substances by filtration washed and finally dried in a steam-oven when it is obtained as a brown powder.

Boron thus obtained is 93.8% pure. It contains silicon, silica as impurities. This impure boron is further purified by treatment with hot hydrofluoric and boron of 98.3% purity is obtained.

Crystalline boron : (*a*) Crystalline boron can be prepared by fusion of amorphous boron followed by cooling. (*b*) Crystalline boron can also be prepared by using amorphous boron with aluminium and then dissolving the latter with hydrochloric acid. Boron thus obtained is generally associated with aluminium in the form of aluminium boride having the composition AlB_{12}.

Other methods : (*a*) For the preparation of amorphous boron, potassium fluoborate instead of boron trioxide may be used. Moreover metallic potassium, sodium can also act as the reductant.

$$B_2O_3 + 6K = 3K_2O + 2B \; ; \; KBF_4 + 3K = B + 4KF$$

(*b*) Pure and crystalline boron can be obtained by reduction of boron trichloride or tribromide vapours with hydrogen in an electric arc or by thermal decomposition of vapours of these boron halides. in contact with electrically heated tungsten or tantalum filament at about 1200°C.

$$2BCl_3 + 3H_2 = 2B + 6HCl$$

Boron cannot be prepared by reduction of B_2O_3 by carbon, as boron is not obtained even at white heat B_2O_3 on reduction with carbon at very high temperature very hard boron carbide (B_4C) is produced.

$$2\,B_2O_3 + 7C = B_4C + 6CO$$

Properties of Boron

Atomic Number	5
Relative Atomic Mass	(^{12}C = 12.000) 10.81
Melting Point	2300C
Boiling Point	3658°C
Density/kg m^{-3}	2340 (0°C)
Ground State Electron Configuration	[He]$2s^22p^1$
Electron Affinity (M-M$^-$)/kJ mol^{-1}	15
Oxidation States	B^{III}

Key Isotopes

Nuclide	^{10}B	^{11}B
Atomic mass	10.013	11.009
Natural abundance	20.0%	80.0%
Half-life	stable	stable

Ionisation Energies/kJ mol^{-1}

$M - M^+$	800.6
$M^+ - M^{2+}$	2427
$M^{2+} - M^{3+}$	3660
$M^{3+} - M^{4+}$	25025
$M^{4+} - M^{5+}$	32822

Thermodynamic Data

Enthalpy of Fusion/kJ mol^{-1}	22.2
Enthalpy of Vapourisation/kJ mol^{-1}	504.5

Covalent Bonds/kJ mol^{-1}

B-H	381
B-H-B	439
B-C	372
B-O	523
B-F	644
B-Cl	444
B-B	335

Chemical Properties : Amorphous boron is chemically reactive.

1. *Reaction with air oxygen*: Boron is not affected by air or oxygen at ordinary temperature but when heated in air at about 700°C, it burns with a reducing flame forming boron trioxide (B_2O_3} and some boron nitride (BN).

2. *Reaction with water*: Ordinarily, boron is without action on water but it is oxidised to boron trioxide by super heated steam and hydrogen is evolved.

$$2B + 3H_2O = B_2O_3 + 3H_2.$$

3. *Action on non-metals*: Many non-metals such as nitrogen, halogens, carbon, sulphur etc. react with boron at elevated temperatures forming binary compounds.

When heated in a current or nitrogen at about 900°C, boron forms boron nitride, which is readily decomposed by steam producing ammonia and boric acid.

$$2B + N_2 = 2BN; BN + 3H_2O = NH_3 + NaBO_3$$

Boron nitride is a hard substance having a very high, melting point (3000°C under high pressure). It is slippery and resembles graphite in its structure. This is why it is called *inorganic graphite.*

Boron readily enters into combination with halogens. Boron ignites spontaneously in fluorine giving boron trifluoride. When heated in chlorine, it forms boron trichloride, which is hydrolysed by water to produce boric acid.

$$2B + 3F = 2BF_3$$

$$2B + 3Cl_2 = 2BCl_3$$

$$BCl_3 + 3H_2O = H_3BO_3 + 3HCl$$

Boron unites with carbon at very high temperatures of the electric furnace forming boron carbide (B_4C) which is nearly as hard as diamond. It also combines with sulphur giving sulphide (B_2S_3) at elevated temperatures.

(4) *Action on metals:* Boron combines with most metals at high temperatures forming metallic borides.

$$3Mg + 2B = Mg_3B_2$$

(5) *Reaction with acids:* Boron does not react with hydrochloric acid or dilute oxyacids. On boiling with concentrated nitric acid, aqua regia or concentrated sulphuric acid, boron is oxidised to boric acid.

$$B + 3HNO_3 = H_3BO_3 + 3NO_2$$

$$2B + 3H_2SO_4 = 2H_3BO_3 + 3SO_2$$

Reaction with alkalis: It reacts with fused alkalis or their concentrated solutions giving alkali metaborates and hydrogen.

$$2B + 2NaOH + 2H_2O = 2NaBO_2 + 3H_2.$$

Boron can replace carbon and silicon from their oxides on strong heating.

$$3SiO_2 + 4B = 2B_2O_3 + 3Si$$

Under similar condition, boron and nitric oxide interact to give boron trioxide and boron nitride.

$$5B + 3NO = B_2O_3 + 3BN$$

Boron can also displace many metals from their oxides, sulphides or chlorides when heated.

It is to be noted here that crystal is almost chemically inert. It is not oxidised even on heating in oxygen at high temperatures. Hot, concentrated nitric acid, sulphuric acid or hydrofluoric acid has no action on it. However 1t is converted into borates fusion with sodium peroxide or a mixture of sodium carbonate and potassium nitrate.

Uses

Amorphous boron is used in pyrotechnic flares to provide a distinctive green colour, and in rockets as an igniter. The most important compounds of boron are boric, or boracic acid, widely used as a mild antiseptic, and borax which serves as a cleansing flux in welding and as a water softener in washing powders. Boron compounds are also extensively used in the manufacture of borosilicate glasses. Other boron compounds show promise in treating arthritis. The isotope boron 10 is used as a control for nuclear reactors, as a shield for nuclear radiation, and in instruments used for detecting neutrons. Demand is increasing for boron filaments for a high-strength, lightweight material chiefly employed for advanced aerospace structures.

Elemental boron is not considered a poison, and indeed is essential to plants, but assimilation of its compounds has a cumulative toxic effect.

Compounds of Boron

There are many compounds of boron, they are as follows :

Boric Acid, H_3BO_3: More accurately the name of the compound is orthoboric acid but it is commonly known as boric acid. Sometimes, it is also called boraric acid.

Preparation : (A) From borax: Boric acid is prepared by adding requisite quantity of concentrated hydrochloric acid (or sulphuric acid) to a hot saturated solution of borax; colourless, pearly white, silky crystals of boric acid separate on cooling. The separated crystals are filtered, washed with cold water and recrystallised from hot water; to give pure crystals of the acid.

$$Na_2B_4O_7 + 2HCl + 5H_2O = 4H_3BO_3 + 2NaCl$$

(B) From mineral colemanite ($Ca2B_6O_{11}.5H_2O$) :

On a commercial scale, boric acid is prepared from the native Calcium borate (known as colemanite). Sulphur dioxide is passed through a suspension of finely powdered colemanite in boiling water when calcium sulphite and boric acid are formed. Calcium sulphite on being reacted with excess of sulphur dioxide, goes into the, solution forming calcium bisulphite. Boric acid crystallises out when the solution is cooled.

$$Ca2B_6O_{11} + 2SO_2 + 9H_2O = 2CaSO_3 + 6H_3BO_3$$

$$2CaSO_3 + 2H_2O + 2SO_2 = 2Ca(HSO_3)_2$$

Net reaction:

$$Ca2B_6O_{11} + 11H_2O + 4SO_2 = 2CaSO_3 + 6H_3BO_3$$

Properties of boric acid

Physical : 1. Boric acid or orthoboric acid is a soft, silky, purely white flaky crystalline substance with specific gravity 1.46. (2) It is greasy to the touch. (3) It is slightly soluble in cold water but the solubility rises considerably with rise of temperatures, (solubility at 12°C = 4.01% and about 28% at 100°C). (4) It is volatile in steam.

Chemical : *Action of heat:* When heated, orthoboric acid loses water in three successive stages. At 100°C, boric acid is converted into metaboric acid (HBO_2) which on further heating at 140°C is transformed into pyroboric acid ($H_2B_4O_7$). At red heat, it loses all water and a hygroscopic, glassy mass of boron trioxide is formed.

$$H_3BO_3 \xrightarrow{100°C} HBO_2 + H_2O$$

$$4HBO_2 \xrightarrow{140°C} H_2B_4O_7 + H_2O$$

$$H_2B_4O_7 \xrightarrow{\text{Red heat}} 2B_2O_3 + H_2O$$

2. *Acidic character :* Boric acid is a very weak acid. It is a monobasic acid ionising in aqueous solution as :

$$H_3BO_3 \Re H^+ + H_2BO_3^-$$

Its aqueous solution turns blue litmus slightly red but it has no action on methyl orange. Alkali metal borates (salts of weak acid and strong base) are extensively hydrolysed in aqueous solutions giving alkaline solutions.

Uses

1. Boric acid is widely used in medicine as an antiseptic and for preparing borated cotton, boracic powder, boracic ointment etc. Very dilute solution of the acid is used as an eye lotion.
2. It is used in large quantities in the manufacture of borax, glass, glazes in enamel industries.
3. It also finds applications in tanning candle and dye manufacture.

Test for Boric Acid and Borates

When boric acid mixed with a little methyl alcohol is gently heated in a test tube and—the issuing gas is ignited, a green-edged flame is produced (see Fig. 11).

In this test, boric acid reacts with methyl alcohol giving volatile methyl borate (easter) which burns with a green aged flame.

$$H_3BO_3 + 3CH_3OH = \underset{\text{(Methyl borate)}}{B(OCH_3)_3} + 3H_2O$$

Ethyl alcohol can also be used instead of methyl alcohol, but the ethyl borate formed is less volatile. Hence, it is always preferred to use methyl alcohol.

Borax or Sodium Pyroborate or Sodium Tetra Decabydrate $Na_2B_4O_7$ $10H_2O$

It is the most important salt of boric acids. It occurs naturally as *tincal* in Tibet, Sri Lanka, South America and California. Native colemanite is also regarded as a source of borax.

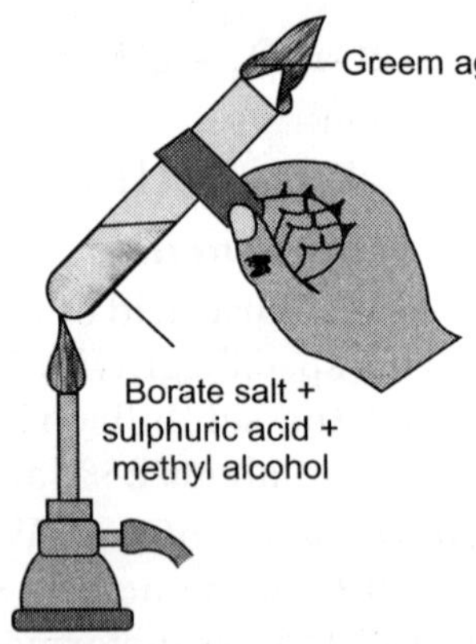

Fig. 11. The testing of Borax.

Preparation

(A) From tincal : Native tincal contains about 55% dried borax. The powdered tincal is treated with boiling water and filtered. On cooling the filtrate, almost pure crystals off borax separate out. The crystals are filtered, washed with water and dried.

(B) From Colemanite : Finely powdered colemanite ($Ca_2B_6O_{11}$ $5H_2O$) is boiled with an aqueous solution of sodium carbonate: when insoluble calcium carbonate is precipitated and a mixture of borax and sodium metaborate remains in solution.

$$Ca_2B_6O_{11} + 2Na_2CO_3 = 2CaCO_3 + +Na_2B_4O_7 + 2NaBO_2$$

The precipitated calcium carbonate is filtered off and the filtrate is concentrated and allowed to cool when crystals of borax separate out. These are filtered and dried. The sodium meta-borate in the mother liquor is converted into borax by passing carbon dioxide through the solution and thus a second crop of crystals of borax is obtained.

$$4NaBO_3 + CO_2 = Na_2B_4O_7 + Na_2CO_3.$$

(C) From boric acid : When boric acid and sodium carbonate solution are boiled, borax is formed. The solution is concentrated and subsequently cooled when crystals of borax separate out.

$$4H_3BO_3 + Na_2CO_3 = Na_2B_4O_7 + CO_2 + 6H_2O$$

Properties : *1. Physical: Ordinary borax, $Na_2B_4O_7.10H_2O$* forms large, colourless, transparent crystals having the density 1.72.

2. It effloresces in dry air.

3. It is fairly soluble in water. The solubility increases with rise of temperature.

Chemical : 1. The aqueous solution of borax reacts alkaline as borax undergoes hydrolysis giving sodium hydroxide (a strong base) and boric acid (a weak acid).

$$Na_2B_4O_7 + 7H_2O = 2NaOH + 4H_3BO_3$$

The aqueous solution of borax gives a pink colour with phenolphthalein. Hydrolysis of the solution is arrested on addition of glycerine.

2. Borax on being heated loses its water of crystallisation, swells up and is ultimately transformed into a colourless transparent (glass-like) mass consisting of sodium metaborate, ($NaBO_2$) and boron trioxide (B_2O_3).

This glassy mass dissolves the oxides of various metals forming coloured metaborates. For example,

$$CuO + Na_2B_4O_7 = 2NaBO_2 + Cu(BO_2)_2$$

(Copper metaborate, blue)

The borax bead test, which is applied in the identification of metallic radicals in the qualitative analysis, is based on these reactions.

Borax bead test : A clean platinum wire attached to a glass rod is taken and a small circular loop is made at the free end of the wire. The loop is heated in the non-luminous Bunsen flame and is then dipped while still red hot into powered borax, which in small quantity adheres to it. The loop with the borax is heated again in the hottest part of the flame when borax swells up and melts to form a transparent glass-like bead on the loop. The hot bead is made to touch the, substance under investigation so that only a minute quantity of the substance adheres to the bead. The bead with the adhering substance is heated again first in the oxidising flame and then in the reducing flame. The colour of the bead is noted in each case in the hot and cold conditions. The following observations are made if the test is performed with the copper and iron compounds.

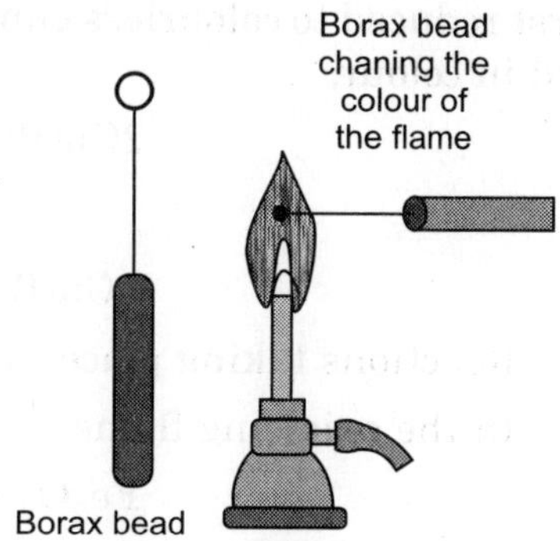

Fig. 12. Borax bead

Compounds of	Observation	
	Colour of the bead in oxidising flame	Colour of the bead in reducing flame
Copper	Green when hot but blue when cold	Colourless when hot and opaque red on
Iron	Reddish-yellow while hot yellow, when cold	Light green when hot or cold

Reactions in Borax Bead Test

Borax when heated loses its water of crystallisation, swells up and finally forms a transparent glass-like mass which is made up, of sodium meta borate and boron trioxide produced upon the decomposition of borax :

$$Na_2B_4O_710H_2O = Na_2B_4O_{11} + 10H_2O$$

$$Na_2B_4O_{11} = 2NaBO_2 + B_2O_3$$

Copper and iron-salts, when heated strongly in the oxidising flame are converted into their oxides which reacts with B_2O_3 and $NaBO_2$ of the bead forming coloured borates. CuO and B_2O_3 interact to form copper (cupric) metaborate [$Cu(BO_2)_2$] which is green when hot and blue when cold.

$$CuO + BeO_3 = \underset{\text{Cupric metaborate}}{Cu(BO_2)_2}$$

Some ortho borate is also formed.

$$CuO + NaBO_2 = \underset{\text{Sodium copper ortho borate, blue.}}{NaCuBO_3}$$

In the reducing flame, the cupric metaborate in contact with carbon of the flame is first reduced to colourless cuprous metaborate and finally to metallic copper which is dull-red in colour.

$$2Cu(BO_2)_2 + C = \underset{\substack{\text{cuprous metaborate}\\ \text{(colourless)}}}{Cu_2(BO_2)_2} + B_2O_3 + CO$$

$$Cu(BO_2)_2 + C = 2Cu + B_2O_3 + CO$$

Reactions taking place in case of an iron compound.

In the oxidising flame,

$$Fe_2O_2 + 3B_2O_3 = 2Fe(BO_2)_3 \text{ (ferric metaborate, yellow)}$$

In the reducing flame

$$2Fe(BO_2)_3 + C = 2Fe(BO_2)_2 + B_2O_3 + CO$$

The colours of the borax bead produced by nickel, cobalt. chromium and manganese salts are given below :

Substance taken	Colour of the borax bead	
	In oxidising flame	In reducing flame
Nickel-compounds	Brownish (cold)	Grey
Cobalt-compounds	Deep blue	Deep blue
Chromium compounds	Green	Green
Manganese compounds	Amethyst	colourless

Uses : (1) Borax is widely used in the preparation of glazes or porcelain enamels and earthenware. (2) It finds its applications in making optical glasses and a special type of high-grade glass, called borosilicate glass. (3) Borax is used as a flux in welding and soldering. (4) It is used as an anti-septic in making medical soaps, in the preservation of food, in the stiffening of candle-wicks and in laundering. In the laboratory, it is used in borax bead test which is carried out for identification of metallic radicals in qualitative-analysis.

N.B. It is to be noted here that ordinary borax crystallises below 62°C and has the composition $Na_2B_4O_7.10H_2O$. If crystallisation is made above 62°C, pentahydrate $Na_2B_4O_7.5H_2O$ separates. It is harder than ordinary borax and is an efflorescent substance. Its density is 1.81.

Preparation of Various Substances Starting from Borax

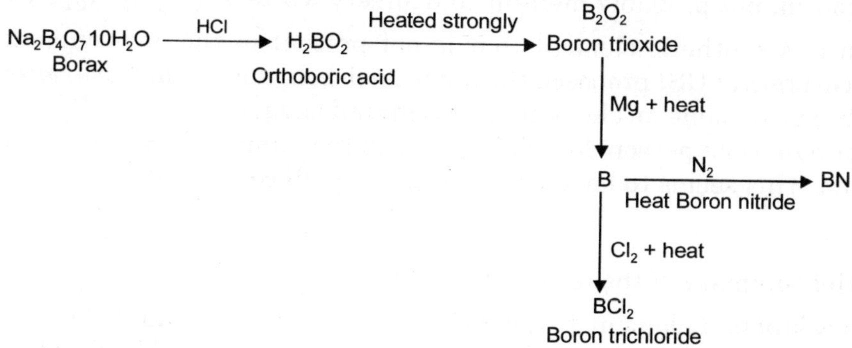

EPITOME–8

Elements 105 to 118

Introduction

This appendix consists of those elements, atomic number 107 to 119, their separation and properties have been explained.

BOHRIUM [107][264.12]

Presumably a solid at 298 K.

Colour: unknown, but probably metallic and silvery white or grey in appearance.

Bohrium is a synthetic element that is not present in the environment at all. The German discoverers at GSI proposed the name Nielsbohrium (symbol Ns) after Niels Bohr. IUPAC are happy to name an element after Bohr but suggest *bohrium (Bh)* on the grounds that the first name of a person does not appear in the names of any other element named after a person. This seems to have been accepted by all concerned.

Isolation

Here is a brief summary of the isolation of bohrium.

Only a few atoms of element 107, bohrium, have ever been made. The first atoms were made through a nuclear reaction involving fusion of an isotope of lead, ^{209}Pb, with one of chromium, ^{54}Cr.

$$^{209}Pb + ^{54}Cr \; ^{262}Bh + ^{1}n$$

Isolation of an observable quantity of bohrium has never been achieved, and may well never be. This is because bohrium decays very rapidly through the emission of α-particles.

More recently, other isotopes have been made at the Paul Scherrer Institute (PSI) in Switzerland.

$$^{249}Bk + ^{22}Ne \; ^{266}Bh + 5^{1}n$$

$$^{249}Bk + ^{22}Ne \; ^{267}Bh + 4^{1}n$$

In this work, it appears the scientists concerned feel bohrium forms the oxychloride $BhClO_3$

DUBNIUM [105][262.11]

Presumably a solid at 298 K.

Colour: unknown, but probably metallic and silvery white or grey in appearance

Isolation

Here is a brief summary of the isolation of dubnium.

Only very small amounts of element 105, dubnium, have ever been made. The first samples were made through nuclear reactions involving fusion of an isotope of californium, ^{249}Cf, with one of nitrogen, ^{14}N. A second route to the same isotope via berkelium is also known

$$^{15}N + {}^{249}Cf \; {}^{261}_{105}Db + 4\ {}^{1}n$$

$$^{16}N + {}^{249}Bk \; {}^{261}_{105}Db + 4\ {}^{1}n$$

Isolation of an observable quantity of dubnium has never been achieved.

HASSIUM [108][269.13]

Presumably a solid at 298 K.

Colour: unknown, but probably metallic and silvery white or grey in appearance.

Here is a brief description of hassium.

Hassium, is a synthetic element that is not present in the environment at all.

Isolation

Here is a brief summary of the isolation of hassium.

Only a few atoms of element 108, hassium, have ever been made. The first atoms were made through a nuclear reaction involving fusion of an isotope of lead, ^{208}Pb, with one of iron, ^{58}Fe.

$$^{208}Bi + {}^{58}Fe \; {}^{265}Hs + {}^{1}n$$

Isolation of an observable quantity of hassium has never been achieved, and may well never be. This is because hassium decays very rapidly through the emission of α-particles.

SEABORGIUM [106] [266.12]

Presumably a solid at 298 K.

Colour: unknown, but probably metallic and silvery white or grey in appearance.

Here is a brief description of seaborgium.

Seaborgium is a synthetic element that is not present in the environment at all. It has no uses.

Isolation

Here is a brief summary of the isolation of seaborgium.

Only very small amount of element 106, seaborgium, have ever been made. The first samples were made through a nuclear reaction involving fusion of an isotope of californium, ^{249}Cf, with one of oxygen, ^{18}O.

$$^{18}O + {}^{249}Cf\ {}^{263}{}_{106}Sg + 4\ {}^{1}n$$

Isolation of an observable quantity of seaborgium has never been achieved.

More recently, other isotopes have been made at the Paul Scherrer Institute (PSI) in Switzerland using neon atoms to bombard californium isotopes.

$$^{248}Cf + {}^{22}Ne\ {}^{266}Sg + 4{}^{1}n$$

UNUNBIUM [112] [277]

Presumably a liquid at 298 K.

Colour: unknown, but probably metallic and silvery white or grey in appearance.

Here is a brief description of ununbium.

Element 112 was discovered on 9th February 1996 at 22:37 at the GSI in Darmstadt, Germany. Further information on element 112 is here (outside WebElements) and in a press release (outside WebElements). The interested reader should consult the on-line version of The Wonderful World of Atoms and Nuclei for a fascinating insight into research on "super-heavy" atoms

Isolation

Here is a brief summary of the isolation of ununbium.

As only a few atoms of element 112 have ever been made (through a nuclear reaction involving fusing a zinc atom with a lead atom) isolation of an observable quantity has never been achieved, and may well never be.

$$^{208}Pb + {}^{70}Zn\ {}^{277}Uub + {}^{1}n$$

This is because atoms of the element decompose through the emission of α-particles with a half life of only about 240 microseconds.

UNUNOCTIUM [118] [293]

Presumably a gas at 298 K

Colour: unknown, but probably a colourless gas

Here is a brief description of ununoctium.

The team of Berkeley Lab scientists that announced in 1999 the observation of what appeared to be Element 118 — heaviest undiscovered transuranic element at the time — has retracted its original paper after several confirmation experiments failed to reproduce the results. This means that the following apparently is wrong. Three atoms of element 118 (ununoctium, Uuo) were synthesized by fusing krypton-86 ($^{86}{}_{36}Kr$) with lead-208 ($^{208}{}_{82}Pb$) in experiments conducted at at The Lawrence Berkeley National Laboratory, Berkeley, (California, USA), The University of Calfornia (USA), and Oregon State University, Corvallis (USA). It is very short-lived and decomposes to element 116.

UNUNQADIUM [114] [285]

Presumably a solid at 298 K.

Colour: unknown, but probably metallic and silvery white or grey in appearance.

Here is a brief description of ununquadium.

Element 114 was reported informally in January 1999 following experiments towards the end of December 1998 involving scientists at Dubna (Joint Institute for Nuclear Research) in Russia apparently using isotopes supplied by scientists at the Lawrence Livermore National Laboratory, USA. Only one atom was identified and the claim has not yet been ratified. The results of calculations suggest that element 114 will not form a tetrafluoride $UuqF_4$, but could be isolated as a water-soluble difluoride $UuqF_2$.

Isolation

Here is a brief summary of the isolation of ununquadium.

Currently, the identification of element 114 is yet to be confirmed. As only about three atoms of element 114 has ever been made (through nuclear reaction involving fusing a calcium atom with a plutonium atom) isolation of an observable quantity has never been achieved, and may well never be.

$$^{244}{}_{94}Pu + {}^{48}{}_{20}Ca \; {}^{289}{}_{114}Uuq + 3\,{}^{1}n$$

$$^{244}{}_{94}Pu + {}^{48}{}_{20}Ca \; {}^{288}{}_{114}Uuq + 4\,{}^{1}n$$

The element decomposes through the emission of an α-particle to form element 112 with a half life of about 30 seconds for $^{289}{}_{114}Uuq$ and 2 seconds for $^{288}{}_{114}Uuq$.

A different isotope of element 114, $^{285}{}_{114}Uuq$, is observed as a decomposition product of the recently observed element 118. Elements 118 and 116 were identified by accelerating a beam of krypton-86 ($^{86}{}_{36}Kr$) ions to an energy of 449 million electron volts and directing the beam onto targets of lead-208 ($^{208}{}_{82}Pb$). After 11 days work, just three atoms of the new element were identified. The production rates for element 118 are approximately one in every 10^{12} interactions.

$$^{208}{}_{82}Pb + {}^{86}{}_{36}Kr \; {}^{293}{}_{118}Uuo + {}^{1}n$$

Element 118 nucleus decays less than a millisecond after its formation by emitting an α-particle. This results in an isotope of element 116 (mass number 289, containing 116 protons and 173 neutrons). This isotope of element 116, is also radioactive and undergoes further α-decay processes to an isotope of element 114 and so on down to at least element 106.

$${}^{293}_{118}\mathrm{Uuo}\ {}^{289}_{116}\mathrm{Uuh} + {}^{4}_{2}\mathrm{He}\ (0.12\ \text{milliseconds})$$
$${}^{289}_{116}\mathrm{Uuh}\ {}^{285}_{114}\mathrm{Uuq} + {}^{4}_{2}\mathrm{He}\ (0.60\ \text{milliseconds})$$
$${}^{285}_{114}\mathrm{Uuq}\ {}^{281}_{112}\mathrm{Uub} + {}^{4}_{2}\mathrm{He}\ (0.58\ \text{milliseconds})$$
$${}^{281}_{112}\mathrm{Uub}\ {}^{277}_{110}\mathrm{Uun} + {}^{4}_{2}\mathrm{He}\ (0.89\ \text{milliseconds})$$
$${}^{277}_{110}\mathrm{Uun}\ {}^{273}_{108}\mathrm{Hs} + {}^{4}_{2}\mathrm{He}\ (3\ \text{milliseconds})$$
$${}^{273}_{108}\mathrm{Hs}\ {}^{269}_{106}\mathrm{Sg} + {}^{4}_{2}\mathrm{He}\ (1200\ \text{milliseconds})$$

UNUNHEXIUM [116] [289]

Presumably a solid at 298 K.

Colour: unknown, but probably metallic and silvery white or grey in appearance.

Here is a brief description of ununhexium.

Element 116 was identified as a decomposition product of element 118 in experiemnts conducted at The Lawrence Berkeley National Laboratory, Berkeley, (California, USA), The University of Calfornia (USA), and Oregon State University, Corvallis (USA). It is very shortlived and decomposes in turn to element 114.

Isolation

Here is a brief summary of the isolation of ununhexium.

Elements 118 and 116 were identified by accelerating a beam of krypton-86 (${}^{86}_{36}\mathrm{Kr}$) ions to an energy of 449 million electron volts and directing the beam onto targets of lead-208 (${}^{208}_{82}\mathrm{Pb}$). After 11 days work, just three atoms of the new element were identified. The production rates for element 118 are approximately one in every 10^{12} interactions.

$${}^{208}_{82}\mathrm{Pb} + {}^{86}_{36}\mathrm{Kr}\ {}^{293}_{118}\mathrm{Uuo} + {}^{1}\mathrm{n}$$

Element 118 nucleus decays less than a millisecond after its formation by emitting an α-particle. This results in an isotope of element 116 (mass number 289, containing 116 protons and 173 neutrons). This isotope of element 116, is also radioactive and undergoes further α-decay processes to an isotope of element 114 and so on down to at least element 106.

$${}^{293}_{118}\mathrm{Uuo}\ {}^{289}_{116}\mathrm{Uuh} + {}^{4}_{2}\mathrm{He}\ (0.12\ \text{milliseconds})$$
$${}^{289}_{116}\mathrm{Uuh}\ {}^{285}_{114}\mathrm{Uuq} + {}^{4}_{2}\mathrm{He}\ (0.60\ \text{milliseconds})$$
$${}^{285}_{114}\mathrm{Uuq}\ {}^{281}_{112}\mathrm{Uub} + {}^{4}_{2}\mathrm{He}\ (0.58\ \text{milliseconds})$$
$${}^{281}_{112}\mathrm{Uub}\ {}^{277}_{110}\mathrm{Uun} + {}^{4}_{2}\mathrm{He}\ (0.89\ \text{milliseconds})$$
$${}^{277}_{110}\mathrm{Uun}\ {}^{273}_{108}\mathrm{Hs} + {}^{4}_{2}\mathrm{He}\ (3\ \text{milliseconds})$$
$${}^{273}_{108}\mathrm{Hs}\ {}^{269}_{106}\mathrm{Sg} + {}^{4}_{2}\mathrm{He}\ (1200\ \text{milliseconds})$$

Results from recent experiments at Dubna (Russia) describe the decay of the isotope ${}^{296}\mathrm{Uuh}$ (produced in the reaction of ${}^{248}\mathrm{Cm}$ with ${}^{48}\mathrm{Ca}$) to ${}^{292}\mathrm{Uuq}$.

$${}^{248}_{96}\mathrm{Cm} + {}^{48}_{20}\mathrm{Ca}\ {}^{292}_{116}\mathrm{Uuh} + 4\ \mathrm{n}$$

UNUNNILIUM [110] [271.15]

Presumably a solid at 298 K.

Colour: unknown, but probably metallic and silvery white or grey in appearance.

Here is a brief description of ununnilium.

Element 110, ununnilium, is a synthetic element that is not present in the environment at all. Further information on element 110 is here (outside Web Elements) and in a press release (outside Web Elements). The interested reader should consult the on-line version of The Wonderful World of Atoms and Nuclei for a fascinating insight into research on "super-heavy" atoms.

Chemically, the element 110 is in the same Group as nickel, palladium, and platinum (Group 10). Unlike these lighter atoms, element 110 decays after a small fraction of a thousandth of a second into lighter elements by emitting α-particles which are the nuclei of helium atoms.

Isolation

Here is a brief summary of the isolation of ununnilium.

Only a few atoms of element 110 have ever been made, initially through a nuclear reaction involving fusion of an isotope of lead, Pb, with one of nickel, Ni.

$$^{208}Pb + {}^{62}Ni \; {}^{269}Uun + {}^{1}n$$

Isolation of an observable quantity has never been achieved, and may well never be. This is because atoms of the element decompose through the emission of α-particles with a half life of only about 270 microseconds. Another isotope was made by using a different isotope of nickel.

$$^{208}Pb + {}^{64}Ni \; {}^{271}Uun + {}^{1}n$$

UNUNUNIUM [111] [272.15]

Presumably a solid at 298 K.

Colour: unknown, but probably metallic and silvery white or grey in appearance.

Here is a brief description of unununium.

Element 111 was discovered on 8th December 1994 at the GSI in Darmstadt, Germany. Further information on element 111 is here (outside WebElements). The information following is an abstract of this source. The interested reader should consult the on-line version of The Wonderful World of Atoms and Nuclei (outside WebElements) for a fascinating insight into research on "super-heavy" atoms. Chemically, element 111 should be in the same group as the elements copper, silver, and gold (Group 11).

Isolation

Here is a brief summary of the isolation of unununium.

Only a few atoms of element 111 have ever been made through a nuclear reaction involving fusion of an isotope of bismuth, ^{209}Bi, with one of nickel, ^{64}Ni.

$$^{209}Bi + {}^{64}Ni \; {}^{272}Uuu + {}^{1}n$$

Isolation of an observable quantity has never been achieved, and may well never be.

PERIODIC TABLES

1 **H** (Hydrogen) 1.00794											
3 **Li** (Lithium) 6.941	4 **Be** (Beryllium) 9.01218										
11 **Na** Sodium 22.9888	12 **Mg** Magnesium 24.3050										
19 **K** Potassium 39.0983	20 **Ca** Calcium 40.078	21 **Sc** Scandium 44.9559	22 **Ti** Titanium 47.88	23 **V** Vanadium 50.9415	24 **Cr** Chromium 51.9941	25 **Mn** Manganese 54.9380	26 **Fe** Iron 55.847	27 **Co** Cobalt 58.9332	28 **Ni** Nickel 58.69	29 **Cu** Copper 63.546	30 **Zn** Zinc 66.39
37 **Rb** Rubidium 85.4678	38 **Sr** Strontium 97.62	39 **Y** Yttrium 88.9059	40 **Zr** Zirconium 91.224	41 **Nb** Niobium 92.9064	42 **Mo** Molybdenum 95.94	43 **Tc** Technetium 98	44 **Ru** Ruthenium 101.07	45 **Rh** Rhodium 102.906	46 **Pd** Palladium 106.42	47 **Ag** Silver 107.868	48 **Cd** Cadmium 112.411
55 **Ce** Cesium 132.905	56 **Ba** Barium 137.327	57 **La** Lanthanum 138.906	72 **Hf** Hafnium 178.49	73 **Ta** Tantalum	74 **W** Tungsten 183.85	75 **Re** Rhenium 186.207	76 **Os** Osmium 190.2	77 **Ir** Iridium 192.22	78 **Pt** Platinum 195.08	79 **Au** Gold 196.97	80 **Hg** Mercury 200.59
87 **Fr** Francium 223	88 **Ra** Radium 226.025	89 **Ac** Actinium 227.028	104 **Ku** Kurchatovium 261	105 **Ha** Hahnium 262	106 **Unh** Unnilhexium 263	107 **Uns** Unnilseptium 264	108 **Uno** Unniloctium 265	109 **Unn** Unnilennium 266			

...(Contd.)

					2 **He** Helium 4.00260
5 **B** Boron 10.11	6 **C** Carbon 12.011	7 **N** Nitrogen 14.0076	8 **O** Oxygen 15.9984	9 **F** Fluorine 18.9984	10 **Ne** Neon 20.1797
13 **Al** Aluminium 26.9815	14 **Si** Silicon 28.05	15 **P** Phosphorous 30.09	16 **S** Sulphur 32.066	17 **Cl** Chlorine 35.4527	18 **Ar** Argon 39.948
31 **Ga** Gallium 69.723	32 **Ge** Germanium 72.59	33 **As** Arsenic 74.9816	34 **Se** Selenium 78.96	35 **Br** Bromine 79.904	36 **Kr** Krypton 83.80
49 **In** Indium 114.82	50 **Sn** Tin 118.710	51 **Sb** Antimony	52 **Te** Tellurium 128.60	53 **I** Iodine 126.905	54 **Xe** Xenon 131.29
81 **Tl** Thallium 204.383	82 **Pb** Lead 207.2	83 **Bi** Bismuth 208.980	84 **Po** Polonium 209	85 **At** Astatine 210	86 **Rn** Radon 222

58 **Ce** Cesium 140.12	50 **Pr** Praseo- dymium 140.908	60 Neodynium **Ce** Cesium 140.12	61 **Pm** Promethium 145	62 **Sm** Samarium 150.36	63 **Eu** Europium 151.965	64 **Gd** Gadolinium 157.25	65 **Tb** Terbium 158.925	66 **Dy** Dysprosium 162.50	67 **Ho** Holmium 164.930	68 **Er** Erbium 167.50	69 **Tm** Thulium 168.930	70 **Yb** Yiterbium 173.04	71 **In** Lutetium 174.967
90 **Th** Thorium	91 **Pa** Protac- tinium 231.038	92 **U** Uranium 238.029	93 **Np** Neptunium 237.048	94 **Pu** Plutonium 244	95 **Am** Americium 243	96 **Cm** Curium 247	97 **Bk** Berkelium 247	98 **Cf** Californium 151	99 **Es** Einsteinium	100 **Fm** Fermium 257	101 **Md** Mende- levium 258	102 **Nb** Nobelium 299	108 **Lr** Lavrenium 260

THE PERIODIC TABLE

Lanthanides *Actinides*

1 18
1A 8A

3b	4b	5b	6b	7b		8b		1b	2b
3	4	5	6	7	8	9	10	11	12

1

2

3

4

5

6

7

KEY

1	Atomic Number
H	Symbol
Hydrogen	Name
1.00794	Atomic Weight

ZENTIL BORDER

Atomic No.	Element	K	L		M		N				O				P				Q			
		1	2		3		4				5				6				7			
	s	s	p	s	p	d	s	p	d	f	s	p	d	f	s	p	d	f	s	p	d	f
1	H	1																				
2	He	2																				
3	Li	2	1																			
4	Be	2	2																			
5	B	2	2	1																		
6	C	2	2	2																		
7	N	2	2	3																		
8	O	2	2	4																		
9	F	2	2	5																		
10	Ne	2	2	6																		
11	Na	2	2	6	1																	
12	Mg	2	2	6	2																	
13	Al	2	2	6	2	1																
14	Si	2	2	6	2	2																
15	P	2	2	6	2	3																
16	S	2	2	6	2	4																
17	Cl	2	2	6	2	5																
18	A	2	2	6	2	6																
19	K	2	2	6	2	6	...	1														
20	Ca	2	2	6	2	6...	2															
21	Sc	2	2	6	2	6	1	2														
22	Ti	2	2	6	2	6	2	2														
23	V	2	2	6	2	6	3	2														
24	Cr	2	2	6	2	6	5*	1														
25	Mn	2	2	6	2	6	5	2														
26	Fe	2	2	6	2	6	6	2														
27	Co	2	2	6	2	6	7	2														
28	Ni	2	2	6	2	6	8	2														
29	Cu	2	2	6	2	6	10*	1														
30	Zn	2	2	6	2	6	10	2														
31	Ga	2	2	6	2	6	10	2	1													
32	Ge	2	2	6	2	6	10	2	2													

...(Contd.)

33	As	2	2	6	2	6	10	2	3							
34	Sc	2	2	6	2	6	10	2	4							
35	Br	2	2	6	2	6	10	2	5							
36	Kr	2	2	6	2	6	10	2	6							
37	Rb	2	2	6	2	6	10	2	6	.	.	1				
38	Sr	2	2	6	2	6	10	2	6	.	.	2				
39	Y	2	2	6	2	6	10	2	6	1	.	2				
40	Zr	2	2	6	2	6	10	2	6	2	.	2				
41	Cb	2	2	6	2	6	10	2	6	4*	.	1				
42	Mo	2	2	6	2	6	10	2	6	5	.	1				
43	Te	2	2	6	2	6	10	2	6	6	.	1				
44	Ru	2	2	6	2	6	10	2	6	7	.	1				
45	Rh	2	2	6	2	6	10	2	6	8	.	1				
46	Pd	2	2	6	2	6	10	2	6	10*	.	.				
47	Ag	2	2	6	2	6	10	2	6	10	.	1				
48	Cd	2	2	6	2	6	10	2	6	10	.	2				
49	In	2	2	6	2	6	10	2	6	10	.	2	1			
50	Sn	2	2	6	2	6	10	2	6	10	.	2	2			
51	Sb	2	2	6	2	6	10	2	6	10	.	2	3			
52	Te	2	2	6	2	6	10	2	6	10	.	2	4			
53	I	2	2	6	2	6	10	2	6	10	.	2	5			
54	Xe	2	2	6	2	6	10	2	6	10	.	2	6			
55	Gs	2	2	6	2	6	10	2	6	10	.	2	6	.	.	1
56	Ba	2	2	6	2	6	10	2	6	10	.	2	6	.	.	2
57	La	2	2	6	2	6	10	2	6	10	.	2	6	1	.	2
58	Ce	2	2	6	2	6	10	2	6	10	2*	2	6	.	.	2
59	Pr	2	2	6	2	6	10	2	6	10	3	2	6	.	.	2
60	Nd	2	2	6	2	6	10	2	6	10	4	2	6	.	.	2
61	Pm	2	2	6	2	6	10	2	6	10	5	2	6	.	.	2
62	Sm	2	2	6	2	6	10	2	6	10	6	2	6	.	.	2
63	En	2	2	6	2	6	10	2	6	10	7	2	6	.	.	2
64	Gd	2	2	6	2	6	10	2	6	10	7	2	6	1	.	2
65	Tb	2	2	6	2	6	10	2	6	10	9*	2	6	.	.	2
66	Dy	2	2	6	2	6	10	2	6	10	10	2	6	.	.	2
67	Ho	2	2	6	2	6	10	2	6	10	11	2	6	.	.	2
68	Fn	2	2	6	2	6	10	2	6	10	12	2	6	.	.	2

...(Contd.)

69	Tm	2	2	6	2	6	10	2	6	10	13	2	6	.	.	2				
70	Yh	2	2	6	2	6	10	2	6	10	14	2	6	.	.	2				
71	Lu	2	2	6	2	6	10	2	6	10	14	2	6	1	.	2				
72	Hf	2	2	6	2	6	10	2	6	10	14	2	6	2	.	2				
73	Ta	2	2	6	2	6	10	2	6	10	14	2	6	3	.	2				
74	W	2	2	6	2	6	10	2	6	10	14	2	6	4	.	2				
75	Re	2	2	6	2	6	10	2	6	10	14	2	6	5	.	2				
76	Os	2	2	6	2	6	10	2	6	10	14	2	6	6	.	2				
77	Ir	2	2	6	2	6	10	2	6	10	14	2	6	7	.	2				
78	Pt	2	2	6	2	6	10	2	6	10	14	2	6	9*	.	1				
79	Au	2	2	6	2	6	10	2	6	10	14	2	6	10	.	1				
80	Hg	2	2	6	2	6	10	2	6	10	14	2	6	10	.	2				
81	Tl	2	2	6	2	6	10	2	6	10	14	2	6	10	.	2	1			
82	Pb	2	2	6	2	6	10	2	6	10	14	2	6	10	.	2	2			
83	Bi	2	2	6	2	6	10	2	6	10	14	2	6	10	.	2	3			
84	Po	2	2	6	2	6	10	2	6	10	14	2	6	10	.	2	4			
85	At	2	2	6	2	6	10	2	6	10	14	2	6	10	.	2	5			
86	Rn	2	2	6	2	6	10	2	6	10	14	2	6	10	.	2	6			
87	Fr	2	2	6	2	6	10	2	6	10	14	2	6	10	.	2	6	.	.	1
88	Ra	2	2	6	2	6	10	2	6	10	14	2	6	10	.	2	6	.	.	2
89	Ae	2	2	6	2	6	10	2	6	10	14	2	6	10	.	2	6	1	.	2
90	Th	2	2	6	2	6	10	2	6	10	14	2	6	10	.	2	6	2	.	2
91	Pa	2	2	6	2	6	10	2	6	10	14	2	6	10	2*	2	6	1	.	2
92	U	2	2	6	2	6	10	2	6	10	14	2	6	10	3	2	6	1	.	2
93	Np	2	2	6	2	6	10	2	6	10	14	2	6	10	4	2	6	1	.	2
94	Pu	2	2	6	2	6	10	2	6	10	14	2	6	10	5	2	6	1	.	2
95	Am	2	2	6	2	6	10	2	6	10	14	2	6	10	6	2	6	1	.	2
96	Cm	2	2	6	2	6	10	2	6	10	14	2	6	10	7	2	6	1	.	2

* Note irregularity.

Index

C

R

T

U